T0296912

# Climate System Dynamics and Modelling

This book presents all aspects of climate system dynamics on all time scales from the Earth's formation to modern human-induced climate change. It discusses the dominant feedbacks and interactions between all the components of the climate system: atmosphere, ocean, land surface and ice sheets. It addresses one of the key challenges for a course on the climate system: students can come from a range of backgrounds. A glossary of key terms is provided for students with little background in the climate sciences, whilst instructors and students with more expertise will appreciate the book's modular nature. Exercises are provided at the end of each chapter for readers to test their understanding. This book will be invaluable for any course on climate system dynamics and modelling and will also be useful for scientists and professionals from other disciplines who want a clear introduction to the topic.

**Hugues Goosse** is a senior research associate with the Fonds National de la Recherche Scientifique (F.R.S.-FNRS-Belgium) and a professor at the Université catholique de Louvain in Belgium. He teaches climate-related topics to students from a wide range of backgrounds, including physics, geography, engineering, bioengineering, biology, Earth and environmental sciences and philosophy. His research is mainly devoted to the development of climate models, model-data comparisons and the application of models to study past, current and future climate change, analysing both natural variability and the response to human-induced perturbations. He is currently editor and former co–chief editor of the journal *Climate of the Past*. He has contributed to several international programs and assessment reports, in particular, to the fourth and fifth assessment reports of the Intergovernmental Panel on Climate Change.

# Climate System Dynamics and Modelling

HUGUES GOOSSE

Université catholique de Louvain

# CAMBRIDGE
## UNIVERSITY PRESS

University Printing House, Cambridge CB2 8BS, United Kingdom

One Liberty Plaza, 20th Floor, New York, NY 10006, USA

477 Williamstown Road, Port Melbourne, VIC 3207, Australia

314-321, 3rd Floor, Plot 3, Splendor Forum, Jasola District Centre, New Delhi - 110025, India

79 Anson Road, #06-04/06, Singapore 079906

Cambridge University Press is part of the University of Cambridge.

It furthers the University's mission by disseminating knowledge in the pursuit of
education, learning and research at the highest international levels of excellence.

www.cambridge.org
Information on this title: www.cambridge.org/9781107445833

© Hugues Goosse 2015

First published 2015

*A catalogue record for this publication is available from the British Library*

*Library of Congress Cataloging in Publication data*
Goosse, Hugues, 1971–
Climate system dynamics and modelling / Hugues Goosse, Université catholique de Louvain.
pages   cm
Includes bibliographical references and index.
ISBN 978-1-107-08389-9 (hardback) – ISBN 978-1-107-44583-3 (pbk.)
1. Climatology–Mathematical models.   2. Climatic changes–Mathematical models.
3. Atmospheric physics–Statistical methods.   I. Title.
QC874.5.G66 2015
551.601´5118–dc23       2015009564

ISBN  978-1-107-08389-9  Hardback
ISBN  978-1-107-44583-3  Paperback

Addition resources for this publication at www.cambridge.org/goosse

To Dominique, Nicolas and Oriane

# Contents

# Preface

The climate has a significant impact on life on Earth as well as on human activities. Temperature and precipitation strongly constrain the type of vegetation that can grow in a particular region. The design and location of houses depend on summer and winter temperatures and also on the probability of flooding. One single, late frost or a heavy hail storm could ruin an entire crop. Since the beginning of humanity, people thus have had to cope with climate and, if possible, to adapt to it. As a consequence, the various human civilisations have observed and tried to understand climate variations. They first provided mythological or religious explanations, often relying on weather lore to obtain forecasts. In parallel, climate has evolved as a science, elaborating more and more sophisticated representations of the observed phenomena. Such a description of climate now involves a very broad range of expertise corresponding to different domains of the sciences, including physics, chemistry, biology and geology.

A comprehensive analysis of all the components of the climate system (i.e., atmosphere, ocean, ice sheets, land surfaces, etc.) and of all the interactions between them is beyond the scope of any course or book. Here I provide only a relatively brief overview of the processes that rule the behaviour of the individual components. More detailed descriptions are provided in meteorology, oceanography and glaciology textbooks, for instance, with some suggestions for reading given in the reference section. *The focus of this book is on the interactions between the different elements of the climate system and on the main feedbacks that govern climate variability on all the time scales. On this basis, the first goal of this book is to analyse the dominant causes of past climate changes and to critically discuss the projections of climate change over the next centuries or millennia.*

Because of the complexity of the system, many analyses devoted to a quantitative estimate of climate change or climate variability rely on the use of comprehensive three-dimensional numerical models. Simple models are also widely applied to underline clearly the fundamental properties of the climate. *The second goal of this book is to give readers the basis on which to develop an understanding of how climate models are built and their specific interests and limitations and to provide key examples of their applications.*

This book is an extended version of an online resource available at www.climate.be/textbook. It was designed originally to support a course proposed to students in their first year of a master's program at the Université catholique de Louvain (Belgium). However, the book has been designed to also be followed by undergraduate students. Because the book covers a wide range of disciplines and is devoted to an audience with different backgrounds, some of the terms or concepts employed may not be familiar to everyone. An extensive glossary thus is provided for readers who feel the need for specific explanations. The corresponding terms are highlighted in **bold** in the text. Some sections include limited mathematical developments, but understanding them in detail is not required to follow the main arguments, which are always developed using words or diagrams. More generally, this book includes an extensive index and many cross-references between the various sections where related topics are discussed. This allows for dynamic navigation inside the text that encourages readers to focus on certain specific parts of interest whilst skipping others or leaving them for a later reading.

The references to textbooks and scientific papers have been chosen to provide up-to-date information that is complementary to the material included herein at an adequate level of complexity. Consequently, they do not necessarily correspond to the historical development of the concepts, but interested readers can consult them to gain more insight into the history of the field. The number of references is also strongly variable between the sections, being larger for subjects in rapid development and smaller for subjects that are only briefly discussed in the present framework. Finally, review exercises are available at the end of each chapter. They include questions that provide an overview of the most important elements covered in the corresponding sections so that readers can directly evaluate their understanding of the text.

A comprehensive understanding of climate modelling requires one to perform simulations on one's own. A very useful exercise thus is to code some of the equations proposed in the various sections, starting from the simplest. Because this may require a significant amount of work, some interactive models are proposed online (www.climate.be/climatebook). They are related to the material covered in this book, but they focus on some specific examples. They offer the opportunity to test the influence on model results of changes in parameters or of forcing. Specific quizzes are also available online that can be answered using those models.

# Acknowledgements

This book is based on a shorter online version that has been publicly available since 2008. I first want to thank Violette Zunz, Marie-France Loutre, Wouter Lefebvre, Pierre-Yves Barriat and Antoine Barthélemy, without whom the online version and thus this current printed version would not have been possible. Their essential contribution includes the production of several figures as well as very useful suggestions and comments. Furthermore, they took in charge all the technical aspects of the online version. Many of my colleagues have carefully read various sections of this book, or previous versions, and have proposed modifications that have improved the quality of the published material significantly. I would like to thank in particular Jean-Marie Beckers, André Berger, Olivier Boucher, Victor Brovkin, Sally Close, Matt Collins, Elisabeth Crespin, Michel Crucifix, Eric Deleersnijder, Anne de Montety, Anne de Vernal, Svetlana Dubinkina, Mike Evans, Thierry Fichefet, Pierre Francus, Pierre Friedlingstein, Yves Godderis, Jonathan Gregory, Joel Guiot, Ed Hawkins, François Klein, Olivier Lecomte, Ralph Lescroart, Gurvan Madec, Aurélien Mairesse, François Massonnet, Pierre Mathiot, Sébastien Moreau, Hans Renssen, Didier Roche, Yoann Sallaz-Damaz, Andrew Schrurer, Ted Shepherd, Benoit Tartinville, Axel Timmermann, Kevin Trenberth, Stephane Vannitsem, Jean-Pascal van Ypsersele, Koffi Warou and Quizhen Yin. I have benefitted from very constructive advice from Françoise Docq, Marcel Lebrun and Denis Smidts of the Institut de Pédagogie universitaire et des Multimédias of the Université catholique de Louvain (IPM, http://www.ipm.ucl.ac.be). The comments of master's students who have followed my lectures at the Université catholique de Louvain and the suggestions of users of the online version also were very valuable. A book such as this is based on the work of many scientists, and I want to thank them for their contributions and apologise for not being able to give more details on important aspects of climatology in the framework of this introductory study. I specifically want to acknowledge the organisations, publishers and scientists who have allowed me to reproduce their work. The online material has been supported by the Fonds de Développement Pédagogique of the Université catholique de Louvain in the framework of the project 'Réalisation de simulations interactives comme support à l'apprentissage dans le cadre du cours d'introduction à la physique du système climatique et à sa modélisation.'

# Main Symbols and Acronyms, Including Typical Values for Constants

| | |
|---|---|
| $\alpha$ | Albedo |
| $\alpha_p$ | Planetary albedo (around 0.3 for present-day conditions) |
| $\alpha_T$ | Thermal expansion coefficient ($\mathrm{kg\,m^{-3}\,K^{-1}}$) |
| $\beta_C$ | Concentration–carbon feedback parameter ($\mathrm{PgC\,ppm^{-1}}$) |
| $\beta_S$ | Haline contraction coefficient ($\mathrm{kg\,m^{-3}\,psu^{-1}}$) |
| $\delta$ | Solar declination (in degrees or radians) |
| $\delta^{13}C$ | Delta value for the relative abundance of $^{13}C$ in a sample (‰) |
| $\delta^{18}O$ | Delta value for the relative abundance of $^{18}O$ in a sample (‰) |
| $\Delta Q$ | Radiative forcing ($\mathrm{W\,m^{-2}}$) |
| $\Delta R$ | Radiative imbalance at the top of the atmosphere ($\mathrm{W\,m^{-2}}$) |
| $\Delta t$ | Time step |
| $\Delta x$ | Spatial step |
| $\varepsilon$ | Emissivity of an object |
| $\varepsilon_{\mathrm{obl}}$ | Obliquity ($=23.45°$ presently) |
| $\gamma$ | True anomaly (in degrees or radians) |
| $\gamma_c$ | Climate–carbon feedback parameter ($\mathrm{PgC\,ppm^{-1}}$) |
| $\kappa_c$ | Ocean heat uptake efficiency ($\mathrm{W\,m^{-2}\,K^{-1}}$) |
| $\phi$ | Latitude (on Earth, in degrees or radians) |
| $\rho$ | Density ($=1{,}000\ \mathrm{kg\,m^{-3}}$ for pure water, $917\ \mathrm{kg\,m^{-3}}$ for ice, around $1\ \mathrm{kg\,m^{-3}}$ for air at sea-level pressure) |
| $\rho_f$ | Climate resistance ($\mathrm{W\,m^{-2}\,K^{-1}}$) |
| $\theta_s$ | Solar zenith distance (in degrees or radians) |
| $\lambda$ | Longitude (on Earth, in degrees or radians) |
| $\lambda_f$ | Climate feedback parameter |
| $\lambda_i$ | Climate feedback parameter for variable $x_i$ |
| $\lambda_t$ | True longitude (in degrees or radians) |
| $\eta$ | Sea-surface elevation |
| $\sigma$ | Stefan-Boltzmann constant ($=5.67\times10^{-8}\ \mathrm{W\,m^{-2}\,K^{-4}}$) |
| $\Gamma$ | Lapse rate ($\mathrm{K\,m^{-1}}$) |
| $\tau_a$ | Infrared transmissivity of the atmosphere |
| $\tilde{\omega}$ | Longitude of the perihelion measured from the vernal equinox (in degrees) |
| $\vec{\Omega}$ | Angular velocity vector of the Earth ($\Omega = \left\|\vec{\Omega}\right\| = 7.292 \times 10^{-5}\ \mathrm{s^{-1}}$) |
| AABW | Antarctic bottom water |

| | |
|---|---|
| AAIW | Antarctic intermediate water |
| ACC | Antarctic circumpolar current |
| A.D. | Anno Domini; year A.D. is the number of years since the beginning of the Christian (or Common) era |
| AGCM | Atmospheric general circulation model |
| AMO | Atlantic Multi-Decadal Oscillation |
| AOGCM | Atmosphere-ocean general circulation model |
| B.C. | Before Christ; year B.C. is the number of years before A.D. 1 |
| B.P. | Before present, that is, before A.D. 1950 |
| $C$ | Condensation |
| CDW | Circumpolar deep water |
| CGCM | Coupled general circulation model |
| $c_m$ | Specific heat capacity of medium $m$ ($J\,K^{-1}\,kg^{-1}$) |
| $C_m$ | Effective heat capacity of medium $m$ ($J\,K^{-1}\,m^{-2}$) |
| CMIP | Coupled Model Inter-Comparison Project |
| $c_p$ | Specific heat at constant pressure ($=1{,}004\,J\,K^{-1}\,kg^{-1}$ for dry air) |
| CRE | Cloud radiative effect |
| $c_v$ | Specific heat at constant volume ($=717\,J\,K^{-1}\,kg^{-1}$ for dry air) |
| $c_w$ | Specific heat of water ($=4{,}180\,J\,K^{-1}\,kg^{-1}$ for pure water at 0°C) |
| DGVM | Dynamic global vegetation model |
| DIC | Dissolved inorganic carbon |
| DJF | December, January and February |
| $e$ | Partial pressure of water vapour (Pa) |
| $E$ | Evapotranspiration |
| EBM | Energy-balance model |
| $ecc$ | Eccentricity ($=0.0167$ for present-day conditions) |
| EMIC | Earth model of intermediate complexity |
| ENSO | El Niño–Southern Oscillation |
| ERF | Effective radiative forcing |
| $e_s$ | Saturation vapour pressure (Pa) |
| ESM | Earth system model |
| $F_{diff}$ | Flux due to diffusion or conduction |
| $f_f$ | Feedback factor |
| $F_{fric}$ | Force due to friction |
| $F_{IR\downarrow}$ | Downward longwave radiation at the surface ($W\,m^{-2}$) |
| $F_{IR\uparrow}$ | Upward longwave radiation at the surface ($W\,m^{-2}$) |
| $F_{LH}$ | Latent heat flux at the surface ($W\,m^{-2}$) |
| $F_{SH}$ | Sensible heat flux at the surface ($W\,m^{-2}$) |
| $F_{SOL}$ | Incoming solar radiation at the surface ($W\,m^{-2}$) |
| $g$ | Acceleration due to gravity at the Earth's surface ($=9.8\,m\,s^{-2}$) |
| GCM | General circulation model |

| | |
|---|---|
| $g_T$ | Total feedback gain |
| GtC | Gigaton of carbon ($10^{15}$ g of carbon) |
| **H** | Observation operator |
| HA | Hour angle |
| IPCC | Intergovernmental Panel on Climate Change |
| IRD | Ice-rafted debris |
| ITCZ | Intertropical Convergence Zone |
| J | Joule |
| JJA | June, July and August |
| K | Kelvin |
| ka | 1000 years |
| $K_H$ | Solubility (of $CO_2$) |
| kyr | 1000 years |
| $L_f$ | Latent heat of fusion of water (= 334 kJ kg$^{-1}$ at 0°C) |
| LGM | Last glacial maximum (around 21 kyr B.P.) |
| $L_v$ | Latent heat of vaporisation of water (= 2250 kJ kg$^{-1}$ at 100°C, 2500 kJ kg$^{-1}$ at 0°C) |
| m | Metre |
| MIP | Model Inter-Comparison Project |
| MOC | Meridional overturning circulation |
| MOS | Model output statistics |
| NADW | North Atlantic deep water |
| NAM | Northern Annular Mode |
| NAO | North Atlantic Oscillation |
| nm | Nanometre ($10^{-9}$ m) |
| NPO | North Pacific Oscillation |
| NPP | Net primary production |
| NPZD | Nutrient-phytoplankton-zooplankton-detritus model |
| OGCM | Ocean general circulation model |
| $p$ | Pressure (Pa) |
| $P$ | Precipitation |
| Pa | Pascal |
| PDE | Partial differential equation |
| PDO | Pacific Decadal Oscillation |
| PERH | Longitude of the perihelion measured from the autumn equinox (= 102.04° in present-day conditions) |
| PETM | Paleocene-Eocene thermal maximum |
| PFT | Plant functional type |
| PgC | Petagrams of carbon ($10^{15}$ g of carbon) |
| PMIP | Paleoclimate Modelling Inter-Comparison Project |
| PNA | Pacific North American pattern |
| ppb | Parts per billion |
| ppm | Parts per million |
| $p_s$ | Surface pressure (Pa) |
| PSA | Pacific South American pattern |

| | |
|---|---|
| psu | Practical salinity unit |
| PW | $10^{15}$ W |
| $q$ | Specific humidity (kg/kg) |
| $R$ | Earth's radius (=6,371 km) |
| $R^*$ | Universal gas constant (=8.3143 J K$^{-1}$ mol$^{-1}$) |
| RF | Radiative forcing |
| $RF_{TOA}$ | Net radiative flux at the top of the atmosphere (W m$^{-2}$) |
| $R_g$ | Gas constant for a gas $g$ (=287.0 J K$^{-1}$ kg$^{-1}$ for dry air) |
| RH | Relative humidity |
| $r_m$ | Mean distance between the Earth and the Sun (=$1.5 \times 10^{11}$ m) |
| RMS | Root mean square |
| $R_{riv}$ | River runoff |
| $R_v$ | Gas constant for water vapour (=461.4 J kg$^{-1}$ K$^{-1}$) |
| $S$ | Ocean salinity |
| $S_0$ | Mean total solar irradiance at mean Earth–Sun distance (~1,360 W m$^{-2}$) |
| SAM | Southern Annular Mode |
| SLP | Sea-level pressure (Pa) |
| $S_m$ | Soil moisture (metres of water) |
| SOI | Southern Oscillation index |
| $S_r$ | Mean total solar irradiance at a distance $r$ from the Sun |
| Sv | Sverdrup (=$10^6$ m$^3$ s$^{-1}$) |
| $T$ | Temperature (K) |
| $t$ | Time |
| $T_a$ | Air temperature (K) |
| TCR | Transient climate response |
| $T_e$ | Effective emission temperature of the Earth (K) |
| $T_s$ | Surface temperature (K) |
| TSI | Total solar irradiance (W m$^{-2}$) |
| $U$ | Internal energy per unit mass (J kg$^{-1}$) |
| $U_a$ | Wind velocity (m s$^{-1}$) |
| $\vec{U}_{ice}$ | Horizontal velocity vector for ice (m s$^{-1}$) |
| $\vec{V}$ | Velocity vector |
| W | Watt |
| WMO | World Meteorological Organisation |
| yr | Year |
| $z$ | Altitude or depth, measured from the bottom upwards |

# Description of the Climate System and Its Components

## OUTLINE

This first chapter describes the main components of the climate system as well as some processes that will be necessary to understand the mechanisms analysed in the chapters that follow. Complementary information is available in the Glossary for readers not familiar with some of the notions introduced here.

## 1.1 Introduction

*Climate* is traditionally defined as a description in terms of the mean and variability of relevant atmospheric variables such as temperature, precipitation and wind. Climate thus can be viewed as a synthesis or aggregate of weather. This implies that portrayal of the climate in a particular region must contain an analysis of mean conditions, of the seasonal cycle and of the probability of extremes such as severe frost, storms and so on. In accordance with the standard of the World Meteorological Organisation (WMO), thirty years is the classic period for performing the **statistics** used to define climate. This is well adapted for studying recent decades because it requires a reasonable amount of data along with a good sample of the different types of weather that can occur in a particular area. However, when analysing the more distant past, such as the last glacial maximum around 21,000 years ago, climatologists are often interested in variables which are characteristic of longer time intervals. As a consequence, the thirty-year period proposed by the WMO should be considered more as a practical indicator than as a norm that must be followed in all cases. This definition of climate as representative of conditions over several decades should not, of course, obscure the fact that climate can change rapidly. Nevertheless, a substantial time interval is needed to observe a difference in climate. In general, the smaller the difference between two periods, the longer is the time required to confidently identify any climate changes between those periods.

We also must take into account the fact that the state of the atmosphere used in the preceding definition of climate is influenced by numerous processes involving not only the atmosphere but also the oceans, sea ice, vegetation and so on. Climate is therefore now defined with increasing frequency in the wider sense of a description of the **climate system**. This includes an analysis of the behaviour of

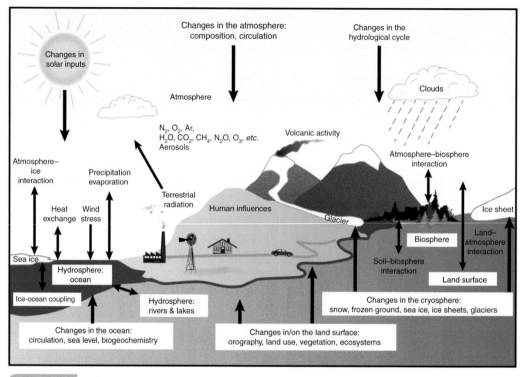

**Fig. 1.1**    Schematic view of the components of the climate system and their potential changes. (Source: IPCC 2007).

its five major components – the atmosphere (the gaseous envelope surrounding the Earth), the **hydrosphere** (liquid water, i.e., oceans, lakes, underground water, etc.), the **cryosphere** (solid water, i.e., **sea ice**, glaciers, ice sheets, etc.), the land surface and the **biosphere** (all the living organisms) – and of the interactions between them (IPCC 2007) (Figure 1.1). Here we will use the word 'climate' to refer to this wider definition. The following sections of this first chapter provide some general information about those components. Note that the climate system itself is often considered to be part of the broader **Earth System**, which includes all the parts of the Earth, not only the elements that are directly or indirectly related to the temperature or precipitation.

## 1.2   The Atmosphere

### 1.2.1   Composition and Temperature

**Dry air** is mainly composed of nitrogen (78.08% by volume), oxygen (20.95% by volume), argon (0.93% by volume) and to a lesser extent carbon dioxide (395 **ppm** or 0.0395% by volume in 2013) (see Section 2.3). The remaining fraction is made up of various trace constituents such as neon, helium, methane and krypton.

In addition, a highly variable amount of water vapour is present in the air. This water content can be measured by the mixing ratio $w$, defined as the ratio between the mass of vapour and the mass of dry air; the **specific humidity** $q$, given by the ratio between the mass of vapour and the mass of air (i.e., including water vapour); or the partial pressure of the water vapour $e$, all these variables being, of course, related. When the water content is high enough to reach condensation in equilibrium conditions, the water vapour pressure is equal to the **saturation vapour pressure** $e_s$, and the mixing ratio is by definition $w_s$. The **relative humidity** is then given by the ratio $w/w_s$. It is nearly identical to $e/e_s$, which is also used directly to define the relative humidity. $e_s$ can be expressed using the **Clausius-Clapeyron equation**, which shows that the amount of water vapour in the air at saturation strongly depends on temperature. For instance, the amount of water vapour that can be present in the atmosphere before saturation at a temperature of 20°C is more than three times higher than at 0°C. Consequently, the relative volume of water vapour in the air is close to 0% in the driest and coldest parts of the atmosphere but can reach 5% in hot regions at or close to saturation. On average, water vapour accounts for 0.25% of the mass of the atmosphere (Wallace and Hobbs 2006).

In nearly all the cases studied in climatology, the dry air and water vapour can be considered in good approximation as ideal gases (also called 'perfect gases'). The density, temperature and pressure thus are related through the **equation of state** of perfect gases, also referred to as the 'ideal gas law'

$$p = \rho R_g T \qquad (1.1)$$

where $p$ is the pressure, $\rho$ is the density, $R_g$ is the gas constant (which is equal to 287.0 J K$^{-1}$ kg$^{-1}$ for dry air) and $T$ is the temperature.

On a large scale, the atmosphere is very close to **hydrostatic equilibrium**, meaning that at a height $z$, the force due to the pressure $p$ on a 1-m$^2$ horizontal surface balances the force due to the weight of the air above $z$. The atmospheric pressure thus is at its maximum at the Earth's surface, and the surface pressure $p_s$ is directly related the mass of the whole air column at a particular location. Pressure then decreases with height, following approximately an exponential law:

$$p \simeq p_s e^{-z/H} \qquad (1.2)$$

where $H$ is a scale height (which is generally between 7 and 8 km for the lowest 100 km of the atmosphere). As a result of this clear and monotonic relationship between height and pressure, pressure is often used as a vertical coordinate for the atmosphere. Indeed, pressure is easier to measure than height, and choosing a pressure coordinate simplifies the formulation of some equations.

The temperature in the **troposphere**, roughly the lowest 10 km of the atmosphere, generally decreases with height. The rate of this decrease is called the 'lapse rate' $\Gamma$:

$$\Gamma = -\frac{\partial T}{\partial z} \qquad (1.3)$$

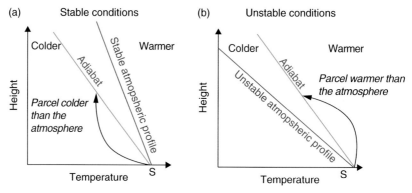

**Fig. 1.2** Schematic illustration of the importance of the adiabatic lapse rate for the stability of the air. (a) If the air parcel following an adiabatic uplift becomes colder and thus denser than the surrounding air, it will tend to go downward, and the atmosphere will be stable. (b) If an air parcel becomes warmer and thus lighter than the surrounding air, it will continue to move upward, and the atmosphere will be unstable.

where $T$ is the temperature. Its global mean value is about 6.5 K km$^{-1}$, or 0.0065 K m$^{-1}$, but $\Gamma$ varies with location and season. The lapse rate depends mainly on the radiative processes in the atmosphere (see Section 2.1) and on the vertical exchanges in the air column (the resulting balance is often referred to as the 'radiative-convective equilibrium') but also on the horizontal heat transport.

The observed lapse rate determines the vertical stability of the atmosphere (Figure 1.2). If an air parcel moves upward because of a perturbation, its temperature does not remain constant: as pressure decreases with height, the parcel expands and thus cools. For dry air, this decrease in temperature with height, which is only due to expansion without any additional exchange of heat with the surrounding air, is referred to as the 'dry **adiabatic** lapse rate'. It has a value of 9.8 K km$^{-1}$.

For an observed lapse rate which is locally smaller than the adiabatic lapse rate (Figure 1.2a), the air parcel after an upward vertical shift will be colder than the environment and thus denser [Eq. (1.1)]. It will tend to move back to its original position, inhibiting vertical movements and leading to a stable atmosphere. Negative lapse rates (i.e., temperature increasing with height), called 'temperature inversions', therefore correspond to highly stable conditions. By contrast, if the observed lapse rate is locally larger than the adiabatic lapse rate (Figure 1.2b), an air parcel displaced upward will be warmer than the environment and less dense. Consequently, it will be further entrained upward, leading to instability, **convection** and mixing between the air parcels at different altitudes until the profile becomes identical to the adiabatic profile, restoring the stability of the air column.

If the air includes water vapour, an additional term has to be taken into account as the cooling during raising motions may induce condensation (and cloud formation) following the Clausius-Clapeyron equation. The **latent heat** released by this condensation partly compensates for the cooling that is due to the expansion. Consequently, the so-called saturated adiabatic lapse rate is lower than the dry adiabatic lapse rate. Its value depends on the temperature and pressure. It is lower

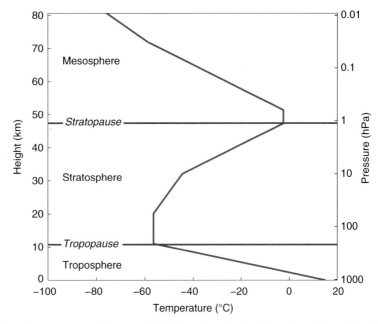

**Fig. 1.3**   Typical vertical temperature profile as given by the International Standard Atmosphere. (Source: From V. Zunz; reproduced with permission.)

at higher temperature because the air may contain more water vapour and is typically between 4 and 7 K km$^{-1}$. Convection leading to condensation is referred to as 'moist convection'.

Although complex mechanisms are involved, an important point to note is that an atmosphere in radiative equilibrium is unstable, in particular, because of the warming at the surface (see Section 2.1.6). Consequently, the air close to the surface is generally less dense than above and tends to rise. The convection processes thus are very important for the vertical structure of the atmosphere, and in many regions, particularly in the tropics, the observed lapse rate is very close to the saturated adiabatic lapse rate. The lapse rate is also involved in **feedbacks**, playing an important role in the response of the climate system to a perturbation (see Section 4.2.1).

At an altitude of about 10 km, a region of weak vertical temperature **gradients**, called the '**tropopause**', separates the troposphere from the **stratosphere**, where the temperature increases with height until the stratopause at around 50 km (Figure 1.3). The stratosphere therefore is generally very stable, being subject to conditions similar to temperature inversions in the troposphere. Above the stratopause, temperature decreases strongly as height increases in the mesosphere, until the mesopause is reached at an altitude of about 80 km, and then increases again in the thermosphere above this height.

The vertical temperature gradients above 10 km are strongly influenced by the absorption of solar **radiation** by different atmospheric constituents and by chemical reactions driven by the incoming light. In particular, the absorption

(a)

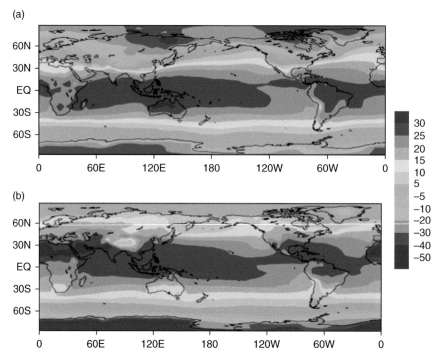

(b)

**Fig. 1.4** Surface air temperature (in °C) averaged over (a) December, January and February and (b) June, July and August. (Source: Data from Brohan et al. 2006.)

of ultraviolet (UV) radiation by stratospheric **ozone**, which protects life on Earth from this dangerous radiation, plays a critical role in stratospheric warming.

Atmospheric **specific humidity** also displays a characteristic vertical profile with maximum values in the lower levels and a marked decrease as height increases. As a consequence, the air above the tropopause is nearly dry. This vertical distribution is mainly due to two processes. First, the major source of atmospheric water vapour is evaporation at the surface. Second, the warmer air close to the surface can contain a much larger quantity of water before saturation occurs than the colder air further aloft; saturation leads to the formation of water or ice droplets, clouds and eventually precipitation.

At the Earth's surface, the temperature reaches its maximum in equatorial regions (Figure 1.4) because of the higher incoming solar radiation in annual mean (see Section 2.1). In those regions, the temperature is relatively constant throughout the year. Given the much stronger seasonal cycle at middle and high latitudes, the temperature gradient between the equator and the polar regions is much larger in winter than in summer. The distribution of surface temperature is influenced by atmospheric and oceanic heat transport as well as by the thermal inertia of the ocean (see Section 2.1.5). Furthermore, the role of topography is important, with a temperature decrease at higher altitudes associated with the positive lapse rate in the troposphere.

## 1.2.2 General Circulation of the Atmosphere

Not only is **convection** important for the vertical structure of the atmosphere (Section 1.2.1), but it is also responsible for horizontal movements on every scale from local or regional to global. The high temperatures at the equator make the air there less dense. It thus tends to rise before being transported poleward at high altitudes in the **troposphere**. This motion is compensated for at the surface by an equator-ward transport of air. On a motionless Earth, this big convection cell would reach the poles, inducing direct exchanges between the warmest and coldest places on Earth. However, owing to the Earth's rotation, such an atmospheric structure would be unstable (see, e.g., Marshall and Plumb 2008). Consequently, the two cells driven by the **ascendance** at the equator, called the '**Hadley cells**', close with a downward branch at about 30° latitude (Figure. 1.5). The poleward boundaries of these cells are marked by strong westerly winds in the upper troposphere called the 'tropospheric jets' or 'tropospheric jet streams'. At the surface, the Earth's rotation is responsible for a deflection of the flow coming from the middle latitudes to the equator towards the right in the northern hemisphere and towards the left in the southern hemisphere (due to the **Coriolis force**). This gives rise to the easterly **trade winds** characteristic of the tropical regions (Figure 1.6).

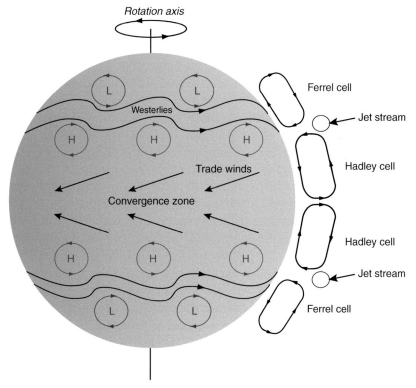

**Fig. 1.5** Schematic representation of the annual mean general atmospheric circulation. H (L) represents high (low) pressure systems.

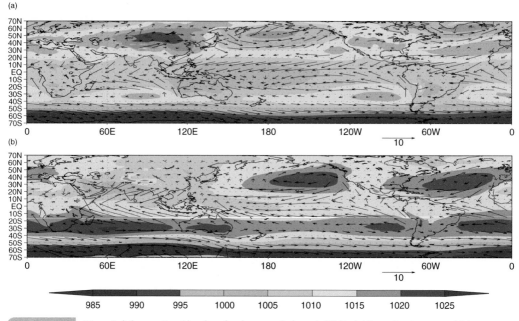

**Fig. 1.6**    10-m winds (arrows, in m/s) and sea-level pressure (colours, in hPa) in (a) December, January and February and (b) June, July and August. (Source: From Wouter Lefebvre using data from NCEP/NCAR **reanalyses**; Kalnay et al. 1996.)

The convergence of these surface winds and the resulting ascendance do not occur exactly at the equator but in a band called the '**Intertropical Convergence Zone**' (ITCZ). Presently, it is located around 5°N on average, with some seasonal shifts.

At the surface, the circulation at middle latitudes is characterised by westerly winds. Their **zonal** symmetry is perturbed by large wavelike patterns caused by an instability in the flow called the '**baroclinic instability**'. Such disturbances are associated with the formation of low- and high-pressure systems that govern the day-to-day variations in the weather in these regions. The dominant feature of the **meridional** circulation at these latitudes is the 'Ferrell cell', which is weaker than the Hadley cell. It is characterised by rising motion in its poleward branch and downward motion in the equator-ward branch. This contrasts with the Hadley cell, which is driven by convection and ascendance in the warmest regions near the equator. As a consequence, the Hadley cell is termed a 'direct cell', whereas the Ferrell cell is termed an 'indirect cell'.

Outside a narrow equatorial band and above the layer that interacts directly with the surface (the so-called **surface boundary layer)**, the large-scale atmospheric circulation is close to **geostrophic equilibrium**. This means that surface pressure and winds are closely related. In the northern hemisphere, the winds rotate clockwise around a high pressure and counter-clockwise around a low pressure, whilst the reverse is true in the southern hemisphere. Consequently, the middle-latitude westerlies are associated with high pressure in the subtropics and low pressure at around 50 to 60° in both hemispheres. Rather than a continuous structure, this subtropical high-pressure belt is characterised by distinct high-pressure centres,

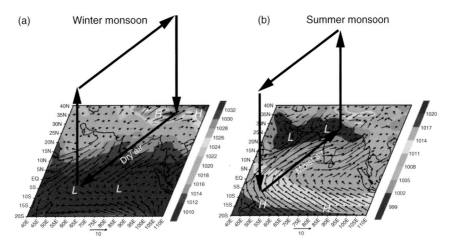

Fig. 1.7 10-m winds (arrows, in m/s) and sea-level pressure (colours, in hPa) in (a) January and (b) July illustrating the wind reversal between the winter and summer monsoons (zoom from Figure 1.6). The arrows represent a schematic view of the three-dimensional overturning circulation. The symbols *L* and *H* indicate the location of low and high pressures. (Source: NCEP/NCAR reanalyses; Kalnay et al. 1996.)

often referred to by the name of a region close to their maximum (e.g., 'Azores high' and 'St. Helena high'). In the northern hemisphere, low pressures at around 50 to 60°N manifest on **climatological** maps as cyclonic centres called, for example, the 'Icelandic low' and the 'Aleutian low'. In the Southern Ocean, because of the absence of large land masses in the corresponding band of latitude, the pressure is more zonally homogeneous, with a minimum surface pressure at around 60°S.

The preceding discussion briefly mentioned the potential impact of the presence of continents on the large-scale circulation, but the role of land surfaces becomes critical in explaining **monsoons** (Figure 1.7). In summer, the continents warm faster than the oceans because of their lower thermal inertia (see Section 2.1.5). The warming of the air close to the surface is associated with a decrease in pressure there whilst the surface pressure is higher over the ocean. This pressure difference between land and sea then induces the transport of moist air from the sea to the land. In winter, the situation reverses, with high pressure over the cold continent and a surface flow generally from land to sea. The monsoon circulation therefore exhibits clear similarities to the Hadley cell, both being driven by thermal differences, and also can be referred to as a 'direct circulation'.

Such a monsoon circulation, with seasonal reversals of wind direction, is present in many tropical areas of Africa, Asia and Australia. Nevertheless, the most famous monsoon is probably the South Asian one, which is a major feature of the Indian sub-continent climate (see the reversal of the winds in this region between Figures 1.7a and 1.7b).

## 1.2.3 Precipitation

Precipitation and temperature are the most important variables in defining the climate of a region. Precipitation is strongly influenced by the large-scale atmospheric circulation that transports water vapour horizontally and vertically. In particular, vertical movements are responsible for large temperature variations that play an important role in the condensation processes and therefore in precipitation.

In the upward branch of the **Hadley cell** along the ITCZ, the cooling of warm and moist surface air as it rises leads to condensation and heavy precipitation. For instance, the Western Tropical Pacific receives more than 3 m of rainfall per year. By contrast, the downward branch of the Hadley cell in the subtropics is associated with the **subsidence** of relatively dry air from the upper levels of the troposphere and thus very low precipitation rates. As a consequence, most of the large deserts on Earth are located in the subtropical belt.

The **monsoon** has a significant impact upon the precipitation over subtropical continents. During the winter monsoon, the inflow of dry continental air is associated with low precipitation. However, the summer brings moist air from the ocean which induces rainfall up to several metres in a few months.

The topography also plays a large role as it can generate significant vertical motion. Where the **ascendance** of moist air is topographically induced, massive precipitation can occur, as it does on the slopes of the Himalaya during the summer monsoon. By contrast, the subsidence of dry air, generated, for instance, because of the presence of mountains nearby, will tend to suppress precipitation, contributing to the occurrence of deserts. Mountains are also barriers to moist air coming from oceanic regions. Within this framework, the distance from the oceanic source also must be taken into account when studying the precipitation regime in a region. This explains why, for example, there is less rainfall in central Asia than in Western Europe at the same latitude.

Notable features are also present over the ocean, for instance, the South Pacific Convergence Zone (SPCZ) associated with the high precipitation rates in a northwest–southeast band from Indonesia towards 30°S, 130°W. In the middle latitudes, precipitation in winter is mainly due to **cyclones**, which tend to follow a common path at about 45°N in the Pacific and the Atlantic. This **storm track** manifests as maximum rainfall in this region. These effects are visible in the precipitation maps reproduced in Figure 1.8.

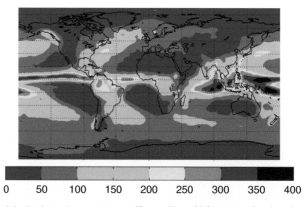

0    50    100    150    200    250    300    350    400

**Fig. 1.8**    Annual mean precipitation in centimetres per year. (Source: Xie and Arkin 1997 and updates.)

## 1.3    The Ocean

### 1.3.1 Composition and Properties

The ocean covers about 71% of the Earth's surface and has an average depth of roughly 3700 m. Seawater is composed of 96.5% water and 3.5% dissolved salts, particles, gases and organic matter. The most important of these components are chloride and sodium, which represent about 85% of the dissolved material. Although the total quantity of dissolved salts varies from place to place, their relative contribution is quite stable in sea water. Rather than specifying each of the components, it is thus very convenient to define a 'bulk salinity' as the total amount of dissolved material (in grams) in a kilogram of sea water. This salinity is then given in parts per thousand (‰) or in grams per kilogram and is a key variable in oceanography. It influences many of the properties of seawater, such as the density or the freezing-point temperature, which decreases at the surface from 0°C for pure freshwater to typically −1.8°C in polar oceans.

In current practice, the salinity measurements are based on the conductivity of sea water. As the correspondence between the total salt content and the conductivity is not easily assessed with a high level of precision, it was recommended that the salinity be given in practical salinity units (psu) based on a ratio of conductivity between the sample analysed and a reference. Additionally, a new standard has been proposed, based on more accurate measurements, in which absolute salinities are given as 'the mass fraction of dissolved solute in so-called Standard Seawater with the same density as that of the sample' (IOC, SCOR and IAPSO 2010). This reference to a 'Standard Seawater' allows small changes in the composition of the salts in different regions to be taken into account. Different choices of unit thus can be seen in the scientific literature, but the resulting salinity values are similar, so no distinction will be made here. The salinity will be expressed in practical salinity units (psu) in the remaining part of this text, following a practice that is still common for databases and model results.

In contrast to the air **equation of state** (see Section 1.2.1), the equation of state for seawater, which gives the density as a function of pressure, temperature and salinity, is a complex non-linear expression (IOC, SCOR and IAPSO 2010). It shows that the density increases with salinity as well as with pressure, whilst it decreases with increasing temperature. In simple terms, temperature is often considered to dominate the density changes at high temperatures, whilst salinity plays a greater role at low temperature. This larger contribution of salinity changes is particularly clear in sea-ice-covered regions as surface-temperature variations are very small, being constrained to be close to the freezing point. Density changes are only of the order of a few percent compared to a mean value of about 1035 kg m$^{-3}$, but those small density differences are very important for the ocean circulation, as we will see in the next section. If the density is known, it is then possible to estimate the pressure as the ocean can also be considered to be in **hydrostatic equilibrium** on a large scale. This results in a nearly linear increase in pressure with depth.

### 1.3.2  Oceanic Circulation

The oceanic circulation is driven by the wind stress at the surface and by density differences. In the top hundreds of metres, the influence of the wind is dominant, as illustrated by the good correspondence between the patterns of surface winds and surface currents. The link between these two variables, however, is more subtle than expected at first sight [see subsequent paragraphs and Marshall and Plumb (2008) for more details]. The atmospheric westerlies at middle latitudes correspond to eastward currents in the ocean (take care to note that the standard convention for the direction of flow is different in oceanography, where the flow is towards the east, and atmospheric science, where the flow comes from the west). In the tropics, westward currents are roughly parallel with the **trade** winds. Owing to the presence of continental barriers, those currents form loops called 'subtropical **gyres**' (Figure 1.9). The surface flow in those gyres intensifies along the western boundaries of the oceans (the east coasts of continents), inducing well-known strong currents such as the **Gulf Stream** off the east coast of the United States and the Kuroshio off the coast of Japan (Figure 1.10). These currents have velocities higher than 1 m s$^{-1}$, whilst the speed in the interior of the subpolar gyres is typically less than 0.1 m s$^{-1}$.

At high latitudes, weaker subpolar gyres are present in each oceanic basin of the northern hemisphere. By contrast, because of the absence of continental barriers in the Southern Ocean, a current which connects all the ocean basins can be maintained: the Antarctic Circumpolar Current (ACC). This is one of the strongest currents on Earth, which transports about 130 Sv eastward (1 Sverdrup = $10^6$ m$^3$ s$^{-1}$). In addition to these currents, which are basically parallel to the surface winds, the equatorial counter-currents run in the opposite direction to the trade winds, at or just below the surface, in all the ocean basins. It should be noted that this description of the surface currents is based on an average of ocean movements over several years to decades. The instantaneous flow is characterised by large departures from this picture, in particular, involving roughly circular structures called 'ocean **eddies**', whose typical scale is of the order of 10 to 100 km.

As a result of the Earth's rotation, the ocean transport induced by the wind is perpendicular to the wind stress (to the right in the northern hemisphere, to the

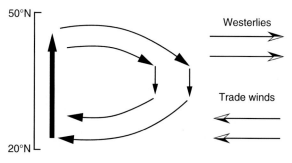

**Fig. 1.9**  Schematic representation of the wind-driven circulation at middle latitudes characterised by gyres and a strong current at the western boundary.

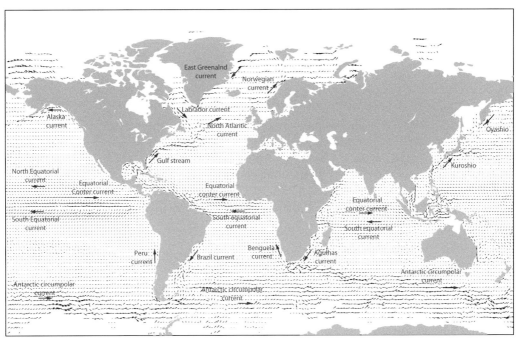

**Fig. 1.10**   Major surface currents as illustrated by the velocity derived from satellite-tracked surface-drifting buoy observations (black arrows). (Source: From V. Zunz using data from Lumpkin and Garraffo 2005; available at: http://www.aoml.noaa.gov/phod/dac/drifter_climatology.html.)

left in the southern hemisphere). This transport is known as the '**Ekman transport**'. Its horizontal variations lead to surface convergence/divergence which has to be compensated by vertical movements in the ocean. An important example of this process is the equatorial **upwelling** (Figure 1.11a). In the northern hemisphere, the Ekman transport is directed to the right of the easterly winds and thus moves northward. By contrast, the wind-driven transport is on the left of the wind and southward in the southern hemisphere. This results in a divergence of the water motion at surface at the equator. At equilibrium, this flow away from the equator in both hemispheres is balanced by an upward motion referred to as the 'equatorial upwelling'. Similar processes may lead to coastal upwelling if the wind stress is parallel to the coast, with the coast on the left when looking in the direction of the wind in the northern hemisphere (e.g., southerly winds along an East Coast oriented north-south). In this configuration, the wind causes an offshore transport and an upwelling to compensate for this transport (Figure 1.11b). The Ekman transport also has much wider consequences as, to a large extent, it explains the path of wind-driven currents. It is outside the scope of this book to describe the processes in detail, but we can briefly state that the vertical movements induced by horizontal variations of the Ekman transport modify the elevation of the surface of the ocean by up to a few tens of centimetres. This alters the pressure in the ocean, leading to wind-driven currents that are in **geostrophic equilibrium,** the sea-surface elevation playing more or less the same role as the surface pressure in the atmosphere (see Section 1.2.2).

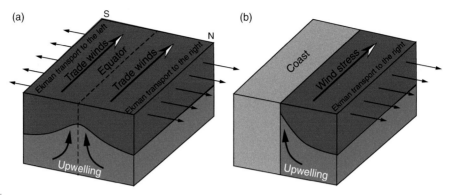

**Fig. 1.11** Schematic representation of (a) the equatorial upwelling and (b) a coastal upwelling in the northern hemisphere.

At high latitudes, because of its low temperature and relatively high salinity, surface water can be dense enough to sink to great depths. This process is only possible in a few places in the world. In the North Atlantic, the salty water transported northward by the Gulf Stream and the North Atlantic current system is cooled by exchanges at the ocean surface and becomes denser, leading to the formation of North Atlantic Deep Water (NADW) in the Labrador and Greenland-Iceland-Norwegian seas. NADW flows southward along the western boundary of the Atlantic towards the Southern Ocean. There it is transported to the other oceanic basins after some mixing with ambient water masses. The deep water then slowly up-wells towards the surface in the different oceanic basins, mixing in the ocean interior also playing a central role in the circulation. This is represented schematically in Figure 1.12 by localised upward fluxes in the Indian, Pacific and Southern oceans. Nevertheless, whilst sinking occurs in very small regions, the upwelling is broadly distributed throughout the ocean. The return flow to the sinking regions is achieved through surface and intermediate-depth circulation.

In winter, dense water is found close to Antarctica on the **continental shelf** of the Southern Ocean, in particular, in the Weddell and Ross seas. Indeed, the temperature is at the freezing point and the salinity is high because of the brine rejection during sea ice formation (see Section 1.4). Given its high density, this dense shelf water sinks along the continental slope, leading to the formation of Antarctic Bottom Water (AABW). AABW is colder and denser than NADW and so flows below it. Note that because of the mixing of **water masses** of different origins in the Southern Ocean, the water that enters the Pacific and Indian basins is generally called 'Circumpolar Deep Water' (CDW).

The **Meridional Overturning Circulation** (MOC), resulting from the formation of dense waters in polar regions and their export, is thus associated with currents at all depths (Figure 1.12). It is often called the oceanic '**thermohaline circulation**' because it is apparently driven by temperature and salinity (and thus density) contrasts. However, the mechanical forcing responsible for the mixing in the ocean interior and the winds play a significant role, in particular, for the energy supply

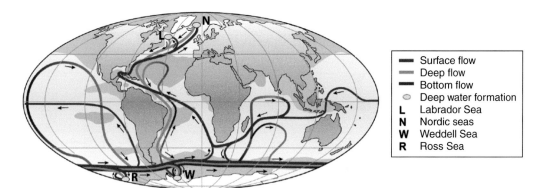

**Fig. 1.12**   Schematic representation of the oceanic thermohaline circulation illustrating the surface (red), deep (blue) and bottom (magenta) flow. The main regions of deep-water formation are indicated by a small yellow circle, whilst the areas characterised by a sea-surface salinity which is higher than 36 are in green and lower than 34 in light blue (see Figure 1.13). (Sources: Rahmstorf 2002 and Kuhlbrodt et al. 2007. Reprinted by permission from Macmillan Publishers, Ltd., *Nature*; copyright 2002.)

for the overturning circulation (see, e.g., Kuhlbrodt et al. 2007). The winds shape the surface circulation and thus the upper branch of the thermohaline circulation which feeds the regions where sinking occurs. Owing to the divergence of the **Ekman transport**, the winds influence the **upwelling** of deep-water masses towards the surface in some regions. This is particularly important in the Southern Ocean. Finally, winds also could act as a local/regional pre-conditioning factor which favours deep **convection** and the deep-water formation.

The thermohaline circulation is quite slow. At the surface, its role is lesser than that of wind-driven circulation but is not negligible. Besides, at depth, where it is the dominant contributor to mean velocities, the circulation is generally quite sluggish, with mean velocities typically 10 to 100 times smaller than at surface. The time needed for water masses formed in the North Atlantic to reach the Southern Ocean is of the order of a century. If the whole cycle is taken into account, the **time scale** is estimated as between several centuries and a few millennia depending on the exact location and mechanism studied. However, this circulation transports huge amounts of water, salts and energy. In particular, the rate of NADW formation is estimated to be around 15 Sv. Uncertainties are larger for the Southern Ocean, but the production rate of AABW is likely close to that of NADW. As a consequence, the thermohaline circulation has an important oceanographic as well as climatological role (see Section 2.1.5).

## 1.3.3 Temperature and Salinity

### 1.3.3.1 Surface Layer

Because of the strong interactions between the ocean and the atmosphere, the sea surface temperature (SST) (Figure 1.13) is close to the temperature of the air above it (Figure 1.4). One exception is the polar regions, where **sea ice** (see

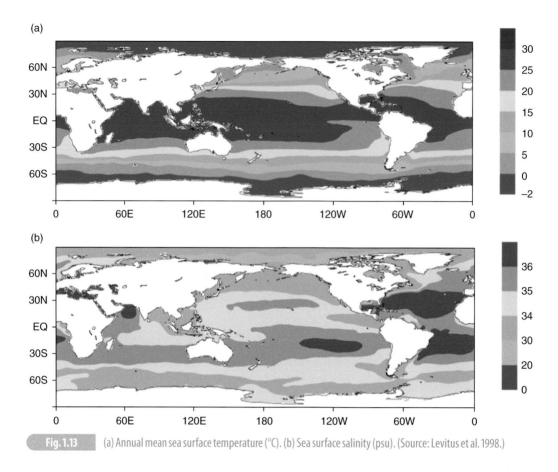

**Fig. 1.13**    (a) Annual mean sea surface temperature (°C). (b) Sea surface salinity (psu). (Source: Levitus et al. 1998.)

Section 1.4) insulates the ocean from the cold atmosphere. Furthermore, the SST has a seasonal cycle at middle and high latitudes which is shifted by one to three months compared to that over land and which has a weaker amplitude because of the large thermal inertia of the ocean (see Section 2.1.5).

The main characteristic of the annual mean SST distribution is a decrease in temperature as latitude increase, but there are also large longitudinal contrasts. In particular, the Eastern Tropical Pacific is colder than the so-called warm pool in the West Tropical Pacific, where temperatures exceed 28°C (see Section 5.2.1); the North East Atlantic is warmer than the North West Atlantic because of the heat transported by ocean currents (Figure 1.10; see also Section 2.1.5.2).

The sea surface salinity is strongly influenced by the freshwater fluxes at the surface: a net inflow of freshwater dilutes the salt present in oceanic waters, leading to a decrease in salinity; a net evaporation removes water from the surface, whilst salt remains in the ocean, increasing the salinity. As a consequence, the salinity reaches a maximum in subtropical areas because of the significant evaporation and low rainfall there. The high precipitation rates induce lower salinity at the equator, whilst weak evaporation is responsible for the lower salinity observed at middle and high latitudes. River input also has a large regional impact. This is clearly seen in Figure 1.13, with low values close to the mouths of the Amazon

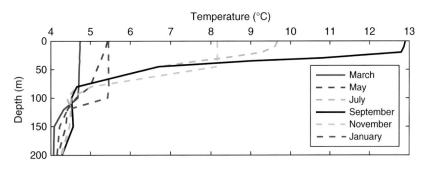

Monthly mean temperature profiles illustrating typical growth and decay of the seasonal thermocline at a middle-latitude site in the northern hemisphere (50°N, 145°W). Note that the mixed layer is not well defined in May for those averaged profiles. (Source: OCS Project Office of NOAA/PMEL; available at: http://www.pmel.noaa.gov/OCS/data/disdel_v2/disdel_v2.html.)

and Mississippi rivers as well as in the Arctic, which receives the freshwater input from several major rivers. Another important characteristic is the higher salinity in the Atlantic compared to the Pacific, explaining why water is dense enough in the North Atlantic to form deep water but not in the North Pacific.

The top tens of metres closest to the surface of the ocean are vertically homogeneous, forming what is called the '**ocean mixed layer**' (Figure 1.14). At middle and high latitudes, the mixed layer is deeper in winter, with depths generally of about 50 to 100 m but which can reach several hundred metres in some regions. This is due first to the mechanical stirring by the winds which generates turbulence and therefore mixing. Second, cooling and salinity increase at the surface destabilise the water column and generate shallow **convection** which induces a homogenisation of the surface and the deeper layers (Figure 1.15). When the temperature rises in spring and summer, the density at the surface decreases, stabilising the water column and inhibiting convection. As the winds also tend to be weaker in spring and summer, generating less turbulence, the mixed layer becomes shallower. The warming is thus concentrated in a shallow layer, whose depth is generally lower than 40 m. Below this summer mixed layer, the temperature is insulated from the surface and so retains the properties it has acquired through contact with the atmosphere in winter (see also Section 1.3.3.2). This seasonal process induces the formation of a region with strong vertical **gradients** at the base of the summer mixed layer referred to as the 'seasonal **thermocline**' (Figure 1.14). In the autumn, the mixed layer deepens again, entraining thermocline waters back in the mixed layer. This sequence of processes is focussed on the contribution of the winds and heat fluxes in the seasonal development of the mixed layer, but density and therefore stability changes associated with salinity variations also should be taken into account (Figure 1.15).

The mixed-layer dynamics, and in particular, the seasonal changes in its depth, have a considerable influence on the surface ocean properties and on the exchanges of heat, water and gases between the ocean and the atmosphere (see Section 2.1). Its development also has a large impact on the growth of phytoplankton, which is the basis of the whole oceanic food web. In order for photosynthesis to occur,

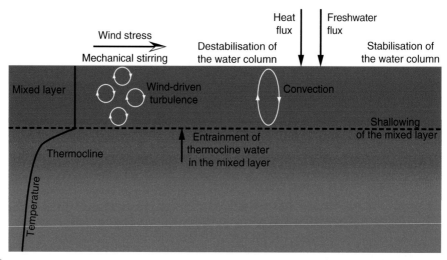

**Fig. 1.15**    Schematic illustration of the processes governing the time development of the mixed-layer depth.

phytoplankton needs light, which is only available close to the surface. If the mixed layer is deep, as it is in winter, the phytoplankton is mixed over a large depth range by surface turbulence and thus spends a large part of its time in the dark, deep levels. Additionally, the fluxes of solar radiation at the surface are relatively low, explaining why biological production is weak in winter. By contrast, the availability of light for phytoplankton is greater in summer given the shallow mixed layer and the large amount of incoming solar radiation. However, such a shallow mixed layer limits the exchanges between the surface and the deep waters, which are rich in the nutrients required for phytoplankton growth (see Section 2.3). The summer concentration of those nutrients is therefore generally too low to sustain a significant biological production. As a consequence, phytoplankton growth often reaches its maximum during spring **blooms**. The mixed layer is also relatively shallow during this period, but the nutrient concentration is high enough, thanks to the exchanges with deeper layers which occurred during the previous winter.

### 1.3.3.2  Intermediate and Deep Layers

The temperature and salinity of sea water are strongly modified at the surface by interactions with the atmosphere, leading to the formation of various **water masses.** The mixed layer is the area where the greatest mixing occurs, diffusion being weaker in the ocean interior. When these waters masses flow beneath the mixed layer, they tend to keep the properties they have acquired close to the surface. This is particularly clear in the deep ocean. As a consequence, the path of important water masses, such as the NADW and AABW, can easily be followed from their region of formation on vertical sections showing temperature and salinity (Figures 1.16 and 1.17). The influence of Antarctic Intermediate Water (AAIW), originating from the Southern Ocean, is also clearly identified as a low-salinity tongue reaching the equator at intermediate depth.

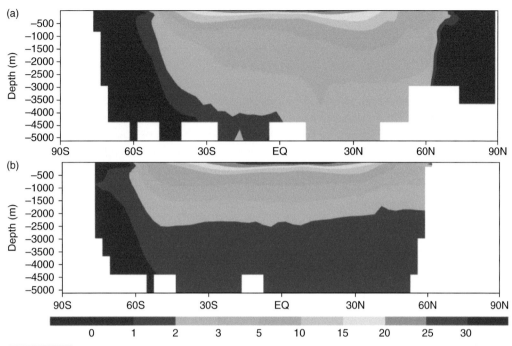

Fig. 1.16 Temperature (°C) averaged over all latitudes (i.e., zonal mean) in (a) the Atlantic and (b) the Pacific. (Source: Levitus et al. 1998.)

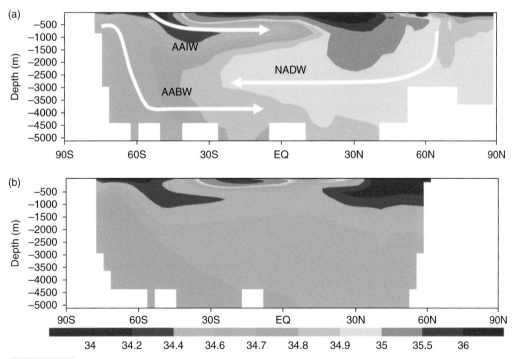

Fig. 1.17 Salinity (psu) averaged over all latitudes (i.e., **zonal** mean) in (a) the Atlantic and (b) the Pacific. The schematic paths of three important water masses are shown for the Atlantic. (Source: Levitus et al. 1998.)

More generally, below the mixed layer and the seasonal **thermocline**, most water properties originate from a nearby location (generally poleward), where surface density in winter is high enough to allow the water to sink to an intermediate depth. This depth depends on the density of the water and thus mainly on its temperature. It results in a strong vertical temperature **gradient** (except in some regions at high latitudes) which defines the permanent thermocline (Figures 1.16 and 1.17). In the deep ocean, the vertical gradients are much weaker. It might be considered surprising that near the equator the temperature difference between water at the surface and at a depth of 1000 m can be more than 20°C, whilst the temperature difference between water at a depth of 1000 m and the bottom of the ocean is only of 3°C. Nevertheless, this illustrates that the whole ocean is 'stratified', meaning that light water sits above dense water, as required by the vertical stability of the water column.

## 1.4 The Cryosphere

### 1.4.1 Components

The **cryosphere** is the portion of the Earth's surface where water is in solid form. It thus includes **sea ice**, lake and river ice, snow cover, **glaciers**, **ice caps** and **ice sheets** and frozen ground. Snow cover has the largest extent, with a maximum area of more than $45 \times 10^6$ km² (Table 1.1). Given the present distribution of the continents, land surfaces at high latitudes are much larger in the northern hemisphere than in the southern hemisphere. As a consequence, a large majority of the snow cover is located in the northern hemisphere (Figure 1.18). The same is true for the freshwater ice that forms on rivers and lakes in winter. Both the snow cover and freshwater ice have a very strong seasonal cycle as they nearly disappear in summer in both hemispheres.

Sea ice, which is a moving medium formed when sea water freezes, generally does not cover the whole oceanic surface in a given region. The 'sea-ice concentration' is defined as the fraction of the surface of interest (pixel from a satellite image, area surrounding a boat, etc.) which is effectively covered by sea ice. A concentration of ice of 1 (or 100%) thus corresponds to a continuous ice pack, whilst a value of 0 corresponds to open ocean. The relatively narrow elongated areas of open water inside the pack are called '**leads**', whilst larger areas of open water are called '**polynyas**'.

Sea ice covers a similar area in both hemispheres (Table 1.1). Its seasonal cycle is larger in the Southern Ocean (Figure 1.19) where the majority of the ice cover is first-year sea ice (i.e., sea ice that has not survived one summer). Owing to the large thermal inertia of the ocean (see Section 2.1.5), the minimum and maximum in sea ice extent are shifted by about two months compared to the snow cover

**Table 1.1**  Areal extent and volume of snow cover and sea ice for the late twentieth century

| Component | Maximum area ($10^6$ km$^2$) | Minimum area ($10^6$ km$^2$) | Maximum ice volume ($10^6$ km$^3$) | Minimum ice volume ($10^6$ km$^3$) |
|---|---|---|---|---|
| Northern hemisphere snow cover | 46.5 (late January) | 3.9 (late August) | 0.002 | Very small |
| Southern hemisphere snow cover | 0.83 (late July) | 0.07 (early May) | Very small | Very small |
| Sea ice in the northern hemisphere | 14.0 (late March) | 6.0 (early September) | 0.05 | 0.02 |
| Sea ice in the southern hemisphere | 15.0 (late September) | 2.0 (late February) | 0.02 | 0.002 |

*Source:* Data compiled in Climate and Cryosphere (CliC) Project Science and Co-ordination Plan (2001).

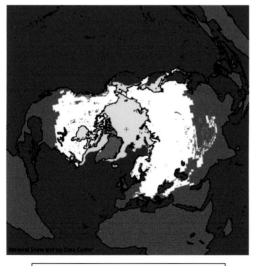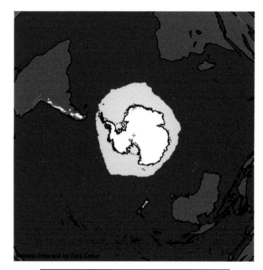

- ■ Glaciers and permanent land ice
- □ Snow extent, January (1967–2005)
- ■ Land
- ▨ Sea ice extent, January (1979–2005)

- ■ Glaciers and permanent land ice
- □ Snow extent, August (1987–2002)
- ■ Land
- ⠿ Antarctic ice shelves
- ▨ Sea ice extent, August (1979–2005)

**Fig. 1.18**  The distribution of sea ice, snow and land ice in January in the northern hemisphere and in August in the southern hemisphere. (Source: Atlas of the Cryosphere, National Snow and Ice Data Center (NSIDC); available at: http://nsidc.org/data/atlas/; Maurer 2007.)

Fig. 1.19 Location of the ice edge in March (green) and September (blue) in both hemispheres. The 'ice edge' is commonly defined as the line where the ice concentration is 15%. (Source: Rayner et al. 2003.)

on land, with maximum/minimum values around March and September in both hemispheres (Figure 1.19). The sea ice is thinner in the southern hemisphere, with a mean thickness of less than 1 m, whilst the mean ice thickness in the Central Arctic is around 3 m.

Like snow, the seasonally frozen ground covers a large fraction of the continents in the northern hemisphere. Where the annual mean temperature is below about −1°C, the ground can be perennially frozen below an active layer which is frozen in winter but melts in summer. This frozen ground is referred to as the '**permafrost**', which is estimated to cover more than 20% of the land area in the northern hemisphere (Table 1.2). The thickness of the frozen layer can exceed 500 m at high latitudes. Further south, this layer thins, and the permafrost becomes discontinuous close to its margins (Figure 1.20).

Most of the ice present on Earth today is located in two big ice sheets: the Greenland and Antarctic ice sheets. The Antarctic ice sheet is itself commonly divided into two parts, East Antarctica and West Antarctica, roughly corresponding to the eastern and western hemispheres relative to the Greenwich meridian. The thickness of the ice in these ice sheets can reach several kilometres (Figure 1.21). Ice sheets are formed by the accumulation of snow layers over tens of thousands of years. As snow falls on the surface, the pressure on the older snow layers increases, transforming them into ice. Ice sheets (like glaciers) are not stagnant and generally flow slowly towards their margins. However, in some regions (called 'ice streams'), the flow is much faster than in other parts of the ice sheet, sometimes reaching several kilometres per year.

The bedrock below ice sheets is depressed because of the weight of the ice and is, in some areas, well below sea level. This is the case for large areas of the West Antarctic ice sheet, with a total volume below sea level estimated at around $1.9 \times 10^6$ km$^3$. Most of the East Antarctic ice sheet is grounded above sea level. The volume of ice below sea level is also small in Greenland.

Antarctica is surrounded by **ice shelves**. These are floating platforms made of ice originating from the continent which has flowed down the coastline into the

| Component | Area (10$^6$ km$^2$) | Ice volume (10$^6$ km$^3$) | Sea-level equivalent (m) |
|---|---|---|---|
| Continuous permafrost | 10.69 | 0.0097–0.0250 | 0.024–0.063 |
| Discontinuous permafrost | 12.10 | 0.0017–0.0115 | 0.004–0.028 |
| East Antarctica | 10.1 | 22.7 | 56.8 |
| West Antarctica and Antarctic Peninsula | 2.3 | 3.0 | 7.5 |
| Greenland | 1.8 | 2.6 | 6.6 |
| Small ice caps and mountain glaciers | 0.68 | 0.18 | 0.5 |
| Ice shelves | 1.5 | 0.66 | – |

**Table 1.2** Areal extent and volume of permafrost and land ice for the late twentieth century

*Note*: The sea-level equivalent is computed as the thickness of a water layer corresponding to the ice mass distributed over the whole world ocean. This is not directly equal to the resulting sea-level rise as parts of the Antarctic and Greenland ice sheets are presently below sea level.

*Source*: Data compiled in the Climate and Cryosphere (Clic) Project Science and Co-ordination Plan (2001).

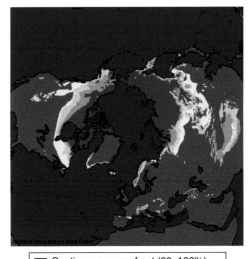

■ Continuous permafrost (90–100%)
■ Discontinuous permafraost (50–90%)
■ Sporadic permafrost (10–50%)
□ Isolated permafrost (0–10%)
■ Land

**Fig. 1.20** Location of permafrost in the northern hemisphere. (Sources: Atlas of the Cryosphere, National Snow and Ice Data Center (NSIDC); available at: http://nsidc.org/data/atlas/; Maurer 2007.)

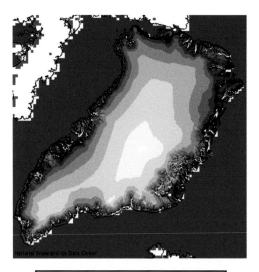

Filchner–
Ronne ice
shelf
Antarctic
Peninsula

West
Antarctica        East Antarctica

Ross ice
shelf

3250–3500 m, Greenland elevation
3000–3250 m, Greenland elevation
2500–3000 m, Greenland elevation
2000–2500 m, Greenland elevation
1500–2000 m, Greenland elevation
1000–1500 m, Greenland elevation
500–1000 m, Greenland elevation
0–500 m, Greenland elevation
Land

4100–5000 m, Antarctica elevation
3500–4150 m, Antarctica elevation
3000–3500 m, Antarctica elevation
2500–3000 m, Antarctica elevation
2000–2500 m, Antarctica elevation
1500–2000 m, Antarctica elevation
1000–1500 m, Antarctica elevation
500–1000 m, Antarctica elevation
100–500 m, Antarctica elevation
<100 m, Antarctica elevation
Land
Antarctic ice shelves

**Fig. 1.21**   Greenland and Antarctica surface elevation. To obtain the ice thickness, the bedrock elevation has to be subtracted from this figure. (Sources: Atlas of the Cryosphere, National Snow and Ice Data Center (NSIDC); available at: http://nsidc.org/data/atlas/; Maurer 2007.)

ocean. The two largest ice shelves are the Ross and Filchner-Ronne ice shelves, which together cover more than 850,000 km². Ice shelves and glaciers which reach the shore are able to release **icebergs**, a process called 'iceberg calving'. Icebergs can drift over long distances, pushed by the ocean currents and winds. They are thus found in the open ocean, but they should not be confused with sea ice. They are usually much thicker (sometimes more than 100 m) and consist of freshwater, whilst sea ice is salty and is formed directly from sea water.

## 1.4.2  Properties

Snow and ice have a large **albedo**; that is, they reflect the majority of the incoming solar radiation. They therefore play a major role in the global heat balance of the Earth. By storing and releasing **latent heat**, they affect the seasonal cycle of the surface temperature (see Section 2.1). They are also good insulators which reduce the heat loss from the underlying surface (land or ocean) towards the cold atmosphere in winter. More generally, the presence of sea ice restricts the exchanges of heat and gases between the ocean and the atmosphere.

When sea ice forms, only a fraction of the salt present in the ocean is trapped in the ice, the remainder being ejected into the ocean (this is called 'brine rejection'). The resulting sea-ice salinity is between 10 in relatively young ice and less than 2 in very old ice (compared to around 35 for the ocean; see Section 1.3). Because of this brine rejection, sea-ice formation increases the salinity at the ocean surface, whilst melting sea ice is associated with surface freshening. Sea-ice drift is also associated with a horizontal freshwater transport. If there is a net convergence of the sea-ice transport and intense ice melting in a region, this will decrease the salinity of surface water there. However, in coastal **polynyas**, strong winter winds continually push the newly formed ice off shore, leading to a strong divergence of the sea-ice transport. This implies high ice formation rates in these polynyas (up to 10 m/year at some locations) and thus large amounts of brine rejection, which can lead to very high local ocean salinities.

**Ice sheets** store large amounts of water on land. Any change in their volume therefore has a considerable effect on the sea level and ocean salinity. It is estimated that if all the ice sheets melted completely, taking into account the fact that some ice sheets are grounded below sea level, the sea level would rise by more than 60 m. However, if we neglect the effect of dilution on sea water density and volume, the melting of sea ice and floating **ice shelves** does not influence sea levels. Indeed, because of Archimedes law, floating ice displaces its own weight of sea water, and the melt water thus simply replaces the volume of ice previously below sea level. Ice sheets are also big mountains that divert the air flow. Because of their height, they help to maintain cold conditions on the surface. Furthermore, the presence of cold air on the ice sheet has a regional influence, cooling the surrounding areas.

## 1.5   The Land Surface and the Terrestrial Biosphere

As discussed earlier, many characteristics of the climate are influenced by the distribution and topography of land surface. For instance, mountain chains such as the Andes and the Rocky mountains (Figure 1.22) are formidable barriers to the westerly winds which influence the climate on a continental scale. Mountains also have an important role in the hemispheric scale by affecting planetary waves and the global atmospheric circulation (see Section 1.2). Distance from the coast influences the temperature and aridity of a region, a feature sometimes referred to as the '**continentality**' of the local climate. The presence of land boundaries to the ocean (and more generally the ocean bathymetry) affects the location of the strong western boundary currents and of the straits that allow water exchanges between the different basins (see Section 1.3). The shape and even the existence of an ice sheet are strongly conditioned by the underlying bedrock (see Section 1.4).

In addition to the influence of the land geometry, the type of vegetation present on land also has a critical influence on climate on all spatial and temporal scales.

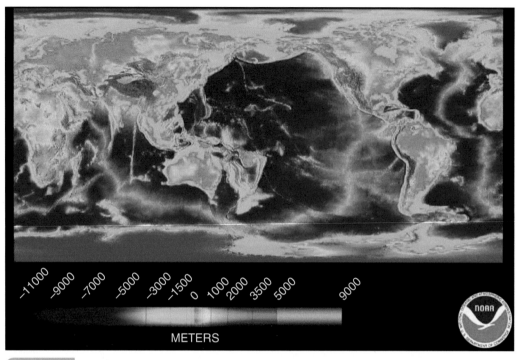

**Fig. 1.22**    Map of the surface topography. (Source: U.S. National Geophysical Data Center, National Ocean and Atmosphere Administration (NOAA); available at: http://www.ngdc.noaa.gov/mgg/image/relief_slides2.html.)

One of the most important roles of terrestrial vegetation is related to its **albedo** (Figure 1.23; see also Section 2.1). Vegetation usually has a lower albedo than soil (Table 1.3) and much smaller than that of deserts. This is why subtropical deserts such as the Sahara appear as regions of particularly high albedo on global maps (Figure 1.23). Maximum levels are observed at high latitudes because of the presence of snow and ice. At these latitudes, the vegetation modulates the influence of the snow. In the absence of vegetation or in the presence of low-growing vegetation such as grass, the snow can cover the whole area, leading to highly reflective white areas with a high albedo. If snow falls on a forest, relatively dark trunks, branches and possibly needles or leaves will partially emerge from the snow, resulting in a much lower albedo than with a homogeneous snow blanket (see Section 4.3.2).

The terrestrial **biosphere** also has a clear impact on the hydrological cycle (see Section 2.2). Water storage is generally greater in soil covered by vegetation than on bare land, where direct runoff often follows precipitation. Stored water can be taken up later by plant roots and transferred back to the atmosphere by **evapotranspiration** (i.e., the sum of evaporation and transpiration). An additional effect of the vegetation cover is related to the surface roughness which influences the stress at the atmosphere-land interface and the turbulent exchanges at the surface (see Section 2.1). Finally, the role of the terrestrial biosphere in the global carbon cycle will be discussed in Section 2.3.

| Table 1.3 Typical range of the albedo of various surfaces ||
| Surface type | Albedo |
| --- | --- |
| Ocean | 0.05–0.10 |
| Fresh snow | 0.75–0.90 |
| Old snow | 0.40–0.75 |
| Bare sea ice | 0.50–0.70 |
| Soil | 0.05–0.35 |
| Desert | 0.20–0.45 |
| Cropland | 0.15–0.25 |
| Grassland | 0.15–0.25 |
| Deciduous forest | 0.15–0.20 |
| Coniferous forest | 0.05–0.15 |
| Snow-covered coniferous forest | 0.15–0.3 |

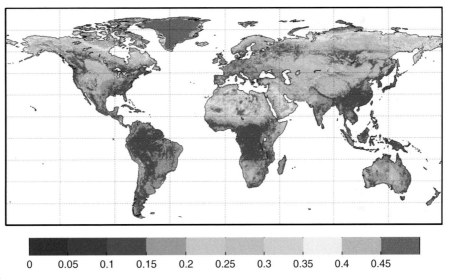

**Fig. 1.23** Annual mean land surface albedo from *Moderate Resolution Imaging Spectroradiometer* (*MODIS*) satellite data. *MODIS* data are provided by the Land Processes Distributed Active Archive Center (LP DAAC), located at the U.S. Geological Survey (USGS) Earth Resources Observation and Science (EROS) Center (lpdaac.usgs.gov), distributed in net CDF format by the Integrated Climate Data Centre (ICDC, http://icdc.zmaw.de), University of Hamburg, Germany. The data have been processed by Stefan Kern.

Given this climatic role of vegetation, it is useful to summarise the variations of its characteristics over the globe by defining different **biomes**, which are regions with distinctive large-scale vegetation systems (Figure 1.24). Their exact definition, as well as the number of important biomes which are considered, differs from one study to another. Nevertheless, it is generally considered that the natural

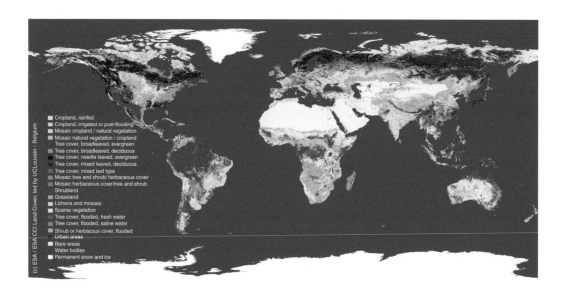

Cropland, rainfed
Cropland, irrigated or post-flooding
Mosaic cropland / natural vegetation
Mosaic natural vegetation / cropland
Tree cover, broadleaved, evergreen
Tree cover, broadleaved, deciduous
Tree cover, needle leaved, evergreen
Tree cover, mixed leaved, deciduous
Tree cover, mixed leaf type
Mosaic tree and shrub/ herbaceous cover
Mosaic herbaceous cover/tree and shrub
Shrubland
Grassland
Lichens and mosses
Sparse vegetation
Tree cover, flooded, fresh water
Tree cover, flooded, saline water
Shrub or herbaceous cover, flooded
Urban areas
Bare areas
Water bodies
Permanent snow and ice

**Fig. 1.24**   Terrestrial land cover. (Source: http://maps.elie.ucl.ac.be/CCI/viewer/index.php. Copyright ESA/ESA CCI land cover, led by UC Louvain, Belgium. Reproduced with permission.)

biomes can be classified based on their typical percentages of grass and trees into five main groups: desert, grassland, shrub land, woodland and forest. Cropland and built-up areas (cities) can be added to take into account the role of land use associated with human activities.

Deserts are characterised by a very small amount of vegetation. Grassland, as indicated by its name, is mainly covered by grass and lichens. It can be found at various latitudes and includes **tundra**, steppe and savannah. In shrub land, low woody plants are present in addition to grass. The fraction of trees is higher in woodland, but there are still significant areas covered by grass and often relatively large distances between trees. Finally, in forests, a dense cover of trees is observed (as in tropical rainforests and boreal conifer forests, also called '**taiga**').

Specifying the biome is not enough to determine all the surface properties as they can change with season or with the age and density of the vegetation cover. Additional variables therefore have been introduced, such as the **leaf-area index**, defined as the total surface covered by the leaves on a surface of $1 \text{ m}^2$, which modulates the evapotranspiration or the photosynthetic activity of the plants.

We discussed earlier how vegetation influences climate, but, of course, climate also affects vegetation through the distribution of incoming solar radiation, temperature and precipitation. This leads to powerful **feedback** loops that will be described in more detail in Section 4.3.2. If precipitation and/or temperature are too low, desert biomes dominate (as in the Sahara or Antarctica). At higher temperatures, forests can be maintained if a sufficient supply of water from rainfall is available. Between those two extremes, different combinations of grass and trees are found (see also Figure 3.13).

## Review Exercises

1. The climate is defined as
   a. the mean and seasonal cycle of relevant atmospheric variables (mainly related to temperature, precipitation and winds).
   b. the mean and variability of relevant atmospheric variables (mainly related to temperature, precipitation and winds).
   c. a description of the whole climate system.
2. The temperature decreases with height in the lowest 10 km of the atmosphere (the troposphere). The lapse rate $\Gamma$, which measures this decrease, has a mean value of about
   a. $2.5 \text{ K km}^{-1}$, corresponding to 25 K over 10 km.
   b. $6.5 \text{ K km}^{-1}$, corresponding to 65 K over 10 km.
   c. $10 \text{ K km}^{-1}$, corresponding to 100 K over 10 km.
3. The stability of an air column depends on its density and thus on its temperature. The limit of neutral stability occurs when
   a. the temperature is homogeneous over the vertical.
   b. the temperature follows the adiabatic lapse rate.
4. Hadley cells are large cells in which the air rises at the equator because of the high temperatures that prevail there and subsides at the poles, the coldest places on Earth.
   a. True
   b. False
5. Precipitations are higher at the equator than in the subtropics because of the ascendance around the equator and the subsidence around a latitude of 30°.
   a. True
   b. False
6. The oceanic circulation close to the surface is influenced by the wind, whilst density contrasts play a role only for the deep circulation.
   a. True
   b. False
7. At middle latitudes, the seasonal evolution of the mixed-layer depth corresponds to
   a. a maximum in winter of typically 500 m and a minimum in summer of less than 40 m.
   b. a maximum in winter of typically 100 m and a minimum in summer of less than 40 m.
   c. a maximum in winter of typically 100 m and a minimum in summer of less than 10 m.
8. The component of the cryosphere that has the largest spatial extent in winter is
   a. sea ice.
   b. frozen soil.
   c. ice sheets.
   d. snow on continents.

# 2 Energy Balance, Hydrological and Carbon Cycles

## OUTLINE

Climate is characterised by large exchanges of energy and various chemical species between its components. Modifications of those exchanges are a major source of climate change. For energy, water and carbon, this chapter describes first the budget of the various reservoirs on a global scale before focussing on the spatial distribution of some key variables.

## 2.1 The Earth's Energy Budget

### 2.1.1 The Radiative Balance at the Top of the Atmosphere: A Global View

Nearly all the energy entering the **climate system** comes from the Sun in the form of **electromagnetic radiation**. Additional sources are present, such as geothermal heating, for instance, but their contribution is so small that their influence can safely be neglected. At the top of the Earth's atmosphere, a surface at the mean Earth–Sun distance perpendicular to the solar rays receives about 1360 W m$^{-2}$ (Kopp and Lean 2011) (see also Figure 4.6). This is called the **'total solar irradiance'** (TSI) or the **'solar constant'** $S_0$. A bit less than half this energy comes in the form of radiation in the visible part of the electromagnetic spectrum, the remaining part occurring mainly in the near-infrared region, with a smaller contribution from the ultraviolet region of the spectrum (Figure 2.1).

On average, the total amount of incoming solar energy per unit of time outside the Earth's atmosphere (Figure 2.2) is the TSI times the cross-sectional surface (i.e., the surface that intercepts the solar rays, which corresponds to a surface equal to $\pi R^2$, where $R$ is the Earth's radius of 6,371 km). For simplicity and because it is a reasonable approximation, we will neglect the thickness of the atmosphere compared to the Earth's radius in our computations of distances or surfaces. Some of the incoming energy is reflected straight back to space by the atmosphere, in particular, by the clouds, and the Earth's surface. This fraction of the energy that is reflected is called the **'albedo** of the Earth' or 'planetary albedo' $\alpha_p$. In present-day conditions, it has a value of about 0.3.

To achieve a balance, the energy coming from the Sun must be compensated for by an equivalent energy loss. If this were not true, the Earth's temperature

30

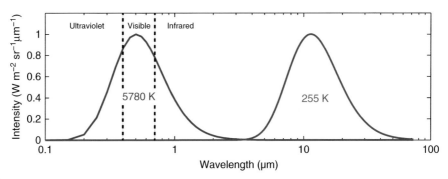

**Fig. 2.1** Normalised blackbody spectra for temperatures representative of the Sun (blue, temperature of 5,780 K) and the Earth (red, temperature of 255 K).

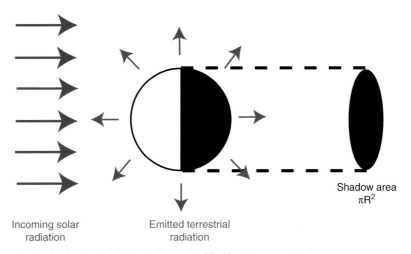

**Fig. 2.2** Energy absorbed and emitted by the Earth. (Source: Modified from Hartmann 1994.)

would rapidly rise or fall. At the Earth's temperature, following **Wien's law**, this is achieved by radiating energy mainly in the infrared region of the electromagnetic spectrum. As the radiation emitted by the Earth has a much longer wavelength than that received from the Sun (see Figure 2.1), it is often termed '**longwave radiation**', whilst that received from the Sun is called '**shortwave radiation**'. Treating the Earth as a **blackbody**, the total amount of energy that is emitted by a 1-m² surface per unit of time ($A\uparrow$) can be computed by the **Stefan-Boltzmann** law:

$$A\uparrow = \sigma T_e^4 \tag{2.1}$$

where $\sigma$ is the Stefan-Boltzmann constant ($\sigma = 5.67 \times 10^{-8}$ W m⁻² K⁻⁴). This equation defines $T_e$, the effective emission temperature of the Earth. $A\uparrow$ is measured in watts per square metre (W m⁻²), as is the TSI. This is also an irradiance, but it is often referred to in the scientific literature as the 'longwave flux' at surface or, more simply, the 'surface radiation'.

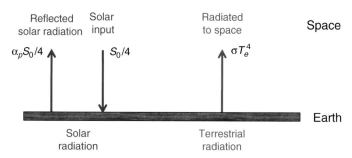

**Fig. 2.3**    Simplified energy balance of the Earth (assuming that it behaves like a perfect blackbody and a transparent atmosphere).

The Earth emits energy in all directions, so the total amount of energy emitted by the Earth per unit of time is $A\uparrow$ times the surface of the Earth, $4\pi R^2$. To achieve equilibrium, therefore, we must have (Figure 2.3).

Absorbed solar energy = emitted terrestrial energy

$$\pi R^2 \left(1-\alpha_p\right)S_0 = 4\pi R^2 \sigma T_e^4 \tag{2.2}$$

This leads to

$$\frac{1}{4}\left(1-\alpha_p\right)S_0 = \sigma T_e^4 \tag{2.3}$$

and finally to

$$T_e = \left[\frac{1}{4\sigma}\left(1-\alpha_p\right)S_0\right]^{1/4} \tag{2.4}$$

This corresponds to $T_e = 255\ \mathrm{K}\ (=-18°\mathrm{C})$. Note that we can interpret Eq. (2.3) as the mean balance between the emitted terrestrial radiation and the absorbed solar radiation for 1 m$^2$ of the Earth's surface. As just shown, the factor 1/4 arises from the spherical geometry of the Earth because, at any time, only part of the Earth's surface receives solar radiation directly.

The temperature $T_e$ is not a real temperature that could be measured somewhere on Earth. It is the equivalent blackbody temperature required to balance the solar energy input. It can be interpreted as the temperature that would occur on the Earth's surface if it were a perfect blackbody, there were no atmosphere and the temperature was the same at every point.

## 2.1.2  The Greenhouse Effect

The atmosphere is nearly transparent to visible light, absorbing about 20% of the incoming solar energy. As a consequence, most of the absorption takes place at the Earth's surface. However, the atmosphere is almost opaque across most of the infrared region of the **electromagnetic spectrum** (see Section 2.1.6). A large fraction of the energy emitted by the Earth's surface thus is absorbed by the atmosphere and re-emitted, significantly increasing the temperature of the system. This is the result of the presence of clouds and the radiative properties of some minor constituents of the atmosphere, such as water vapour, carbon dioxide ($CO_2$), methane ($CH_4$), nitrous oxide ($N_2O$) and ozone ($O_3$). These gases constitute only a small fraction of the atmospheric composition, whilst the two dominant components (molecular nitrogen and oxygen; see Section 1.2) play nearly no part in this opacity. For present-day conditions, water vapour is the main contributor, accounting for about 50% of the total effect according to the computation of Schmidt et al. (2010), whilst clouds and carbon dioxide are responsible of about 25 and 20% of the net amount of **longwave radiation** absorbed in the atmosphere, respectively.

In a garden greenhouse, panes of glass are transparent to visible light but nearly opaque to infrared radiation, 'trapping' part of the energy emitted by the surface and resulting in a warming of the air. By analogy, the alteration of the energy budget by some of the minor atmospheric constituents just described is called the 'greenhouse effect', and those minor constituents are called 'greenhouse gases'. However, in a garden greenhouse, a significant fraction of the warming is related to reduction of the turbulent heat exchanges with the atmosphere, not the modification of the radiative fluxes. Furthermore, absorption and re-emission of infrared radiation at different altitudes in the atmosphere are important for the radiative balance of the Earth in contrast to the single glass layer of a garden greenhouse. The analogy thus should be used with great caution.

The greenhouse effect can be illustrated by a very simple model in which the atmosphere is represented by a single homogeneous layer of temperature $T_a$, totally transparent to the solar radiation and totally opaque to the infrared radiation emitted by the Earth's surface (Figure 2.4). Because of this assumed opacity of the atmosphere to surface radiation, all the energy radiated to space is from the atmosphere. Using Eq. (2.3), the balance at the top of the atmosphere is thus

$$\sigma T_a^4 = \frac{1}{4}\left(1 - \alpha_p\right)S_0 = \sigma T_e^4 \tag{2.5}$$

In this simple model, $T_a$ is thus equal to $T_e$, the effective emission temperature of the Earth. At the Earth's surface, the balance among the flux emitted by the surface, the incoming solar fluxes and the infrared flux coming from the atmosphere gives

$$\sigma T_s^{\,4} = \frac{1}{4}(1 - \alpha_p)S_0 + \sigma T_a^{\,4} \tag{2.6}$$

where $T_s$ is the surface temperature. Combining Eqs. (2.5) and (2.6) leads to

$$\sigma T_s^{\,4} = \sigma T_e^{\,4} + \sigma T_e^{\,4} \tag{2.7}$$

and

$$T_s = 2^{1/4}\,T_e = 1.19 T_e \tag{2.8}$$

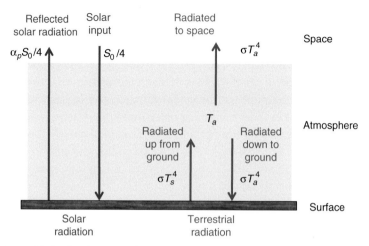

**Fig. 2.4** Energy balance of the Earth with an atmosphere represented by a single layer which is totally transparent to solar radiation and opaque to infrared radiation. (Source: Modified from Marshall and Plumb 2008.)

Because of the greenhouse effect, the surface temperature is much higher than $T_e$, reaching 303 K (30°C) in this example. This temperature is actually higher that the observed mean surface temperature of 288 K (15°C) because of the crude approximations applied in this simple model.

We can improve the model by taking into account the fact that the atmosphere is not a perfect blackbody (Figure 2.5). Using the 'emissivity' of an object $\varepsilon$ (which is defined as the ratio of energy radiated by this object to energy radiated by a blackbody at the same temperature), we can write the balance at the surface as

$$\sigma T_s^4 = \frac{1}{4}\left(1-\alpha_p\right)S_0 + \varepsilon\sigma T_a^4 \tag{2.9}$$

The emissivity is also equal to the fraction of the radiation that is absorbed by the object. The fraction being transmitted through the object is thus equal to $(1-\varepsilon)$, and the balance at the top of the atmosphere is

$$\frac{1}{4}\left(1-\alpha_p\right)S_0 = \varepsilon\sigma T_a^4 + \left(1-\varepsilon\right)\sigma T_s^4 = \sigma T_e^4 \tag{2.10}$$

Equations (2.9) and (2.10) lead to

$$\sigma T_s^4 = \frac{2}{2-\varepsilon}\frac{1}{4}\left(1-\alpha_p\right)S_0 = \frac{2}{2-\varepsilon}\sigma T_e^4 \tag{2.11}$$

and

$$T_s = \left(\frac{2}{2-\varepsilon}\right)^{1/4} T_e \tag{2.12}$$

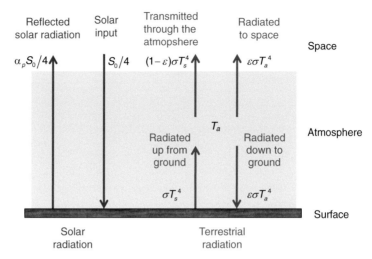

**Fig. 2.5**   Energy balance of the Earth with an atmosphere represented by a single layer totally transparent to solar radiation and with an infrared emissivity $\varepsilon$.

We can also compute $T_a$ as

$$T_a = \left(\frac{1}{2-\varepsilon}\right)^{1/4} T_e = \left(\frac{1}{2}\right)^{1/4} T_s \tag{2.13}$$

For $\varepsilon = 0$, corresponding to an atmosphere which is totally transparent to infrared radiation, Eq. (2.12) leads to $T_s = T_e$, which is well in agreement with the result of Section 2.1.1. For a perfect blackbody, we get a result identical to Eq. (2.8), as expected. A typical value of 0.97 for the atmosphere provides a value of $T_s = 1.18 T_e$, that is, 301 K (28 °C). This is still far away from the observed surface temperature. To get a more precise estimate of the radiative balance of the Earth, it is necessary to take into account the impact of the vertical structure of the atmosphere on the radiative fluxes with multiple absorption by the various atmospheric layers and re-emission at a lower intensity as the temperature decreases with height. Furthermore, the radiative properties of the greenhouse gases are not homogeneous over the infrared region of the electromagnetic spectrum, with strong absorption only in some specific ranges of frequencies called 'bands' which are characteristic of each component. Consequently, the atmosphere is nearly totally opaque in some parts of the infrared spectrum, whilst it is nearly transparent in some narrow bands (Wallace and Hobbs 2006; Archer 2011; Boucher 2012). Finally, the contribution of non-radiative exchanges has to be included to close the surface energy balance (see Section 2.1.6).

### 2.1.3 Present-Day Insolation at the Top of the Atmosphere

The instantaneous '**insolation**' is defined as the energy received per unit time and unit surface on a horizontal plane at the top of the atmosphere (or on a horizontal

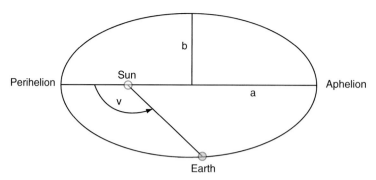

**Fig. 2.6** Schematic representation of the Earth's orbit around the Sun showing the true anomaly $v$. The eccentricity has been exaggerated for clarity.

plane at the Earth's surface, if we neglect the influence of the atmosphere). It thus has the same unit as the irradiance. It depends on the geographical position on Earth as well as on the position of the Earth relative to the Sun. The influence of those factors is described in the following sub-sections.

### 2.1.3.1 Earth's Orbit around the Sun

According to Kepler's first law, the Earth's trajectory around the Sun is an ellipse with the Sun at one focus. The point of the Earth's orbit which is closest to the Sun is called the '**perihelion**', whilst the '**aphelion**' is the point farthest from the Sun (Figure 2.6). Here $a$ is half the length of the major axis, and $b$ is half the length of the minor axis. The shape of the ellipse is then characterised by its **eccentricity** $ecc$, defined by

$$ecc = \frac{\sqrt{a^2 - b^2}}{a} \qquad (2.14)$$

The parameters of the Earth's orbit vary with time (see Section 5.5.1), but at present, $ecc = 0.0167$, meaning that the Earth's orbit is very close to a circle (which corresponds to an eccentricity of zero).

The amount of incoming solar energy per unit of area and time at the top of the atmosphere is a function of $r$, the distance from the Sun to the Earth. $r$ can be computed as a function of $v$, the true anomaly, according to the formula for an ellipse

$$r = \frac{a(1 - ecc^2)}{1 + ecc \cos v} \qquad (2.15)$$

We can also define $r_m$ as the mean distance between the Earth and the Sun by

$$r_m = \sqrt{ab} = a\sqrt{(1 - ecc^2)} \qquad (2.16)$$

This means that a circle with radius $r_m$ would have the same area as the ellipse corresponding to the Earth's orbit.

$S_r$, the solar irradiance measured on the outer surface of the Earth's atmosphere in a plane perpendicular to the rays at a distance $r$ from the Sun, then can be computed as a function of the TSI, $S_0$, estimated at a distance $r_m$. This is achieved by considering that the total energy emitted per unit time by the Sun is equal to the total energy received on the surface of a sphere of radius $r$, centred on the Sun, as well as to that received on the surface of a sphere of radius $r_m$; that is, total energy emitted per unit time by the Sun measured at a distance $r_m$ = total energy emitted per unit of time by the Sun measured at a distance $r$:

$$4\pi r_m^2 S_0 = 4\pi r^2 S_r \tag{2.17}$$

$$S_r = \frac{r_m^2}{r^2} S_0 \tag{2.18}$$

### 2.1.3.2 Computation of the Zenith Distance

Once we have obtained $S_r$, the solar irradiance measured in a plane perpendicular to the rays, we can estimate $S_h$, the amount of solar energy received per unit of time on a unit horizontal surface at the top of the atmosphere. $S_h$ is proportional to the cosine of $\theta_s$, the solar **zenith distance**, which is defined as the angle between the solar rays and the normal to the Earth's surface at any particular point:

$$S_h = S_r \cos\theta_s \tag{2.19}$$

$S_h$ rises as $\cos\theta_s$ becomes closer to 1, that is, as the horizontal surface becomes more normal to the Sun's rays. When the surface is inclined at an oblique angle to the solar rays, the amount of energy received by the surface per square metre and unit of time is lower because the total amount of energy received by the perpendicular surface ($S_r A_1$ in Figure 2.7) is distributed across a larger surface ($S_r A_1 = S_h A_2 = S_h A_1/\cos\theta_s$).

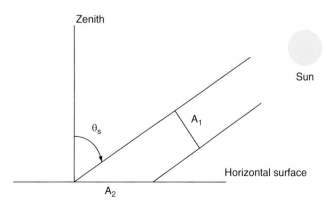

**Fig. 2.7** Influence of the zenith distance $\theta_s$ on the amount of radiation received on a horizontal surface. $A_1$ is the surface perpendicular to the solar beam, whilst $A_2$ is the horizontal surface illuminated by the rays crossing $A_1$.

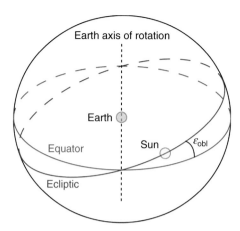

**Fig. 2.8**    Representation of the ecliptic and the obliquity $\varepsilon_{obl}$ in a geocentric system.

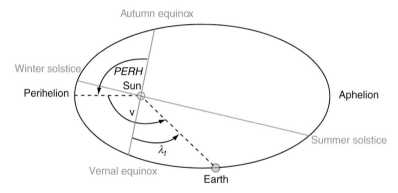

**Fig. 2.9**    Representation of the longitude of the perihelion *PERH*, the true anomaly $v$, the true longitude $\lambda_t$, and the seasons in the ecliptic plane.

In the computation of the zenith distance, we must account for the angle between the rotation axis of the Earth and its orbital plane. Indeed, the **ecliptic plane**, which is the geometrical plane containing the mean orbit of the Earth around the Sun, is inclined relative to the '**celestial equatorial plane**', which is the projection of the Earth's equator into space (and thus is perpendicular to the Earth's rotation axis). The angle between those two planes is called the '**obliquity**' of the ecliptic $\varepsilon_{obl}$ (Figure 2.8). At present, it is about 23°27′ (or 23.45° as used later).

The intersections of the ecliptic and orbital planes are used to define the seasons. In particular, the **vernal equinox**, which is often used as a reference in the coordinate system to define the **true longitude** $\lambda_t$ (or ecliptic longitude; Figure 2.9), corresponds to the intersection of the ecliptic plane with the celestial equator when the Sun moves from the austral to the boreal hemisphere in its apparent movement around the Earth. This occurs around March 20–21 and is often called the 'spring equinox'. However, this term could be misleading because this date corresponds to the beginning of autumn in the southern hemisphere, and 'vernal equinox' is preferred.

By definition, the vernal equinox corresponds to a true longitude equal to zero, the solstices to the true longitudes equal to 90 and 270°, and the autumn equinox to a true longitude equal to 180°. This can be used to compute the length of the different seasons using Kepler's second law, which states that as the Earth moves in its orbit, a line from the Sun to the planet sweeps out equal areas in equal time. If we define $PERH$ as the longitude of the perihelion measured from the autumn equinox ($PERH = 102.04°$ in present-day conditions, corresponding to a true longitude of $180° + PERH = 282.04$), we can write

$$\lambda_t = 180 + PERH + v \qquad (2.20)$$

The true longitude $\lambda_t$ is related to day of the year. Additionally, $\cos\theta_s$ is a function of $\phi$, the latitude of the point on Earth, and $HA$, the hour angle. It can be computed using the standard astronomical formula

$$\cos\theta_s = \sin\phi\sin\delta + \cos\phi\cos\delta\cos HA \qquad (2.21)$$

where $\delta$ is the 'solar declination', which is defined as the angle between a line from the centre of the Earth towards the Sun and the celestial equator (Figure 2.10). It varies from $+\varepsilon_{obl}$ at the summer solstice in the northern hemisphere to $-\varepsilon_{obl}$ at the winter solstice and zero at the equinoxes. During the day, its value is constant to a very good approximation. Knowing the true longitude and the obliquity, the declination $\delta$ can be estimated using the formula

$$\sin\delta = \sin\lambda_t \sin\varepsilon_{obl} \qquad (2.22)$$

Furthermore, if we denote the number of the day, starting on the first of January, by $NDAY$, the value of $\delta$ also can be estimated by using the raw approximation (zero order to eccentricity)

$$\delta = 23.45°\sin\left[360°\left(NDAY-80\right)/365\right] \qquad (2.23)$$

The hour angle $HA$ indicates the time since the Sun was at local meridian, measured from the observer's meridian westward. $HA$ thus is zero at the local solar noon. It is generally measured in radians or hours ($2\pi$ rad = 24 hours). It is more formally defined as the angle between the half-plane determined by the Earth's axis and the zenith (local meridian half-plane) and the half-plane determined by the Earth's axis and the Sun.

### 2.1.3.3  Daily Insolation at the Top of the Atmosphere

If the Sun rises above the horizon on a particular day, we can compute the times of sunrise ($HA_{sr}$) and sunset ($HA_{ss}$) using Eq. (2.21) because both events correspond to a solar zenith angle of 90° ($\cos\theta_s = 0$); that is,

$$HA_{sr,ss} = \pm\arccos\left(-\tan\phi\tan\delta\right) \qquad (2.24)$$

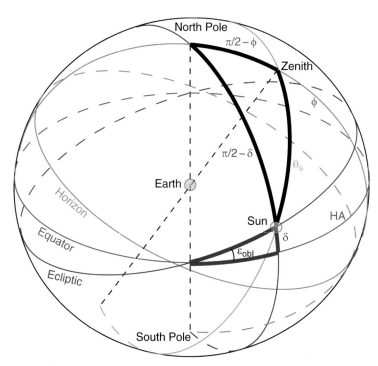

**Fig. 2.10** Representation of the declination $\delta$, the hour angle $HA$, the latitude $\phi$ and the zenith distance $\theta_s$ in a geocentric system. In this figure, the equator and the ecliptic are fixed. The apparent motion of the Sun corresponds to a complete revolution along the ecliptic in one year. The hour angle [and the location of the local meridian (brown circle) through the North Pole and the zenith] changes as the Earth makes a complete rotation around its axis in one day. Applying the standard spherical trigonometric rules to the bold black triangle leads to Eq. (2.21). This figure also can be used to demonstrate Eq. (2.22) using the blue bold triangle.

The length of the day ($LOD$) then is given by

$$LOD = \frac{24}{\pi} \arccos\left(-\tan\phi\tan\delta\right) \tag{2.25}$$

The coefficient $24/\pi$ comes from the factor $24/2\pi$ used to convert the result from radians for $HA$ to hours for $LOD$ and the factor 2 corresponding to the difference between the two times given by Eq. (2.24) for sunrise and sunset when using the + or – in the solution. The coefficient would be $86{,}400/\pi$ if $HA$ is measured is seconds [see Eq. (2.26)].

At the equator, since $\phi$ is equal to zero, $LOD$ is always equal to 12 hours. At the equinoxes, $\delta$ is equal to zero, so $LOD$ is equal to 12 hours everywhere. We also can estimate from Eq. (2.25) that in the polar regions of the northern hemisphere, where $\phi + \delta \geq 90°$, $\tan\phi\tan\delta \geq 1$ in summer, and the Sun is visible during the whole day (midnight Sun). Where $\phi - \delta \geq 90°$, $\tan\phi\tan\delta \leq -1$ in winter, and the Sun is always below the horizon (polar night). Similar formulae can be obtained for the southern hemisphere.

Using Eqs. (2.18), (2.19), (2.21) and (2.24), we can integrate $S_h$ over time to compute $S_{h,\text{day}}$, the daily insolation on a horizontal surface (in J m$^{-2}$). The mean

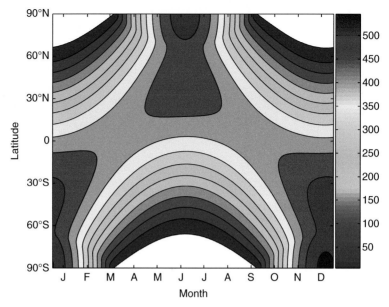

**Fig. 2.11** Daily mean insolation (in W m$^{-2}$) on a horizontal surface at the top of the atmosphere as a function of the day of the year and the latitude. White areas correspond to the polar night. (Courtesy of M. F. Loutre.)

insolation over one day (in W m$^{-2}$) is also often used. It is simply $S_{h,\text{day}}$ divided by 24 hours

$$
\begin{aligned}
S_{h,\text{day}} &= S_0 \frac{r_m^2}{r^2} \int_{\text{Sunrise}}^{\text{Sunset}} \left( \sin \phi \sin \delta + \cos \phi \cos \delta \cos HA \right) dt \\
&= S_0 \frac{r_m^2}{r^2} \frac{86{,}400}{2\pi} \int_{HA_{sr}}^{HA_{ss}} \left( \sin \phi \sin \delta + \cos \phi \cos \delta \cos HA \right) dHA \qquad (2.26) \\
&= S_0 \frac{r_m^2}{r^2} \frac{86{,}400}{\pi} \left( HA_{ss} \sin \phi \sin \delta + \cos \phi \cos \delta \sin HA_{ss} \right)
\end{aligned}
$$

As expected, the daily insolation (Figure 2.11) is higher in the summer hemisphere than in winter because of the lower zenith distance (i.e., the Sun higher above the horizon) and longer duration of the day. This is mainly due to the obliquity of the Earth's axis of rotation, which implies that a different hemisphere is 'facing the Sun' in each half of the orbit. The obliquity is thus the main reason for the existence of seasons on Earth (see also Section 5.5.1). If $\varepsilon_{\text{obl}}$ were equal to zero, night and day would be 12 hours long everywhere [Eqs. (2.24) and (2.25)], and if, additionally, *ecc* also were equal to zero, each point on Earth would have the same daily mean insolation throughout the year [Eqs. (2.22) and (2.26)].

The daily insolation is also a strong function of the latitude because of the influence of the latter on the zenith distance [Eq. (2.21)]. Averaged over one day, the maximum solar insolation occurs at the poles on the summer solstice, whilst averaged over one year, the solar energy received at the top of the atmosphere at the equator is about twice that received at the poles.

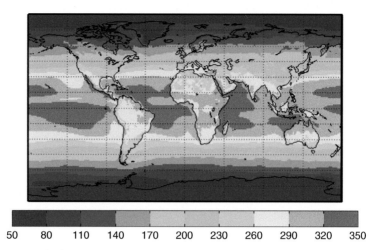

50   80   110   140   170   200   230   260   290   320   350

**Fig. 2.12**   Annual mean net solar flux at the top of the atmosphere absorbed by the Earth (in W m$^{-2}$). (Source: Data from the Clouds and the Earth's Radiant Energy System (CERES); available at: http://ceres.larc.nasa.gov/; Loeb et al. 2009.)

### 2.1.4 The Radiative Balance at the Top of the Atmosphere: Geographical Distribution

The geographical distribution of the net incoming solar radiative flux at the top of the atmosphere (i.e., the incoming minus the reflected solar radiative flux) is a function of the insolation distribution described in the preceding sub-section as well as the regional variations in the planetary **albedo** (Figure 2.12). The latter is influenced by several factors, including the albedo of the surface (see Section 1.5) and the presence of clouds which reflect a significant fraction of the incoming solar radiation back to space. The influence of clouds is particularly evident in tropical regions, where it explains, for instance, why the absorbed solar flux is larger in the relatively cloud-free eastern equatorial Pacific than in the cloudier western Pacific. At high latitudes, the surface albedo is high because of the high zenith distance (the Sun low above the horizon) and the high reflectance of snow and ice (see Section 1.4). This high surface albedo at high latitudes amplifies the latitudinal variations in solar radiation associated with the Earth's geometry (see Figure 2.11), resulting in a difference of about a factor of 5 in annual mean absorbed solar flux at the poles compared to the equator.

The **Stefan-Boltzmann law** says that the radiative flux emitted is a function of the temperature of the emitting surface. A difference of about 50°C between the equator and the poles roughly corresponds to a variation in the emitted thermal flux of about 50 W m$^{-2}$, which is in reasonable agreement with the estimated values (Figure 2.13). The presence of clouds and water vapour also has a large influence. They absorb part of the infrared radiation emitted by the surface before re-emitting radiation, generally at a lower temperature higher in the atmosphere (see Section 2.1.2). This results in a lower outgoing **longwave** flux. As a consequence, the maximum outgoing longwave flux is found above warm, dry areas

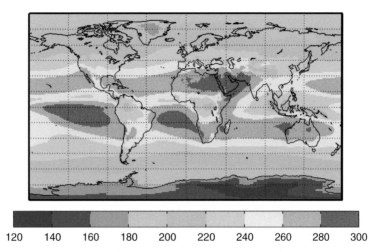

120    140    160    180    200    220    240    260    280    300

**Fig. 2.13**  Net annual mean outgoing longwave flux at the top of the atmosphere (in W m$^{-2}$). (Source: Data from the Clouds and the Earth's Radiant Energy System (CERES); available at: http://ceres.larc.nasa.gov/; Loeb et al. 2009.)

such as the subtropical deserts. More generally, wet equatorial areas emit less radiation than dry tropical areas (Figure 2.13).

When averaged over longitude, the outgoing longwave flux clearly shows fewer latitudinal variations than the net incoming solar flux absorbed by the Earth. As a consequence, the absorbed solar flux outbalances the outgoing flux in regions located between roughly 40°S and 40°N, whilst a net deficit in the radiative flux at the top of the atmosphere ($RF_{\mathrm{TOA}}$) is observed poleward of 40°N and 40°S (Figure 2.14). $RF_{\mathrm{TOA}}$ also displays some longitudinal variations, the most spectacular probably being the net negative flux over the Sahara because of the dry conditions there and the high albedo of its sand.

## 2.1.5  Heat Storage and Transport

In an area delimited by latitudes $\phi_1$, $\phi_2$ and longitudes $\lambda_1$, $\lambda_2$ (Figure 2.15), the net radiative flux at the top of the atmosphere $RF_{\mathrm{TOA}}$ must be balanced by the sum of the net horizontal energy transport, the heat exchanges with the deep ground and the contribution to the energy budget associated with changes in the heat storage $E_s$ in the atmosphere, the ocean and the ground. Because the ground has a low thermal conductivity, only the top few metres interact with the surface on seasonal to decadal **time scales**. In most climatic applications, it is sufficient to take the top few metres of the ground (typically 10 m) into account and assume that exchanges with the deep ground or the Earth's interior can be represented by a geothermal heat flux. The value of this flux is weak in most regions (~0.08 W m$^{-2}$). It is thus often neglected in energy balance computations. This can be expressed by

$$\frac{\partial E_s}{\partial t} = RF_{\mathrm{TOA}} - \Delta F_{\mathrm{transp}} \tag{2.27}$$

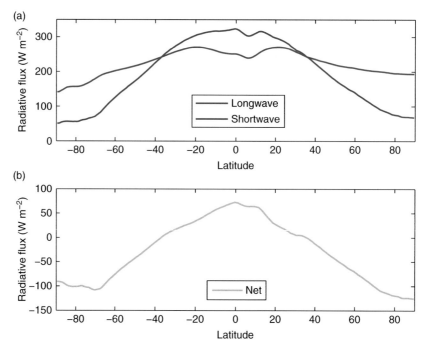

(a) **Zonal mean** of the absorbed solar flux (blue) and the outgoing longwave flux (red) at the top of the atmosphere in annual mean (in W m$^{-2}$). (b) Zonal mean of the difference between the absorbed solar flux and the outgoing longwave flux at the top of the atmosphere in annual mean (in W m$^{-2}$). (Source: Data from the Clouds and the Earth's Radiant Energy System (CERES); available at: http://ceres.larc.nasa.gov/; Loeb et al. 2009.)

where $\Delta F_{transp}$ is the net local effect of the oceanic and atmospheric heat transport, that is, the divergence of the horizontal heat flux.

### 2.1.5.1 Heat Storage

On daily and seasonal **time scales**, the heat storage by the climate system plays a large role in mitigating the influence of the changes in radiative flux at the top of the atmosphere. These variations in heat storage for the ocean, atmosphere and ground can be estimated by

$$\text{Rate of change in heat storage} = \frac{\partial E_s}{\partial t} = \int_V \rho c_m \frac{\partial T}{\partial t} dV \qquad (2.28)$$

where $\rho$, $c_m$ and $T$ are the density, specific heat capacity (in J kg$^{-1}$ K$^{-1}$) and temperature of the medium (i.e., atmosphere, sea or ground) included in the volume $V$, which corresponds to an area $S$ on the Earth's surface. This term can be approximated by

$$\text{Rate of change in heat storage} \simeq m_m c_m \frac{\partial T_m}{\partial t} = C_m S \frac{\partial T_m}{\partial t} \qquad (2.29)$$

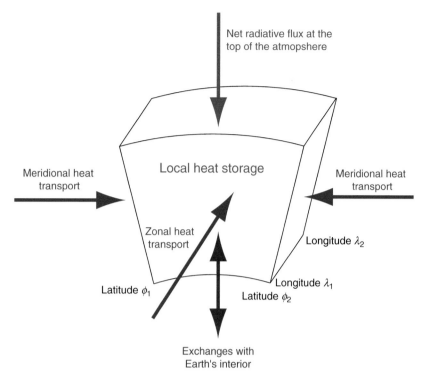

Net radiative flux at the
top of the atmopshere

Meridional heat
transport

Local heat storage

Meridional heat
transport

Zonal heat
transport

Longitude $\lambda_2$

Latitude $\phi_1$

Longitude $\lambda_1$
Latitude $\phi_2$

Exchanges with
Earth's interior

**Fig. 2.15** Schematic representation of the heat balance for the atmosphere, ocean, land surface and ground in a volume lying between latitudes $\phi_1$, $\phi_2$ and longitudes $\lambda_1$, $\lambda_2$.

where $m_m$ and $T_m$ are the characteristic mass and temperature of the medium that is storing heat. $C_m$ is the effective heat capacity of the medium per surface unit (measured in J K$^{-1}$ m$^{-2}$). The value of $m_m$ strongly depends on the volume that displays significant changes in heat content on the time scale of interest.

On the seasonal time scale, the heat content of the whole atmosphere changes. The mass of the atmosphere is about $10^4$ kg m$^{-2}$. Assuming **hydrostatic equilibrium**, this corresponds roughly to a pressure of $10^5$ Pa. $c_m$ for the atmosphere is the specific heat at constant pressure $c_p = 1004$ J K$^{-1}$ kg$^{-1}$. The effective heat capacity of the atmosphere is thus

$$C_{\text{atmosphere}} \simeq 1000 \times 10^4 = 10^7 \, \text{J K}^{-1}\text{m}^{-2} \tag{2.30}$$

Using a specific heat capacity of water $c_w = 4180$ J K$^{-1}$ kg$^{-1}$, this thermal capacity of the whole atmosphere is equivalent to one of the top 2.5 m of the ocean only. Because the top 50 to 100 m of the ocean displays a significant seasonal cycle in temperature, if we take a mass of $7.5 \times 10^4$ kg m$^{-2}$ (i.e., 75 m × 1,000 kg m$^{-3}$), we have

$$C_{\text{ocean}} \simeq 4,180 \times 7.5 \times 10^4 \simeq 3.10^8 \, \text{J K}^{-1}\text{m}^{-2} \tag{2.31}$$

The ground has a specific heat capacity similar to that of the ocean, but only a few metres are affected by the seasonal cycle. As a consequence, the effective

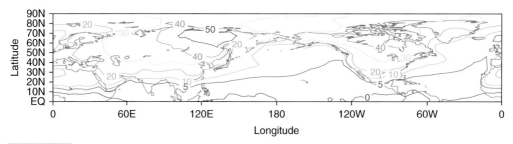

**Fig. 2.16**   Amplitude of the seasonal cycle in surface temperature (in °C) in the northern hemisphere measured as the difference between July and January monthly mean temperatures. (Source: Data from the HadCRUT2 data set; Rayner et al. 2003.)

heat capacity of the ground is much lower than that of the ocean on this time scale.

This rough comparison clearly shows that the effective thermal capacity of the ocean is an order of magnitude larger than that of the atmosphere and the ground on a seasonal time scale. Thus, the ocean stores much more energy during summer than the other media – energy that is released during winter. This moderates the amplitude of the seasonal cycle over the ocean by comparison with the land. A strong difference in the amplitude of the seasonal cycle is also seen in land areas that are directly influenced by the sea compared to regions characterised by a larger **continentality** (Figure 2.16). It is particularly clear at middle latitudes because of the westerly winds which transport air previously in contact with the sea water to land masses located east of the oceans. This explains why Western Europe displays a much smaller amplitude of the seasonal cycle than central or eastern Asia. A similar analysis on a daily time scale shows that heat storage by land, sea and atmosphere is all important in that case.

Energy also can be stored in the form of **latent heat** in water vapour (positive storage as heat is released when the vapour condensates) and in snow and ice (negative storage compared to liquid water). For instance, melting 10 cm of ice requires a net heat flux of 25 W m$^{-2}$ over about two weeks and thus may delay significantly any surface warming (without taking into account the **feedbacks** associated with the cryosphere; see Section 4.2.2).

For decadal to centennial variations, such as the warming observed since the mid-nineteenth century, thermal heat storage in the first hundred metres of the ocean (and at greater depths in regions of **deep-water formation**) also moderates the transient temperature changes (see Sections 4.1.5 and 6.2.3). On much longer time scales, such as the glacial-interglacial cycles, we have to take into account the full depth of the ocean (~4000 m). For deglaciation, which is faster than the **glacial inception** (see Section 5.5), we can estimate the order of magnitude of the mean ocean temperature as a 1°C change in 5000 years. This corresponds to a mean heat flux at the ocean surface of 0.1 W m$^{-2}$[$= (4000$ m $\times$ 1000 kg m$^{-3}$ $\times$ 4000 J K$^{-1}$ kg$^{-1}$ $\times$ 1°C)/(5000 $\times$ 365 $\times$ 24 $\times$ 3600 s)]. This demonstrates that the change in oceanic heat storage plays a small role on these time scales.

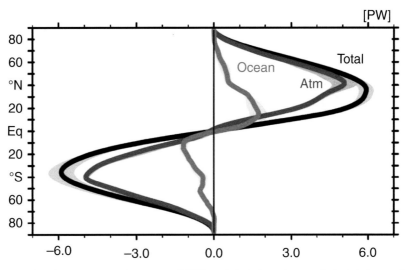

**Fig. 2.17**   The required total energy transport in petawatts ($10^{15}$ W) needed to balance the net radiative imbalance at the top of the atmosphere (in black) and the repartition of this transport in oceanic (blue) and atmospheric (red) contributions, accompanied by the associated uncertainty range (shaded). A positive value of the transport on the x-axis corresponds to a northward transport. (Source: Fasullo, J. T., and K. E. Trenberth (2008). The annual cycle of the energy budget: II. Meridional structures and poleward transports. *Journal of Climate* 21, 2313–25. © American Meteorological Society. Used with permission.)

### 2.1.5.2  Energy Transport

Heat storage in the **climate system** cannot compensate locally for the net radiative flux imbalance at the top of the atmosphere, and annually, the balance is nearly entirely achieved by energy transport from regions with a positive net radiative flux to regions with a negative net radiative flux. When the balance is averaged over latitudinal circles (**zonal** mean), this corresponds to a **meridional** energy transport from equatorial to polar regions (Figure 2.17). If in Eq. (2.27) we neglect the contribution from heat storage, the poleward energy transport $F_{transp}$ at a latitude $\phi$ can be estimated by integrating the net radiative balance at the top of the atmosphere $RF_{TOA}$ from the South Pole (latitude $-\pi/2$) to latitude $\phi$

$$F_{transp}(\phi) = \int_{-\pi/2}^{\phi} \int_{0}^{2\pi} RF_{TOA}\left(\lambda, \phi'\right)R^2 \cos\phi' d\lambda d\phi' \qquad (2.32)$$

The energy transport obtained is close to zero at the equator, rising to more than 5 PW at latitudes of about 35°, before declining again towards zero at the poles by construction (see Figure 2.17). It can be divided into oceanic and atmospheric contributions, the horizontal transport on the continental surface being negligible. This shows that except in tropical areas, the atmospheric transport is much larger than the oceanic transport.

The energy can be transported as **sensible heat** $c_p T$, potential energy $gz$, **latent heat** $L_v q$ and kinetic energy $0.5u^2$ and is expressed per unit of mass as

$$E = c_p T + gz + L_v q + 0.5u^2 \qquad (2.33)$$

where $z$ is the altitude (or depth), $L_v$ is the latent heat of vaporisation of water, $q$ is the **specific humidity** and $u$ is the velocity of the medium. The terms $c_pT + gz$ is called 'dry static energy', and $c_pT + gz + L_vq$ is the 'moist static energy' [see, e.g., Wallace and Hobbs (2006) or Marshall and Plumb (2008) for more details].

The transport of kinetic energy is much weaker that the other transports and is generally neglected. In the atmosphere, the three remaining terms must be taken into account. In the ocean, the transport of sensible heat is clearly dominant. Moreover, an additional term representing the transport of latent heat by sea ice and icebergs must be considered for local or regional analyses at high latitudes.

In the tropics, most of the atmospheric poleward energy transport is achieved by the Hadley circulation. By contrast, the mean circulation plays a much weaker role at middle to high latitudes, where nearly all the transport is effected by the disturbances corresponding to high and low pressure systems (see Section 1.2.2). In the ocean, both the wind-driven and deep-oceanic circulation are responsible for a significant part of the oceanic poleward energy transport. The role of the oceanic **eddies** is less well known, but it can be significant in at least some regions (such as the Southern Ocean).

In addition to its dominant role in reduction of the temperature contrast between the equator and the poles on Earth (compared to a planet without an ocean and an atmosphere), horizontal energy transport is also responsible for some temperature differences on a regional scale. This can be illustrated by analysing the departure of the local temperature from the **zonal** mean temperature. At first sight, Figure 2.18 emphasises the mountainous area such as the Tibetan Plateau and Greenland, where the temperature is much lower than at other locations at the same latitude. However, the influence of the atmospheric circulation is also clearly apparent, for instance, in cold areas such as northeast Canada. This is so because the dominant winds have a strong northerly component in this region, whilst the western North Atlantic is warmer partly because of the south-westerly winds in this region (see Section 1.2). Over the ocean, the influence

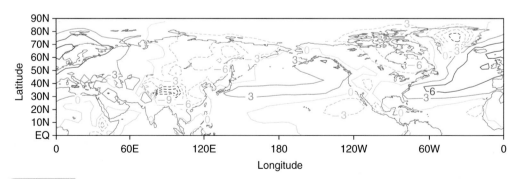

**Fig. 2.18**   Difference between the annual mean surface temperature and the zonal mean temperature (computed as the annual mean temperature measured at one particular point minus the mean temperature obtained at the same latitude but averaged over all possible longitudes). (Source: Data from the HadCRUT2 data set; Rayner et al. 2003.)

of the northward western boundary currents (see Section 1.3) results in a generally higher surface oceanic temperature at about 30−40°N in the western part of the basin than in the eastern part (where the oceanic currents are generally bringing colder water from the North).

The **thermohaline circulation** is an additional source of longitudinal asymmetry because, in the northern hemisphere, **deep-water formation** only occurs in the North Atlantic and not in the North Pacific (see Section 1.3). The associated circulation transports cold water southward at great depths, with the mass balance ensured by a corresponding northward transport of warmer water in the surface layer. This results in a net oceanic transport in the North Atlantic of about 0.8 PW at 30°N (i.e., more than twice the estimated oceanic heat transport in the wider Pacific at the same latitude). The thermohaline circulation is also responsible for the oceanic heat transport being northward at all latitudes in the Atlantic, even in the southern hemisphere.

This oceanic heat transport contributes to the fact that higher temperatures are observed in the North Atlantic than in other oceanic basins. Its influence is particularly large in the Barents Sea, north of Norway. Thanks to the oceanic heat transport, this area located north of 70°N (i.e., at the same latitude as the northern part of Alaska) remains free of sea ice all year long. Climate model calculations have shown that if deep-water formation were suppressed in the North Atlantic, the temperature in the North Atlantic and in western Europe would be reduced by about 3°C at 45°N, whilst the annual mean temperature would decrease by more than 15°C in northern Norway and the Barents Sea. More generally, the net northward heat transport by the thermohaline circulation at the equator partly explains the larger surface temperatures of the northern hemisphere compared to the southern hemisphere in present-day conditions and the position of the Intertropical Convergence Zone (ITCZ) slightly north of the equator (see, e.g., Feulner et al. 2013; Marshall et al. 2014).

## 2.1.6  Energy Balance at the Surface

As discussed in Section 2.1.1, the incoming solar flux on a horizontal surface at the top of the atmosphere is about 340 W m$^{-2}$, with roughly 30% of this being reflected back into space. An analysis of the Earth's global energy balance (Figure 2.19) shows that about 75% of the reflection takes place in the atmosphere, mainly because of the presence of clouds and **aerosols**. The remaining 25% is reflected by the surface. By contrast, most of the absorption of solar radiation occurs at the surface, which absorbs twice as much solar energy as the whole atmosphere. This shows clearly that most atmospheric warming occurs from below and not by direct absorption of solar radiation. This important property of the system explains the major characteristics of the Earth's atmosphere, including the vertical temperature profile and the large-scale circulation of the atmosphere (see Section 1.2).

The outgoing **longwave** flux required to balance the Earth's energy budget at the top of the atmosphere is mainly emitted by the atmosphere. Among the 398 W m$^{-2}$ emitted by the surface, only about 40 W m$^{-2}$ can exit the **climate system** directly in

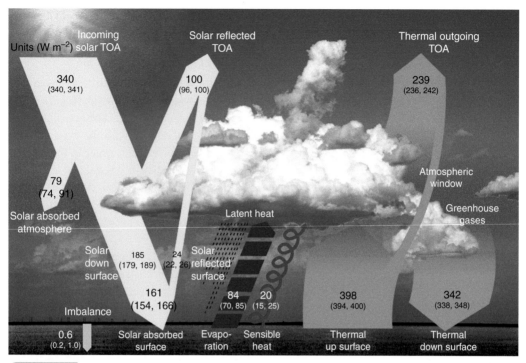

**Fig. 2.19**    Estimate of the Earth's annual and global mean energy balance in the first decade of the twenty-first century. The numbers represent estimates of each individual energy flux, whose uncertainty is given in the parentheses using smaller fonts. The uncertainties are larger at the surface than at the top of the atmosphere because for the latter, the fluxes can be observed directly from satellites. (Source: Hartmann et al. 2013, as adapted from Wild et al. 2013.)

the part of the infrared spectrum where the atmosphere is relatively transparent (i.e., the so-called atmospheric window). Most of the surface longwave radiation is absorbed by the atmospheric greenhouse gases. The atmospheric re-emission towards the surface makes the downwards longwave flux (342 W m$^{-2}$) the largest term in the surface energy balance.

The net radiative balance of the atmosphere is negative, whilst that of the surface is positive. To close the budget, the surface and the atmosphere exchange heat through direct contact between the surface and the air (**sensible heat** flux or thermals) as well as through evaporation and transpiration. Indeed, when evaporation (or sublimation) takes place at the surface, the **latent heat** required for the phase transition is taken out of the surface and results in surface cooling. Later, mainly during the formation of clouds, the water vapour condenses, and the latent heat is released into the atmosphere. This leads to a net mass and heat transfer from the surface into the atmosphere, which is one of the main drivers of the general atmospheric circulation. Actually, because the system is not presently at equilibrium, an imbalance equivalent to a mean flux of 0.6 W m$^{-2}$ is estimated both at the surface and at the top of the atmosphere corresponding to the heat storage in the ocean (see Section 5.7).

The fluxes of sensible and latent heat are generally estimated as a function of the wind speed at a reference level and the difference in temperature (for the sensible heat flux $F_{SH}$) or specific humidity (for the latent heat flux $F_{LH}$) between the surface and the air at this reference level using classical bulk aerodynamic formulae:

$$F_{SH} = \rho c_p c_h U_a \left( T_s - T_a \right) \tag{2.34}$$

$$F_{LH} = \rho L_v c_L U_a \left( q_s - q_a \right) \tag{2.35}$$

where $U_a$, $T_a$ and $q_a$ are the wind velocity, air temperature and specific humidity at the reference level (generally 2 or 10 m), $T_s$ and $q_s$ are the surface temperature and **specific humidity** at the surface and $c_h$ and $c_L$ are the aerodynamic (bulk) coefficients. In general, they are function of the stability of the **atmospheric boundary layer**, the roughness of the surface, the wind speed and the reference height. In most cases, $c_h$ and $c_L$ are not too different from each other, and their value ranges from $1 \times 10^{-3}$ to $5 \times 10^{-5}$. The highest values occur with unstable boundary layers and very rough surfaces, which tend to generate strong turbulent motions and thus higher exchanges between the surface and the air than quieter situations.

The **specific humidity** $q_s$ above a wet surface is generally very close to saturation and thus strongly depends on the temperature (see Section 1.2). As a consequence, the evaporation rate and the latent heat flux are much larger at low latitudes than at high latitudes. The latent heat flux thus is larger than the sensible heat flux at low latitudes, whilst the two fluxes are generally of the same order of magnitude over the ocean at high latitudes. The ratio between the sensible heat and latent heat fluxes is usually expressed as the Bowen ratio $B_o$; that is,

$$B_o = \frac{F_{SH}}{F_{LH}} \tag{2.36}$$

Over land surfaces, the latent heat flux is a function of water availability, and $B_o$ can be much higher than unity over dry areas (see Section 2.2).

The energy balance shown in Figure 2.19 for the whole Earth also can be computed for any particular surface on Earth. This is generally the method used to compute the surface temperature $T_s$. Let us consider a unit volume at the Earth's surface with an area of 1 m² and a thickness $h_{su}$ (Figure 2.20). $h_{su}$ is supposed to be sufficiently small to safely make the approximation that the temperature is constant over $h_{su}$ and equal to $T_s$. The energy balance of this volume then can be expressed as

$$\rho c_{\text{ground}} h_{su} \frac{\partial T_s}{\partial t} = \left( 1 - \alpha \right) F_{\text{solar}} + F_{IR\downarrow} + F_{IR\uparrow} + F_{SH} + F_{LH} + F_{\text{cond}} \tag{2.37}$$

The left-hand side of the Eq. (2.37) represents the heat storage in the layer $h_{su}$, with $c_{\text{ground}}$ being the specific heat capacity of the ground (see Section 2.1.5). $F_{\text{solar}}$ is the incoming solar flux at the surface, which is a function of the incoming

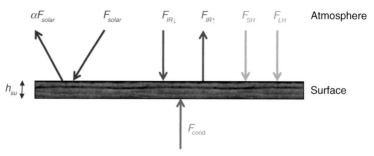

Fig. 2.20 The energy balance of a surface.

solar flux at the top of the atmosphere and the transmissivity of the atmosphere (related to the presence of clouds and aerosols, the humidity of the air, etc.). A fraction $\alpha$ of $F_{solar}$ is reflected by the surface and not absorbed, and we assume in this equation that no solar radiation is transmitted below $h_{su}$. $F_{IR\downarrow}$ is the downwards **longwave** flux at the surface. This flux is caused by the emission of infrared radiation at various levels in the atmosphere. It is thus a complex function of the temperature and humidity profiles in the atmosphere, the cloud cover and the height of the clouds, the presence of various greenhouse gases (in addition to water vapour), and so on. The longwave upward flux $F_{IR\uparrow}$ can be computed using the **Stefan-Boltzmann law**, whilst the expressions for $F_{SH}$ and $F_{LH}$ are given by Eqs. (2.34) and (2.35). $F_{cond}$, the flux from below the surface, is a conduction flux for solid surfaces (e.g., ground and ice), which can be represented following **Fourier's law.** It is generally associated with small vertical heat transfer fluxes because of the low conductivity of the ground. For the ocean, this flux is related to the dynamics of the **ocean's mixed layer**. Because of the turbulence that induces efficient vertical exchanges over a depth reaching several tens of metres (see Section 1.3.3), this leads to higher vertical heat fluxes than over land. Additionally, if the medium at the surface is (partly) transparent, a fraction of the flux is not absorbed in the layer of thickness $h_{su}$ and must be subtracted from the term $(1-\alpha F_{solar})$ in Eq. (2.37). For the other fluxes, the exchanges take place in a very shallow layer and reasonably can be considered as purely surface processes.

Figure 2.20 displays a relatively simple situation where the surface (i.e., the interface between the atmosphere and the material below) is clearly defined. In complex terrain with very rough topography, for instance, over forests or urban areas, defining the lower limit of the atmosphere is less straightforward. Computing the surface fluxes in these regions is a very complex issue which is currently the subject of intense research (see Section 3.3.4).

When snow or ice is present at the surface, the temperature $T_s$ cannot be higher than the freezing point of water. As a consequence, Eq. (2.37) remains valid as long as $T_s$ is below the freezing point. When surface melting occurs (i.e., when $T_s$ equals the freezing point of water), an additional term corresponding to the latent heat of fusion required to keep the temperature unchanged must be added to the right-hand side of Eq. (2.37).

## 2.2 **The Hydrological Cycle**

### 2.2.1 Global Water Balance

As discussed in Section 2.1, the water – or hydrological – cycle plays an important role in the energy cycle on Earth. It has a considerable impact on the radiative balance: water vapour is the most important greenhouse gas in the atmosphere (see Section 2.1.2); the presence of snow and ice strongly modifies the **albedo** of the surface (see Table 1.3 and Sections 2.1.4 and 4.2.2); clouds influence both the **longwave** and **shortwave** fluxes (see Sections 2.1.4, 2.1.6 and 4.2.3). Moreover, water is an essential vehicle for energy: the **latent heat** released during the condensation of water is a dominant heat source for the atmosphere (see Section 2.1.6); the transport of water vapour in the atmosphere and of water at different temperatures in the ocean is an essential term in the horizontal heat transport (see Section 2.1.5.2).

The hydrological cycle is also essential in shaping the Earth's environment, the availability of water being a critical factor for life as well as for many chemical reactions and transformations affecting the physical environment. Describing the various components of the hydrological cycle and analysing the mechanisms responsible for the exchanges of water between the different reservoirs thus are important elements of climate science.

By far the largest reservoir of water on Earth is located in the crust, with estimates on the order of $10^{22}$ kg of water (equivalent to $10^{19}$ m$^3$ at surface pressure, i.e., about ten times the amount of water in the oceans, the second largest reservoir). However, exchanges between deep Earth and other reservoirs are so slow that they have only a very weak impact on the hydrological cycle at the surface and thus generally are not taken into account in estimates of the global hydrological cycle (Figure 2.21).

A large amount of water is also stored in groundwater and in the form of ice, mainly in the Greenland and Antarctic ice sheets (see Section 1.4). By contrast, the stock of water in the soils and the atmosphere is low. The estimate of soil water is relatively uncertain, but a value of $122 \times 10^3$ km$^3$ proposed in Figure 2.21, distributed over the whole Earth, would correspond to a layer of about 20 cm $[= 122 \times 10^3$ km$^3/(4\pi R^2)$, where $R$ is the Earth's radius]. If the $12.7 \times 10^3$ km$^3$ of atmospheric water estimated in Figure 2.21 all precipitated, it would lead to a mean rainfall of about 2.5 cm. As the actual precipitation on the Earth's surface is on the order of $500 \times 10^3$ km$^3$ yr$^{-1}$ or 1 myr$^{-1}$ on average, the water in the atmosphere must be replaced very quickly. This is achieved by evaporation over the oceans and other water bodies as well as by **evapotranspiration** over land. Most of the water that evaporates over the oceans falls back over the oceans (and similarly, water that evaporates over land falls back over land), but there is also water transfer by the atmosphere from the oceanic area to the land area. This net transfer corresponds to roughly 35% of the total precipitation over land and is compensated by a surface flow of water (mainly in rivers) from the land to the sea.

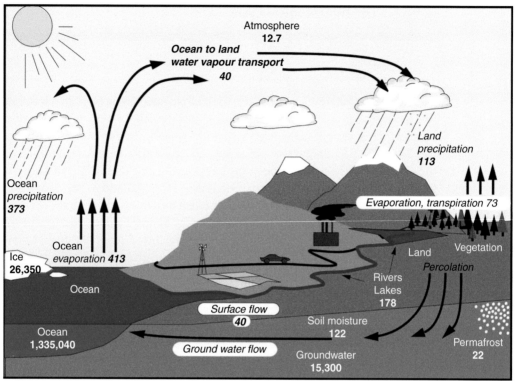

Units: Thousand cubic km for storage, and *thousand cubic km per year* for exchanges

**Fig. 2.21**   The long-term mean global hydrological cycle. Estimates of the main water reservoirs in regular type (e.g., 'Soil moisture') are given in $10^3$ km$^3$, and estimates of the flows between the reservoirs in italic type (e.g. *Surface flow*) are given in $10^3$ km$^3$ yr$^{-1}$. (Source: Trenberth, K. E., L. Smith, T. T. Qian, A. G. Dai and J. Fasullo (2007). Estimates of the global water budget and its annual cycle using observational and model data. *Journal of Hydrometeorology* 8, 758–69, who provide information about the sources used to estimate the magnitude of the elements of the cycle and about the uncertainties of the various terms. © American Meteorological Society. Used with permission.)

## 2.2.2  Water Balance on Land

The soil water content per volume unit $\theta_{sw}$, also referred to as 'volumetric soil moisture', is defined as the volume of water contained in a specified volume $V$ divided by this volume

$$\theta_{sw} = \left( \frac{\text{volume of water in } V}{V} \right) \tag{2.38}$$

When the ratio is performed for a small volume, typically a few cubic centimetres, $\theta_{sw}$ represents the local conditions in soils at a particular depth. Alternatively, $V$ could be chosen as the volume between the surface and a specified depth or as the total volume that can interact directly with the surface, generally corresponding to the zone that contains all the plants roots. $\theta_{sw}$ then can be related to

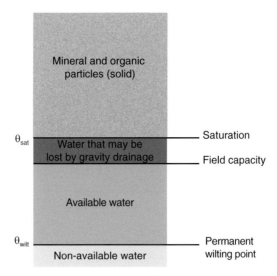

Fig. 2.22 Schematic representation of the fraction of the soil volume occupied by solid particles, unavailable water, the water available for plants and the water that can be rapidly lost by gravity drainage.

$S_m$, the absolute 'soil moisture content' or simply 'moisture content', defined as the water content per unit surface of a soil layer of thickness $d$ (in metres)

$$S_m = \theta_{sw} d \qquad (2.39)$$

Note that $S_m$ is sometimes used to represent the total water content at the surface. Consequently, in addition to the soil moisture, it has to take into account snow, ice and the water present at surface, in ponds, for instance.

The soils are composed of inorganic (mineral) and organic particles that, because of their various shapes and sizes, do not fill the full available space, allowing the presence of **pores** which can contain air and water (Figure 2.22). When all the pores are filled with water ($\theta_{sw} = \theta_{sat}$), the soil is said to be saturated, and the soil reaches its maximal water-holding capacity. $\theta_{sat}$ depends on the composition and type of soil but is generally on the order of 50% (see, e.g., Bonan 2008). Except in wetlands, the saturation is observed only during or immediately after heavy rains because a fraction of the water is rapidly drained by gravity. The water remaining after this drainage, which thus leaves the soil unsaturated, is called the 'field capacity'. A part of this water is too strongly bound with soil particles to be extracted by the plants. When the soil water content reaches the corresponding level because of drying, the plants wilt. As a consequence, it is called the 'wilting point' or 'permanent wilting point' ($\theta_{sw} = \theta_{wilt}$).

In order to estimate the time development of $S$, the water balance of the soil can be obtained from

$$\frac{dS_m}{dt} = P - E - R_s - R_g \qquad (2.40)$$

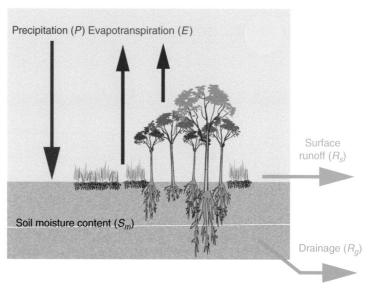

Precipitation ($P$) Evapotranspiration ($E$)

Surface runoff ($R_s$)

Soil moisture content ($S_m$)

Drainage ($R_g$)

**Fig. 2.23**  Schematic representation of the soil water balance. (Source: Modified from Seneviratne et al. 2010.)

knowing the precipitation $P$, the **evapotranspiration** $E$, the surface flow $R_s$ and the drainage $R_g$ that feeds the groundwater reservoir or the deeper soil levels depending on the choice of the depth over which the balance is performed (Figure 2.23). On bare soils, all the precipitation reaches the ground surface directly, but this is not the case when the vegetation cover is dense. In particular, trees can intercept a fraction of the incoming precipitation which depends on cover type and the intensity of precipitation. This interception is larger for light rain, leading to no precipitation reaching the ground, for most of the forest types, if it does not exceed 1 mm (Crockford and Richardson 2000; Bonan 2008). This water holding on branches and leaves then can be rapidly evaporated, significantly affecting the water balance and the renewal of soil moisture.

Evapotranspiration makes the link between the energy balance [Eq. (2.37)] and the water balance [Eq. (2.40)] of the surface because the latent heat flux $F_{LH} = -L_v E$ is a dominant contributor of the energy and freshwater cycles ($L_v$ being the latent heat of vaporisation of water). Besides, the evaporation is controlled by both the availability of water and the energy budget of the land surface. If the soil is wet, evapotranspiration is restrained only by the heat required to induce the shift from liquid to vapour phase. This situation is found at high latitudes because of the cold temperatures but also in some humid tropical forests (Seneviratne et al. 2010). If the soil is dry, the soil moisture is the limiting factor, as observed in many tropical and subtropical regions.

Because of this influence of several factors, it is convenient to define the 'potential evapotranspiration' as the evapotranspiration that would occur in the absence of limitation by water availability. This often corresponds to the evaporation at the surface of a lake or pond in similar weather conditions, but some definitions of the potential evapotranspiration take into account the influence of the vegetation cover on surface characteristics such as the roughness. The observed evapotranspiration then is given as a fraction of this maximum value given by the potential evapotranspiration.

### 2.2.3  Local Water Balance and Water Transport

In agreement with the **Clausius-Clapeyron equation**, intense evaporation occurs in the warm equatorial areas and in the tropics. In equatorial areas, because of the convergence at the surface and upward motions, the moist air at low levels rises, reaching a colder level. This induces condensation, the formation of clouds and high precipitation rates (see Figure 1.8). Despite the high temperature and high evaporation rate, equatorial regions thus have more precipitation than evaporation at the surface (negative $E - P$; Figure 2.24). In the subtropics, evaporation minus precipitation $(E - P)$ is clearly positive because of the general **subsidence** at these latitudes. At middle to high latitudes, $E - P$ is again negative on **zonal** average because of the net moisture transport from tropical areas.

The imbalance of the water budget over land is generally small and is compensated by runoff (in the long-term mean, the total runoff including surface river runoff and groundwater flow nearly equals $P - E$) which returns water to the sea. Because of the land topography, this river runoff is an important element of

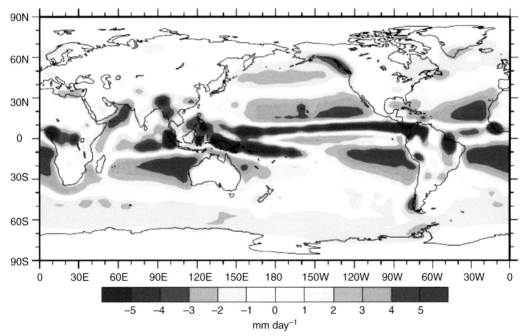

**Fig. 2.24**   Long-term annual mean evaporation minus precipitation $(E - P)$ budget based on ERA-40 **reanalyses**. Units are millimetres per day. The major features of the $E - P$ budget are represented in this figure, but because of large uncertainties, the values displayed should not be considered to be quantitatively exact. There are also some clear discrepancies over land, in particular, where net positive evaporation is shown (negative values generally should occur over land as river runoff is positive). (Source: Trenberth, K. E., L. Smith, T. T. Qian, A. G. Dai and J. Fasullo (2007). Estimates of the global water budget and its annual cycle using observational and model data. *Journal of Hydrometeorology* 8, 758–69. © American Meteorological Society. Used with permission.)

the water balance for some ocean basins. For instance, the Arctic Ocean receives about 10% of the total river runoff (mainly from the Russian rivers), although it only constitutes about 3% of the World Ocean. This partly explains why surface water in the Arctic is relatively fresh (see Figure 1.13).

Over the oceans, the imbalance of the water budget at the surface is larger than over land because it can be compensated for by a net oceanic water transport which is much more efficient (and much larger) than that associated with river runoff. This net oceanic transport counter-balances the large atmospheric moisture transport out of the subtropics towards the equatorial regions and middle and high latitudes. In a way, the net meridional water transport and the associated energy transport in the atmosphere are only possible because the ocean transport is able to compensate for the imbalance at the surface due to the $E - P$ fluxes.

Net oceanic water transport also can counter-balance the **zonal** transport by the atmosphere. In particular, because of high $E - P$ rates, the Atlantic Ocean is a net evaporative basin and thus is more saline at the surface than the Pacific, where the $E - P$ balance integrated over the whole basin is negative (see Figure 1.13). As a consequence, the global oceanic circulation must induce a net water transport from the Pacific to the Atlantic to achieve a water balance in both basins.

## 2.3  The Carbon Cycle

### 2.3.1 Overview

The transfers between the different components of the climate system include exchanges of energy (Section 2.1), exchanges of single molecules such as water ($H_2O$) with phase changes (Section 2.2) and exchanges involving chemical transformations. This latter type of exchange implies elements such as oxygen, nitrogen, phosphate and sulphur, but one of the most important cycles from a climatic point of view is the carbon cycle because it leads to changes in the atmospheric concentration of two important greenhouse gases (Section 2.1.2): carbon dioxide ($CO_2$) and methane ($CH_4$).

One of the major changes brought about by human activity is the large increase in the atmospheric concentration of those two gases. The concentration of carbon dioxide has increased from 278 ppm around 1750 to 395 ppm in 2013. Because $CO_2$ is relatively stable chemically, it is well mixed, and its concentration in the atmosphere is nearly homogeneous away from zones where strong exchanges with the biosphere occur. Those exchanges with the biosphere are also responsible for the weak seasonal cycle observed in the long Mauna Loa record (Figure 2.25). $CH_4$ is more reactive than $CO_2$ (see Section 2.3.5). Its concentration is lower than that of $CO_2$, but it has increased quickly from 722 ppb to more than 1800 ppb over the last 250 years (Ciais et al. 2013).

The atmosphere is a relatively small reservoir of carbon compared to sedimentary rocks, the oceans and the terrestrial biosphere (which includes in common

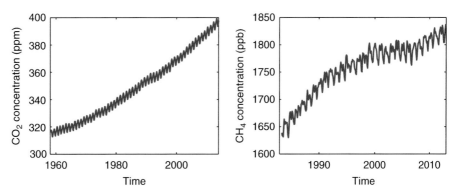

**Fig. 2.25**    (a) Monthly mean atmospheric carbon dioxide concentration (in ppm) at Mauna Loa Observatory, Hawaii (in red). (Sources: Dr. Pieter Tans, NOAA/ESRL (www.esrl.noaa.gov/gmd/ccgg/trends/), and Dr. Ralph Keeling, Scripps Institution of Oceanography (scrippsco2.ucsd.edu/).) (b) Monthly mean methane concentration (in ppb) at Mauna Loa (in red). (Source: NOAA Annual Greenhouse Gas Index (AGGI); available at: http://www.esrl.noaa .gov/gmd/aggi/; Dlugokencky et al. 2013.)

practice non-living organic material such as soil carbon) (Figure 2.26). In particular, more than $50 \times 10^6$ PgC (petagrams of carbon, $10^{15}$ gC or, equivalently, gigatons of carbon, GtC) is stored in the Earth's crust. This is more than 1000 times the stock in the ocean, more than 30,000 times the stock in soils and more than 80,000 times the stock in the atmosphere. However, the changes in the carbon concentration in sedimentary rocks are very small, and the associated fluxes are much lower than those between the ocean, the atmosphere and the soil.

Before the industrial era (i.e., before 1750), the exchanges between the various reservoirs were close to equilibrium. However, because of anthropogenic carbon release mainly related to fossil-fuel burning and cement production (~375 PgC) and changes in land use (e.g., deforestation and agriculture processes, about 180 PgC), the flux of carbon into the atmosphere has increased dramatically during the last 150 years. Roughly 43% of the anthropogenic carbon released up to now has remained in the atmosphere (240 PgC), which explains the observed rise in atmospheric $CO_2$. This is often referred to as the 'airborne fraction'. The remaining part has been absorbed by the ocean (~28%, 155 PgC) or the terrestrial biosphere (~29%, 160 PgC) (Ciais et al. 2013). The net contribution of the terrestrial biosphere thus includes emissions due to land-use changes and a partial compensation due to the land accumulation (Figure 2.26).

The processes responsible for repartition of the emitted carbon among the atmosphere, the ocean and the biosphere will be discussed later, mainly in Sections 4.3.1 and 6.2.9. Nevertheless, it should be stressed at this stage that the fraction stored in each reservoir is a function of the history of the forcing and the state of the system and thus is not constant in time.

### 2.3.2  Oceanic Carbon Cycle

A flux of $CO_2$ between the ocean and the atmosphere $\Phi^{CO_2}$ occurs when the $CO_2$ content of the ocean surface is not in equilibrium with the atmospheric

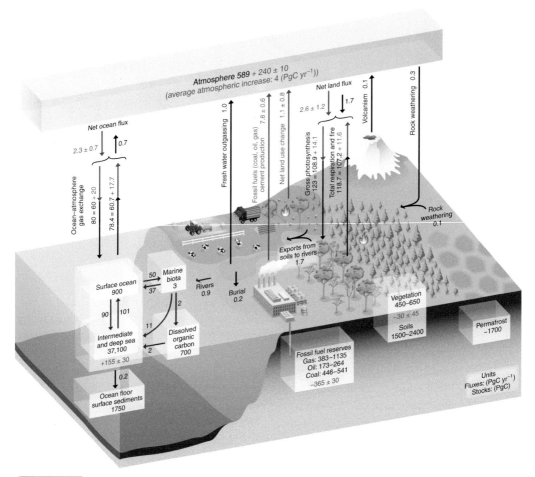

**Fig. 2.26** The global carbon cycle showing the main annual fluxes in petagrams of carbon (PgC) per year. Pre-industrial 'natural' fluxes, corresponding to conditions around 1750, are shown in black, and 'anthropogenic' fluxes averaged over 2000–2009 are shown in red. The carbon stocks in the different reservoirs are given in PgC. The carbon stored in deep sediments and in the Earth's crust is estimated at around $50 \times 10^6$ PgC. The atmospheric inventories have been calculated using a conversion factor of 2.12 PgC/ppm. (Source: Ciais et al. 2013.)

concentration. The magnitude of $\Phi^{CO_2}$ is proportional to the imbalance, and it can be computed as a function of the difference in the partial pressure $p^{CO_2}$ between the two media

$$\Phi^{CO_2} = k^{CO_2} \left( p_W^{CO_2} - p_A^{CO_2} \right) \qquad (2.41)$$

where the subscripts $A$ and $W$ refer to air and water, respectively. $k^{CO_2}$ is a transfer coefficient which strongly depends on the wind velocity. At equilibrium, $p_W^{CO_2}$ is obviously equal to $p_A^{CO_2}$.

As the surface $CO_2$ concentration in the atmosphere is nearly homogeneous, the repartition of the flux mainly depends on the oceanic $p^{CO_2}$. Supersaturated zones, where the partial pressure of $CO_2$ in sea water is higher than in the air, have a positive

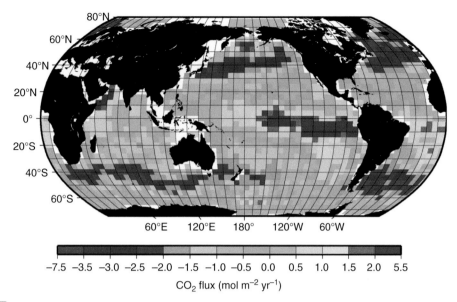

Fig. 2.27 Estimates of sea-to-air flux of $CO_2$. (Source: Denman et al. (2007), based on the work of T. Takahashi; available at: http://www.ldeo.columbia.edu/res/pi/CO2/carbondioxide/pages/air_sea_flux_rev1.html.)

flux from the ocean to the atmosphere. In present-day conditions, this occurs in tropical regions, particularly in the eastern equatorial Pacific (Figure 2.27). However, under saturated areas, such as the middle to high latitudes around 40 to 60° in both hemispheres (except the northern Pacific), display $CO_2$ fluxes from the atmosphere to the ocean, that is, an uptake of $CO_2$ from the atmosphere.

### 2.3.2.1 Inorganic Carbon Cycle

When gaseous $CO_2$ is transferred from the atmosphere to the ocean, it is hydrated to form aqueous $CO_2$ and reacts with water to form carbonic acid ($H_2CO_3$). As it is difficult to distinguish aqueous $CO_2$ and $H_2CO_3$, they are generally combined, and for simplicity, we will keep here the notation $H_2CO_3$ for the sum of those two species. $H_2CO_3$ dissociates, leading to the formation of bicarbonate ($HCO_3^-$) and carbonate ($CO_3^{2-}$) ions (see, e.g., Sarmiento and Gruber 2006)

$$CO_{2(gas)} + H_2O \rightleftharpoons H_2CO_3 \tag{2.42}$$

$$H_2CO_3 \rightleftharpoons H^+ + HCO_3^- \tag{2.43}$$

$$HCO_3^- \rightleftharpoons H^+ + CO_3^{2-} \tag{2.44}$$

These three equations are sometimes referred to as the 'carbonate equilibrium system', and the sum of the concentration of the three forms of carbon is defined as **'dissolved inorganic carbon'** (DIC)

$$DIC = \left[H_2CO_3\right] + \left[HCO_3^-\right] + \left[CO_3^{2-}\right] \tag{2.45}$$

These reactions [Eqs. (2.42)–(2.44)] are so fast that, to a good approximation, the three components are always in equilibrium. The equilibrium relationship between the different molecules involved in reaction (2.41) then can be used to define the solubility $K_H$ of $CO_2$, which relates the concentration of carbonic acid to the partial pressure of carbon dioxide ($p^{CO_2}$)

$$K_H = \frac{[H_2CO_3]}{p^{CO_2}} \qquad (2.46)$$

By definition, for the same atmospheric $p^{CO_2}$, the amount of carbonic acid in the ocean at equilibrium will be larger for a high solubility than for a low solubility. The transfer of $CO_2$ between the ocean and the atmosphere [Eq. (2.41)] then can be easily expressed as a function of $[H_2CO_3]$ using Eq. (2.46).

The solubility of $CO_2$ strongly depends on temperature. A water parcel that is cooled as it flows northward, for instance, will take up atmospheric $CO_2$, whilst a water parcel that is warmed will release $CO_2$ to the atmosphere. This generally leads to positive ocean–atmosphere fluxes in tropical regions and negative ones at high latitudes, as shown in Figure 2.27.

In the current conditions, the equilibrium between the different forms of carbon occurs at sea when nearly 90% of the DIC is in the form of bicarbonate, around 10% is carbonate and only 0.5% is carbonic acid. This predominance of carbonate and bicarbonate ions explains why the ocean is able to store much more carbon than the atmosphere, whilst this is not true for other gases (e.g., oxygen) which have similar solubility to $CO_2$. The atmospheric $CO_2$ must balance the whole pool of DIC, not just $H_2CO_3$, as would be the case if only reaction (2.42) were present. Specifically, it can be estimated that only one molecule in twenty of the $CO_2$ entering the ocean stays as $H_2CO_3$, the large majority reacting with $CO_3^{2-}$ to form $HCO_3^-$, the dominant species in DIC:

$$H_2CO_3 + CO_3^{2-} \rightleftharpoons 2HCO_3^- \qquad (2.47)$$

The ocean–atmosphere fluxes thus are strongly influenced by the availability of $CO_3^{2-}$. Additionally, because of the small fraction of incoming $CO_2$ staying as $H_2CO_3$ and thus influencing the $p^{CO_2}$ in the ocean, the time taken to reach equilibrium between the ocean and the atmosphere is about six months – around twenty times longer than if this reaction were not active. Reactions (2.42)–(2.44) also influence the dependency of equilibrium $CO_2$ concentration in sea water as a function of the temperature, amplifying the direct contribution via the solubility $K_H$ (Sarmiento and Gruber 2006).

As shown in reactions (2.43) and (2.44), the dissociation of $H_2CO_3$ causes the water to become more acidic. This effect is commonly measured via the 'alkalinity' (alk), defined as the excess of bases over acids in water

$$Alk = [HCO_3^-] + 2[CO_3^{2-}] + [OH^-] - [H^+] + \left[B(OH)_4^-\right] + \text{minor bases} \qquad (2.48)$$

where $[B(OH)_4^-]$ is the concentration of the borate ion. The total alkalinity is dominated by the influence of bicarbonate and carbonate ions, meaning that a tight link exists between alkalinity and the concentration of the three forms of carbon. Conversely, changes in total alkalinity or in the acidity of the ocean can have a strong influence on the equilibrium of reactions (2.43) and (2.44). For instance, if alkalinity decreases (or, equivalently, if the system become more acidic), the equilibrium of reactions (2.43) and (2.44) will be pushed towards the formation of more $H_2CO_3$ and $HCO_3^-$, increasing the concentration of $H_2CO_3$ and thus the $p^{CO_2}$ with a potential influence on air-sea fluxes. The estimates of this effect suggest an increase in 10% of the $p^{CO_2}$ for a 1% decrease in alkalinity (Sarmiento and Gruber 2006).

This influence of the alkalinity or the $CO_3^{2-}$ concentration on the equilibrium of the carbonate system also can be described using Le Châtelier's principle, which states that in response to a perturbation, a system in chemical equilibrium will react in a way that mitigates the influence of the initial perturbation in order to restore the equilibrium. For instance, reducing the $CO_3^{2-}$ concentration (corresponding to a decrease in alkalinity) would induce a shift of the equilibrium of reaction (2.47), leading to more production of $CO_3^{2-}$ and $H_2CO_3$ from $HCO_3^-$. This would reduce the magnitude of the changes in $CO_3^{2-}$ but increase the concentration of $H_2CO_3$ and thus $p^{CO_2}$, as mentioned earlier.

### 2.3.2.2 Biological Pumps

In addition to the effect of temperature on solubility, biological processes also play a significant role in the distribution of surface fluxes of $CO_2$ by affecting DIC and alkalinity. A first important reaction is photosynthesis, in which phytoplankton use solar radiation to form organic matter from $CO_2$ and water

$$6CO_2 + 6H_2O \rightleftharpoons C_6H_{12}O_6 + 6O_2 \tag{2.49}$$

Conversely, organic matter can be dissociated to form inorganic carbon (the reverse process of photosynthesis) by respiration and **remineralisation** of dead phytoplankton and detritus.

Reaction (2.49) is a highly simplified representation of the complex biological processes associated with photosynthesis. In particular, it hides the fact that in order to produce organic matter, phytoplankton need nutrients (mainly nitrates, phosphates and silicates) as well as minor elements such as iron. As those nutrients generally have low concentrations in surface water where light is available for photosynthesis, their concentration is often the limiting factor for biological production.

Because particles whose density is more than that of seawater settle out, and some particles are transported by ocean currents, a fraction of the organic matter produced in the surface layer is exported downwards. The net flux of carbon associated with this downwards transport of organic matter is called the 'soft-tissue pump'. A significant part of the remineralisation thus occurs in the deep layers, where it produces an increase in DIC and the release of nutrients. The deep waters are thus rich in nutrients. Where they **upwell** towards the surface, the surface concentration of nutrients increases, generally leading to high biological production, such as that observed off the coasts of Peru and Mauritania.

A second important biological process is related to the production of calcium carbonate (in form of **calcite** or **aragonite**) by different species, in particular, to form their shells

$$Ca^{2+} + CO_3^{2-} \rightleftharpoons CaCO_3 \qquad (2.50)$$

This production influences both the DIC and the alkalinity and thus can have a large influence on the carbon cycle. In particular, $CaCO_3$ production implies a reduction in alkalinity [see Eq. (2.48)], which, in turn, leads to an increase in oceanic $p^{CO_2}$ and reduces the uptake of atmospheric $CO_2$ by the ocean. An alternative way to view this mechanism is to say that $CaCO_3$ production reduces the concentration of $CO_3^{2-}$ in the ocean and thus the availability of this ion to combine with $H_2CO_3$ to produce $HCO_3^-$ [reaction (2.47)], so increasing $[H_2CO_3]$ and $p^{CO_2}$.

The dissolution of calcite and aragonite mainly occurs at great depth (see Section 4.3.3), following the precipitation of particles and dead organisms. This leads to the alkalinity and DIC being transported downwards, a system called the 'carbonate pump'. A third pump, called the 'solubility pump', is associated with the sinking of cold surface water, characterised by a relatively high solubility of $CO_2$ and thus high DIC, to great depths at high latitudes (see Section 1.3.2). All these downwards transports have to be compensated for at equilibrium by an upwards flux of inorganic carbon by the oceanic circulation.

Because of these three pumps, DIC is about 15% higher in the deep ocean than at the surface. The soft tissue pump plays the largest role in the observed vertical gradient. This distribution has a profound influence on the atmospheric $CO_2$ concentration. Indeed, if DIC were perfectly homogeneous in the water column (i.e., had higher surface values and lower depth values than currently observed), the concentration of atmospheric $CO_2$ would be much higher. More realistically, when deep water upwells to the surface, $CO_2$ will tend to escape from the ocean because of the high DIC. However, as the deep waters are rich in nutrients, the biological uptake associated with photosynthesis can compensate for the influence of a higher DIC, the net effect depending on the region.

### 2.3.3 Terrestrial Carbon Cycle

The uptake of carbon through photosynthesis by land plants is larger than the corresponding uptake by phytoplankton, in particular, in spring because of the greening of forest at middle and high latitudes and the growth of herbaceous plants (Figure 2.28). About half of this primary production is transferred back to the atmosphere by the respiration of the land plants themselves (Bonan 2008), the remaining part being incorporated into leaves, woods and roots [this fraction is defined as the '**net primary production**' (NPP)].

A large fraction of the carbon fixed by NPP is transferred as dead organic matter to the soil carbon pool, which is a larger carbon reservoir than the living biomass for many ecosystems (see Figure 2.26). This soil organic carbon then can be

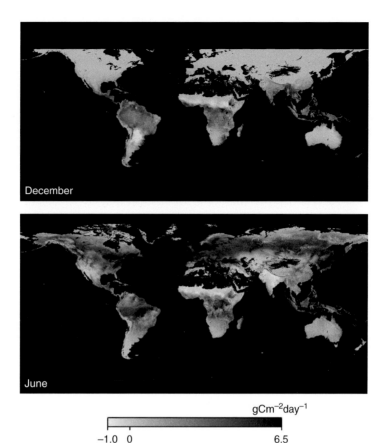

December

June

gCm$^{-2}$day$^{-1}$

−1.0  0                    6.5

**Fig. 2.28**  Net productivity over land in December 2013 and June 2014 based on Terra/MODIS satellite data. (Source: NASA Earth Observations; available at: http://neo.sci.gsfc.nasa.gov/view.php?datasetId=MOD17A2_M_PSN.)

decomposed and returned to the atmosphere. The decomposition is related to a large number of relatively complex processes, and its rate depends strongly on the chemical composition of the organic material, the temperature and the humidity of the soil. In particular, it strongly increases as temperature rises, leading to potential feedbacks with climate change (see Section 4.3.1). The decomposition is relatively fast for some elements such as leaves at the soil surface. Besides, the carbon can be stored for several decades in tree trunks or roots and at depth in the form of **humus**.

The preservation of carbon in soils is particularly good in wetlands as the water saturation reduces the exchanges of oxygen with the atmosphere, inducing lower decomposition rates. These conditions are also favourable for the formation of methane (see Section 2.3.5). If the sink due to decomposition is lower than the source of carbon from plant detritus, this leads to accumulation, mainly as peat. These peat deposits can reach thicknesses of several metres because the older layers are buried below the new ones that are formed by input at the surface. The associated storage at depth is estimated to be around 300 to 700 PgC (Ciais et al.

2013) and is generally not included in soil inventories, which take into account the top 1 m only.

Because of the low temperatures, the decomposition is also very slow in the **permafrost** region. This leads to a carbon storage estimated to be about 1700 PgC (see Figure 2.26) (Tarnocai et al. 2009), that is, about two times the current amount in the atmosphere. As this carbon is mainly derived from surface primary production, most of the carbon storage occurs in the top layer of the soil, with still a significant amount below 1 m originating in peats or because of processes related to freezing of soil that transport carbon at depth (Schuur et al. 2008). Additionally, more than a third of the permafrost carbon pool occurs in deeper deposits that were, for instance, buried by aeolian or alluvial sedimentation at the surface.

In addition to the decomposition, a part of the carbon fixed by land plants is released to the atmosphere by the respiration of herbivores and carnivores and by fires. Fires are widespread on continents, affecting the surface and the canopy of forests, but they also can take place underground, in particular, in peat bogs. They are associated with significant carbon fluxes (Figure 2.29), leading to a global value of about 2 PgC yr$^{-1}$(van der Werf et al. 2010). Locally, the magnitude of the carbon fluxes depends on the area burned and the fuel consumption per unit area. There are wide geographical variations in these quantities as fires in savannas and grasslands are more frequent but are characterised by much less biomass burned per square metre than in major forest fires. Fires are directly influenced by climate, dryer conditions favouring fire development more than wetter ones. Furthermore, climate controls the availability of fuel delivered by the NPP. In equilibrium conditions, the carbon released by fires is balanced by the following recovery of ecosystems, but it can lead to substantial inter-annual variations as well as some longer-term perturbations in transient conditions.

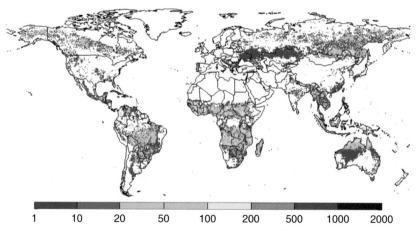

1    10    20    50    100    200    500    1000    2000

**Fig. 2.29** Mean annual carbon emissions by fire (gC m$^{-2}$ yr$^{-1}$) averaged over the period 1997–2009. (Source: van der Werf et al. 2010; available at: http://www.globalfiredata.org/.)

## 2.3.4 Geological Reservoirs

Most of the organic carbon that is exported downwards from the surface layer is remineralised in the water column. In particular, the ocean is under-saturated with respect to **calcite (aragonite)** below 4500 m (3000 m) in the Atlantic and below 800 m (600 m) in the Pacific. As a consequence, the long-term burial of $CaCO_3$ in the sediments to produce **limestone** mainly occurs in shallow seas (e.g., in coral reefs). Averaged over the whole ocean, this long-term burial corresponds to 13% of the export of $CaCO_3$ out of the surface layer (see, e.g., Sarmiento and Gruber 2006). On short **time scales**, this is a small fraction of the whole carbon cycle, but it becomes a crucial component on time scales longer than a century.

Because the sea floor spreads as a result of plate tectonics, sediments are transported horizontally and eventually are incorporated within the mantle through **subduction** along plate boundaries. At higher temperatures and pressure, limestone is transformed during subduction into calcium silicate rocks (this is called '**metamorphism**') by the reaction

$$CaCO_3 + SiO_2 \rightarrow CaSiO_3 + CO_2 \tag{2.51}$$

The $CO_2$ that is released in this reaction can return to the atmosphere, in particular, through volcanic eruptions.

The plate motion also allows the calcium silicate rocks to be uplifted to the continental surface, where they are affected by physical and chemical **weathering**. In particular, the carbonic acid contained in rain water [the same process as reaction (2.42)] can interact with the calcium silicate rocks; thus,

$$CaSiO_3 + H_2CO_3 \rightarrow CaCO_3 + SiO_2 + H_2O \tag{2.52}$$

The products of this reaction are transported by rivers to the sea, where they can compensate for the net export of $CaCO_3$ by sedimentation. Weathering thus tends to reduce atmospheric $CO_2$ by taking up carbonic acid to make $CaCO_3$ and increasing ocean alkalinity, whilst metamorphism and sedimentation tend to increase atmospheric $CO_2$. Overall, carbonate sedimentation, subduction and metamorphism and weathering form a closed loop that takes place over millions of years and is sometimes referred to as the 'long-term inorganic carbon cycle' (Figure 2.30). Additionally, the burial of organic matter, in particular, in the form of natural gas, oil and coal, also should be taken into account in the long-term evolution of the carbon cycle. A part of this buried material is then oxidised when it is brought back to surface by tectonic activity [for more details, see, e.g., Archer 2010).

## 2.3.5 The Methane Cycle

Methane is naturally produced by the decomposition of organic matter in the absence of oxygen by specific micro-organisms called 'archaea' using various biochemical pathways (Whiticar et al. 1986). The favourable anaerobic settings are

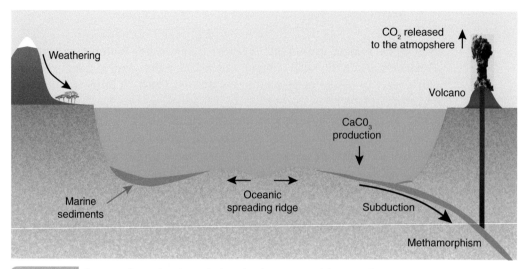

**Fig. 2.30**    Long-term inorganic carbon cycle through sedimentation, subduction and metamorphism and weathering.

mainly found in wetlands as the presence of water at the surface or close to it strongly reduces the diffusion of oxygen from the air into the soil. This is the main source of methane for the atmosphere (Figure 2.31), but some animals, such as the ruminants and the termites, produce significant amount of methane during their digestion process which occurs partly in anoxic conditions. Furthermore, methane is produced because of incomplete combustion during biomass burning. Finally, in addition to the biogenic sources, methane can be obtained in the crust at a depth where temperature is sufficiently high [for more details, see, e.g., Archer (2010) and Schlesinger and Bernhardt (2013)].

Human activities have directly influenced the methane sources because of the extensive rice cultivation occurring over flooded surfaces, through the decomposition in wastes and by strongly increasing the population of ruminants. Furthermore, some gas is released into the atmosphere by mining, for instance, because of leakage during the extraction and transport of methane and through human influence on biomass burning (see Figure 2.31).

The observed atmospheric methane concentration results from the balance of those sources with the sinks due to methane oxidation that ultimately leads to the formation of two other greenhouse gases, $CO_2$ and $H_2O$

$$CH_4 + 2O_2 \rightarrow CO_2 + 2H_2O \tag{2.53}$$

Reaction (2.53) actually represents a simplification of a series of reactions because $O_2$ cannot directly oxidise methane in the atmosphere. This requires the presence of highly reactive constituents such as the hydroxyl radical (OH·), which can be produced, for instance, through the reaction of **ozone** ($O_3$), obtained by photochemical reactions, with water ($H_2O$). The rate of methane oxidation thus is a function of the concentration of the hydroxyl radical in the atmosphere. The latter depends on the production mechanisms of OH (which is influenced by the presence of some pollutants such as NO and $NO_2$, often referred to as $NO_x$) but

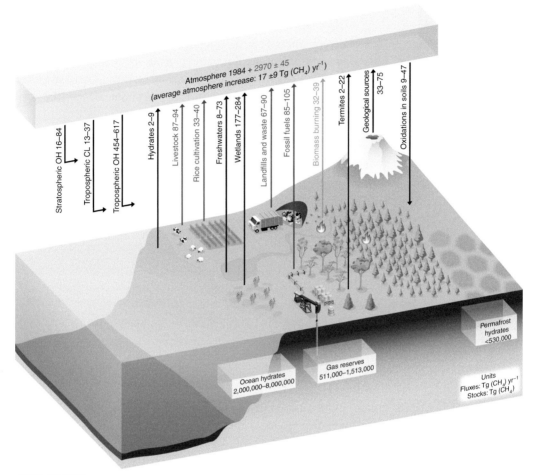

**Fig. 2.31** Schematic of the global cycle of $CH_4$. Numbers represent the range of estimated annual fluxes in $Tg(CH_4)$ $yr^{-1}$ for the period 2000–2009 and $CH_4$ reservoirs in $Tg(CH_4)$: the atmosphere and three geological reservoirs (hydrates on land and in the ocean floor and gas reserves). The atmospheric inventories have been calculated using a conversion factor of 2.75 $Tg(CH_4)$ $ppb^{-1}$. Black arrows denote the natural fluxes, which are not directly related to human activities; red arrows, the anthropogenic fluxes; and light brown arrow, combined natural + anthropogenic flux. (Source: Ciais et al. 2013.)

also on the concentration of other molecules such as CO, whose oxidation and that of methane are the main consumers of the hydroxyl radical. Additionally, a small amount of methane is oxidised in soils by bacteria.

As a result of this interplay between sources and sinks, the lifetime of methane in the atmosphere has been estimated to be around 10 years in present-day conditions. The atmosphere itself is a small reservoir of methane (Figure 2.31). The content of the oceans is low, too, but a large quantity of methane is stored in ocean sediments in the form of **hydrates**, in which frozen water molecules form a kind of cage that traps the gas. To be stable, the hydrates require high pressure and low temperatures, so they are generally found below 500 m, except in cold environments such as north of Siberia, where they can be present below 200 m.

Most of these hydrates originate from the gas locally produced by anaerobic decomposition, so they are mainly located close to sources of organic matter on the continental shelves and slopes. Among the methane that is released every year from those sediments, only a small fraction reaches the ocean surface in the form of methane because of oxidation in sediments and seawater. Large stocks of methane are also stored in **permafrost** and as gas reserves, methane being the most important constituent of natural gas.

## Review Exercises

1. The Sun emits radiation at lower wavelengths than the Earth. The balance is obtained if the Earth emits radiation at a temperature of around (emission temperature)
   a. 210 K (−63°C).
   b. 255 K (−18°C).
   c. 288 K (+15°C).
2. The surface temperature is higher than the emission temperature because of
   a. the low albedo of the Earth.
   b. the heat flow from the interior of the Earth.
   c. the presence of some minor constituent in the atmosphere which could absorb infrared radiation.
3. The temperature in the northern hemisphere is higher during June, July, and August (boreal summer) because
   a. the Earth is closest to the Sun.
   b. the solar rays are closer to the vertical.
4. Integrated over the whole year, the solar flux received at the surface is higher at the equator than at high latitudes.
   a. This is nearly balanced on annual mean by the higher emitted radiative flux in equatorial regions.
   b. The local imbalance between received and emitted flux on an annual mean basis is compensated for by the heat transport by the ocean and the atmosphere.
   c. The local imbalance between received and emitted flux on annual mean basis is compensated for by heat storage.
5. The amplitude of the seasonal cycle (summer minus winter temperatures) can reach 50 K on land compared to 5 to 10 K over the ocean because of different heat storages.
   a. True
   b. False
6. At middle latitudes, around 40°N, the atmosphere and the ocean transport roughly the same amount of heat poleward.
   a. True
   b. False

7. The oceans transport roughly the same amount of heat northward in the North Pacific as in the North Atlantic.
   a. True
   b. False

8. Over the ocean, the latent heat flux is generally larger than the sensible heat flux.
   a. True
   b. False

9. Most of the water that falls as rain over land comes from evaporation over the oceans.
   a. True
   b. False

10. To avoid a strong surface salinity drift, precipitation should be locally more or less equal to evaporation over oceanic surfaces.
    a. True
    b. False

11. Over land, evapotranspiration is always limited by soil moisture availability.
    a. True
    b. False

12. The oceans are a much larger reservoir of carbon than the atmosphere because
    a. the solubility of $CO_2$ is very high in sea water.
    b. most of the stock of carbon is stored as bicarbonate and carbonate ions.

13. The air-sea flux of $CO_2$ is a function of the difference in partial pressure of $CO_2$ between the two media. The regional variations of this flux are mainly due to
    a. variations in the $p^{CO_2}$ of the air due to photosynthesis by the land biosphere.
    b. variations in the $p^{CO_2}$ of the ocean due to biological activities and oceanic circulation.

14. The alkalinity is a measure of the excess of bases in sea water. If alkalinity increases, $p^{CO_2}$ of the water
    a. increases, and there is a tendency to transfer $CO_2$ from the ocean to the atmosphere.
    b. remains the same.
    c. decreases, and there is a tendency to transfer $CO_2$ from the atmosphere to the ocean.

15. Photosynthesis in the ocean tends to lower the surface DIC.
    a. Yes, because this fixes $CO_2$ in form of organic matter.
    b. No, because photosynthesis is compensated by respiration, and thus it has a neutral effect on surface DIC.

16. Formation of calcium carbonate in the ocean induces
    a. a decrease in $p^{CO_2}$.
    b. no effect on $p^{CO_2}$.
    c. an increase in $p^{CO_2}$.

17. The subduction of oceanic sediments induces
    a. a release of $CO_2$.
    b. no release or uptake of $CO_2$.
    c. an uptake of $CO_2$.
18. Weathering of calcium silicate rocks because of the action of the carbonic acid contained in water tends to induce
    a. an uptake of carbon from the atmosphere.
    b. a release of carbon.
19. The transport of weathering products by rivers to the ocean helps to compensate for the loss of calcium carbonate by sedimentation.
    a. True
    b. False
20. The current methane concentration in the atmosphere is controlled by
    a. the natural emissions by wetlands and biomass burning.
    b. the production of wheat.
    c. methane oxidation and thus the oxygen concentration in the atmosphere.
    d. methane oxidation and thus the concentration in hydroxyl radical in the atmosphere.

# 3 | Modelling the Climate System

## OUTLINE

Models are essential tools in climatology to understand the processes responsible for observed changes and to make predictions. After defining what is traditionally meant by climate modelling, this chapter presents the different types of models. Then the chapter briefly reviews the main characteristics of the components of those models. Finally, a discussion is provided of the methodology applied to test the performance of the models and to interpret their results in conjunction with observations.

## 3.1 Introduction

### 3.1.1 What Is a Climate Model?

In general terms, a 'mathematical model' of the climate system is defined as a mathematical representation of the system based on physical, biological and chemical principles or laws. The equations derived from these laws are so complex that they must be solved numerically, leading to the formulation of a numerical or computer model (Figure 3.1). Such a model is used to perform simulations corresponding to realistic past or future conditions and idealised experiments. This is the reason why the term 'climate simulator' has been proposed to make a distinction between various types of models, such as physical, statistical or conceptual models (e.g., Crucifix 2012a). Nevertheless, we will keep here the classical denomination of climate model, which includes the following model equations (mathematical model) and the way these are solved (numerical model). This approach is largely practised in the field, keeping in mind that the word 'model' can have different meanings in different communities.

Numerical resolution of the equations provides an approximation of the true solution, which is, in practise, discrete in space and time. This can be interpreted as if the numerical models give estimates of climate variables averaged over regions, whose size depends on model spatial **resolution**, and over specific time intervals, which are related to the time step. For instance, some models provide globally or **zonally** averaged values, whilst others have a **numerical grid**, whose spatial resolution could be less than 100 km (Figure 3.2). The time step varies

Model development

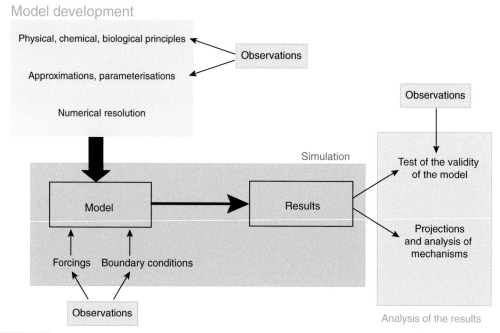

Schematic representation of the development and use of a climate model.

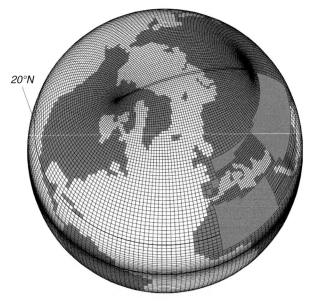

Example of a numerical grid for the ocean model NEMO (http://www.lodyc.jussieu.fr/NEMO/, ORCA2 configuration). Variables such as temperature are obtained at the centre of all the oceanic cells whose sides are defined by the grid coordinates (i.e., the lines in the figure). The resolution of the grid is variable but is on the order of 2°, meaning that the distance between two points where variables are computed is on the order of 200 km. The continents are in grey, with some continental zones not represented on this specific grid. (Source: Copyright 2008, NEMO team. Reproduced with permission.)

between seconds and several years depending on the process studied. This also implies that the boundary conditions, such as the type of vegetation cover and the topography or positions of the coastlines, have to be interpolated on the numerical grid to approximate the real conditions, leading to significant uncertainties at small scales (see Section 3.2.4).

Even for models with the highest resolution, the numerical grid is still much too coarse to represent small-scale processes such as turbulence in the **atmospheric** and oceanic boundary layers and interactions of the circulation with small-scale topographical features, thunderstorms, and **cloud microphysics** processes. Furthermore, many processes are still not sufficiently well known to include their detailed behaviour in models. As a consequence, **parameterisations** have to be designed based on empirical evidence and/or theoretical arguments to account for the large-scale influence of these processes not included explicitly (see Sections 3.3.1 and 3.3.2). Because these parameterisations reproduce only the first-order effects and are usually not valid for all possible conditions, they are often a source of considerable uncertainty in models.

In addition to the physical, biological and chemical knowledge included in the model equations, climate models require some input from observations or other model studies. For a climate model describing nearly all the components of the system, only a relatively small amount of data is required: the solar irradiance, the Earth's radius and period of rotation, the land topography and bathymetry of the ocean, and some properties of rocks and soils. However, for a model that represents only the physics of the atmosphere, the ocean and the sea ice, information in the form of boundary conditions should be provided for all sub-systems of the climate system not interactively simulated in the model: the distribution of vegetation, the topography of the ice sheets, and so on.

These model inputs are often separated into boundary conditions (which are generally fixed during the course of the simulation) and external '**forcings**' (e.g., changes in solar irradiance) which drive the changes in climate. However, such definitions sometimes can be misleading. The forcing of one model could be a key **state variable** of another. For instance, changes in $CO_2$ concentration could be prescribed in some models, but it is computed directly in models including a representation of the carbon cycle. Furthermore, a fixed boundary in some models, such as the topography of the ice sheet, can evolve interactively in a model designed to study climate variations on a longer time scale.

The importance of data is arguably even greater during the development phase of a model because data provide essential information on the properties of the system being modelled (see Figure 3.1). In addition, a large number of observations is needed to test the validity of a model so as to gain confidence in the conclusions derived from model results (see Section 3.5).

Many climate models have been developed to study the response of the system to perturbations, in particular, to perform '**projections**', that is, to simulate and understand future climate changes in response to the anthropogenic emission of greenhouse gases and aerosols (see Chapter 6). In addition, models can be formidable tools to improve our knowledge of the most important characteristics of the climate system and the causes of climate variations. Obviously, climatologists cannot perform experiments on the real climate system to identify the role of a particular process or to test a hypothesis. However, this can be done in the virtual world of climate models. For highly non-linear systems, the design of such tests, often called 'sensitivity experiments', has to be very carefully planned. However, in simple experiments, neglecting a process or an element of the modelled system (e.g., the influence of an increase in $CO_2$ concentration on the radiative properties of the atmosphere) often can provide a first estimate of the role of this process or this element in the system.

## 3.1.2  Types of Models

Simplifications and approximations are unavoidable when designing a climate model as the processes that should be taken into account range from the spatial scale of centimetres (e.g., for atmospheric turbulence) to that of the Earth itself. The involved **time scales** also vary widely from the order of seconds for some waves to hundreds of millions of years when analysing the climatic influence of the carbon exchanges between the Earth's interior and the atmosphere (see Chapter 5). It is thus an important skill for a modeller to be able to select the processes that must be explicitly included compared to those which can be neglected or represented in a simplified way. This choice, of course, is based on the scientific goal of the study. However, it also depends on technical issues since the most sophisticated models require a lot of computational power: even on the largest computer presently available, high-resolution models cannot be used routinely for periods longer than a few centuries to millennia. For longer time scales or when quite a large number of experiments are needed, it is thus necessary to use simpler and faster models. Furthermore, it is often very illuminating to deliberately design a model that includes only the most important properties so as to understand in depth the nature of a feedback or the complex interaction between some components of the system. This is also the reason why simple models are often used to analyse the results of more complex models in which the fundamental characteristics of the system may be hidden by the number of processes represented and the details provided.

Modellers first have to decide the variables or processes to be taken into account and those which will be prescribed from observations. This provides a method of classifying the model as a function of the components which are represented interactively. In most climate studies, at least the physical behaviours of the atmosphere, ocean and sea ice must be represented. In addition, the terrestrial and marine carbon cycles, the atmospheric chemistry, the dynamic vegetation and the ice sheet components are more and more regularly included, leading to what are called '**Earth system models**' (ESMs).

A second way of differentiating between models is related to the complexity of the processes that are included (Figure 3.3). At one end of the spectrum, general circulation models (GCMs) attempt to account for all the important properties of the system at the highest affordable resolution. The term 'general circulation model' was introduced because one of the first goals of these models is to

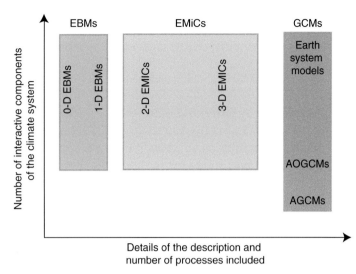

Types of climate models. 0-D, 1-D, 2-D and 3-D correspond to the number of space dimensions explicitly accounted for in the energy-balance models (EBMs) and Earth models of intermediate complexity (EMICs), whilst general circulation models (GCMs) always have a 3-D representation of the system.

simulate the three-dimensional (3-D) structure of winds and currents realistically. They have classically been divided into atmospheric general circulation models (AGCMs) and ocean general circulation models (OGCMs). For climate studies using interactive atmospheric and oceanic components, the acronyms 'AOGCM' (atmosphere-ocean general circulation model) and the broader 'CGCM' (coupled general circulation model) are also used.

At the other end of the spectrum, simple climate models [such as energy-balance models (EBMs); see Section 3.2.1] propose a highly simplified version of the dynamics of the climate system. The variables are averaged over large regions, sometimes over the whole Earth, and many processes are not represented or accounted for by **parameterisations**. EBMs thus include a relatively small number of degrees of freedom.

Earth models of intermediate complexity (EMICs) are located between those two extremes (Claussen et al. 2002). They are based on a more complex representation of the system than EBMs but include simplifications and parameterisations for some processes that are explicitly accounted for in GCMs. EMICs form the broadest category of models. Some of them are relatively close to simple EBMs, whilst others are slightly degraded GCMs.

When employed correctly, all model types can produce useful information on the behaviour of the climate system. There is, however, no perfect model which is suitable for all purposes. This is why a wide range of climate models exists, forming what is called the 'spectrum' or 'hierarchy' of models which will be described in Section 3.2. Depending on the objective or question, an appropriate type of model should be selected. However, combining the results from various types of models is often the best way to gain a deep understanding of the dominant processes in action.

## 3.2 A Hierarchy of Models

### 3.2.1 Energy-Balance Models and Simple Dynamic Systems

As indicated by their name, energy-balance models (EBMs) estimate the changes in the climate system from an analysis of the energy budget of the Earth. In their simplest form, they do not include any explicit spatial dimension, providing only globally averaged values for the computed variables. They are thus referred to as 'zero-dimensional' (0-D) EBMs. The basis for these EBMs was introduced by both Budyko (1969) and Sellers (1969). Their fundamental equation is very similar to those analysed in Sections 2.1.1 and 2.1.5:

Changes in heat storage = absorbed solar radiation – emitted terrestrial radiation

$$C_m \frac{\partial T_s}{\partial t} = \left[ \left(1 - \alpha_p\right) \frac{S_0}{4} - A\uparrow \right] \tag{3.1}$$

where, as in Chapter 2, $C_m$ is the effective heat capacity of the medium (measured in J m$^{-2}$ K$^{-1}$), $T_s$ is the surface temperature, $t$ is time, $\alpha_p$ is the planetary albedo, $S_0$ is the **total solar irradiance** (TSI) and $A\uparrow$ is the total amount of energy emitted by a 1-m$^2$ surface of the Earth. $A\uparrow$ could be represented on the basis of the **Stefan-Boltzmann law** using a factor $\tau_a$ to represent the infrared transmissivity of the atmosphere (including the greenhouse effect) as

$$A\uparrow = \varepsilon \sigma T_s^4 \tau_a \tag{3.2}$$

where $\varepsilon$ is the emissivity of the surface. Using an albedo of 0.3, an emissivity of 0.97 and a value of $\tau_a$ of 0.64 leads to an equilibrium temperature $T_s = 287$K, which is close to the observed one. In some EBMs, Eq. (3.2) is linearised to give an even simpler formulation of the model. $\tau_a$ and $\alpha_p$ are often parameterised as a function of the temperature, in particular, to take into account the fact that cooling increases the surface area covered by ice and snow and thus increases the planetary albedo.

To take the geographical distribution of temperature at the Earth's surface into account, 0-D EBMs can be extended to include one (generally the latitude) or two horizontal dimensions (Figure 3.4). An additional term $\Delta F_{\text{transp}}$ then is included in Eq. (3.1) representing the net effect of heat input and output associated with horizontal transport:

$$C_m \frac{\partial T_i}{\partial t} = \left[ \left(1 - \alpha_p\right) S_i - A_i\uparrow \right] + \Delta F_{\text{transp}} \tag{3.3}$$

An index $i$ has been added to the temperature to indicate that the variable corresponds to the region $i$. A simple form for the transport is to treat it as a diffusion term, but more sophisticated parameterisations are also used.

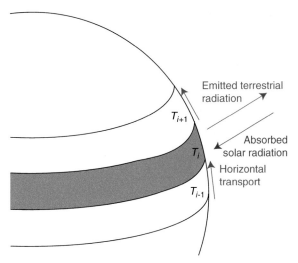

Representation of a 1-D EBM for which the temperature $T_i$ is averaged over a band of longitude.

Besides, instead of a horizontal dimension, some models have several layers along the vertical to represent more precisely the vertical exchanges. This leads to what are called 'radiative-convective models' (see, e.g., Ramanathan and Coakley 1978) because of the focus on those two eminent processes which are responsible to a large extent for the vertical structure of the atmosphere (see Section 1.2.1).

Box models have clear similarities to EBMs as they represent large areas or an entire component of the system by an average which describes the mean over one 'box'. The exchanges between the compartments are then parameterised as a function of the characteristics of the different boxes (see an example in Section 4.2.5). The exact definitions of the boxes depend on the purpose of the model. For instance, some box models have a separate compartment for the atmosphere, the land surface, the ocean surface layers and the deep ocean, possibly also making a distinction between the two hemispheres. Others include additional components which allow a description of the carbon cycle and thus have boxes corresponding to the various reservoirs described in Section 2.3.

The low-order models are also based on a simplified representation of the dynamics of the system (see, e.g., Lorenz 1984; Vannitsem and De Cruz 2014). For instance, the velocity and/or temperature fields can be expressed as the product of only a few time-dependant variables (the unknowns) which are multiplied by fixed spatial modes [similar to Eq. (3.41)]. They have some similarities to EMICs, but the main goal is different. Low-order models generally are developed to study particular processes using a limited number of equations to identify precisely some mechanisms which are important for climate. As a consequence, they use idealised geometries and cannot be applied to realistic conditions.

### 3.2.2  Intermediate-Complexity Models

Like EBMs, EMICs involve some simplifications, but they always include a representation of the Earth's geography; that is, they provide more than averages over

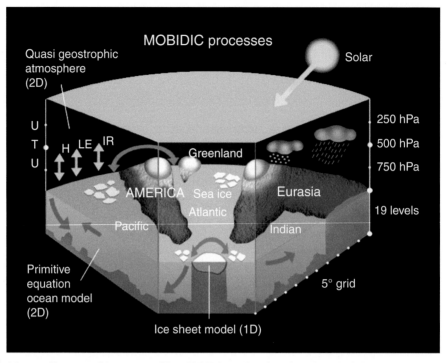

**Fig. 3.5** Schematic illustration of the structure of the MOBIDIC climate EMIC (Crucifix et al. 2002) which includes a zonally averaged atmosphere, a three-basin **zonal** oceanic model (corresponding to the Atlantic, Pacific and Indian oceans) and simplified ice sheets.

the whole Earth or 'large boxes' (Claussen et al. 2002). In addition, they include many more degrees of freedom than EBMs. As a consequence, the parameters of EMICs cannot be easily adjusted to reproduce the observed characteristics of the climate system, the dimension of the **state vector** being much larger than the number of available parameters.

The level of approximation involved in the development of the model varies widely among different EMICs. Some models use a very simple representation of the geography, with a **zonally** averaged representation of the atmosphere and ocean. A distinction is always made between the Atlantic, Pacific and Indian basins (Figure 3.5) because of the strong differences between them in the circulation (see Section 1.3.2). As the atmospheric and oceanic circulations are fundamentally three-dimensional, some **parameterisations** of the **meridional** transport are required. Those developed for EMICs are generally more complex and physically based than those employed in 1-D EBMs.

However, some EMICs include components that are very similar to those developed for GCMs, although a coarser numerical grid is used so that the computations proceed fast enough to allow a large number of relatively long simulations to be run. Some other components are simplified, usually including the atmosphere because this is the component which is most demanding of computer time in coupled climate models.

### 3.2.3  General Circulation Models

General circulation models (GCMs) provide the most precise and complex description of the climate system. Currently, their **grid** resolution is typically on the order of 50 to 200 km. As a consequence, compared to EMICs (which have a grid resolution between 300 km and thousands of kilometres), they provide much more detailed information on a regional scale. Originally, GCMs applied in climate studies only included a representation of the atmosphere, the land surface, the ocean circulation and a very simplified version of the sea ice (see, e.g., Manabe and Bryan 1969), but ESMs potentially now include many more components, as mentioned in Section 3.1.2 (Figure 3.6). The way those components are modelled will be discussed in Section 3.3.

Because of the large number of processes included and their relatively high resolution, GCM simulations require a large amount of computer time. For instance, an experiment covering one century typically takes several days or even weeks to run on the fastest computers, running on hundreds of processors. As computing

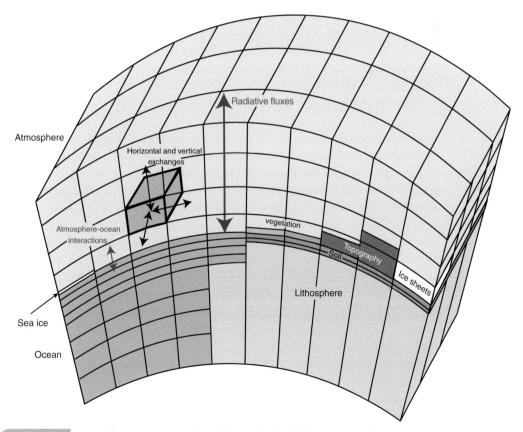

**Fig. 3.6**   A simplified representation of part of the domain of a GCM illustrating some important components and processes. For clarity, the curvature of the Earth has been amplified, the horizontal and vertical coordinates are not to scale and the number of grid points has been reduced compared to state-of-the-art models.

power increases, longer simulations with a higher resolution become affordable, providing more regional details than the previous generation of models.

## 3.2.4   Regional Climate Models

As discussed earlier, the resolution of current global climate models is too coarse to accurately simulate local or regional climate fluctuations. For instance, countries such as Belgium generally are covered by only a few grid points at best. Furthermore, a small number of grid points cannot represent adequately the observed climate averaged over the region corresponding to those points. This is first due to the necessary smoothing of the topography or of the coastline compared to reality to represent them on the model grid (e.g., Figure 3.2). Additionally, the accuracy of the approximations applied in the development of the mathematical model and in the numerical schemes used to solve the equations (see Section 3.4.2) is lower at the grid scale than for averages over wider regions. Consider the clear example of an oceanic strait between two islands where two ocean currents moving in opposite directions are observed. If a model computes only two velocities in this strait, each of them may at best display a current in a different direction, but there is no hope to obtain a reasonable estimate of the magnitude or extent of the two currents. To do so, a minimum of ten grid points in the strait is likely needed. Consequently, models provide skilful estimates of the climate only at a scale that is generally much larger than the grid spacing itself (e.g., Masson and Knutti 2011).

Nevertheless, the study of some specific regional features requires information at a smaller scale than that resolved by GCMs, both for a detailed understanding of the dynamics of the system and for practical applications such as estimation of the impact of future climate changes in a specific area. One option is to perform global simulations at higher resolution over short periods or to use only one component of the system to reduce the cost of the experiment in computer time. Some models also have variable grid resolution, allowing the model to have a coarse resolution globally and a much finer one over a region of interest. This provides a kind of zoom over the region whilst the remaining part of the globe is represented with less detail (see, e.g., Markovic et al. 2010). This option is very promising as it is modular, and the configuration can be adapted to many research and practical questions. It presents, however, several technical and numerical challenges. Furthermore, the design of **parameterisations** suitable for all the grid scales included in models with variable resolution is not straightforward.

Alternatively, a regional climate model (RCM) can be applied over a fraction of the Earth only, a continent, for instance. For a similar number of grid points compared to a global model, the resolution is then much higher, but the time step also has to be reduced to ensure the stability of the numerical scheme used to solve the model equations [e.g., Eqs. (3.37) and (3.40)]. This implies that RCMs are generally expensive to run over long periods. At the boundaries of their domain, the RCMs are constrained by the results of a global model or by

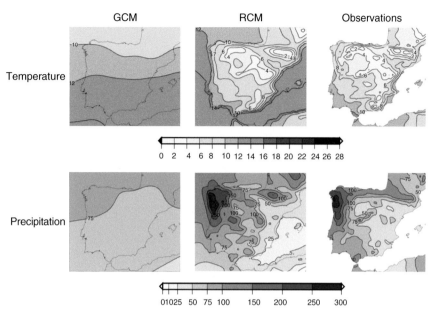

Fig. 3.7 Comparison of the temperature (°C) and precipitation (mm month$^{-1}$) in winter (December, January, and February) over Spain and Portugal simulated by a coarse-resolution GCM with grid spacing on the order of 400 km, a RCM with a resolution of 30 km and an estimate deduced from observations. Because of the too different resolution between the GCMs and the high-resolution RCM, a configuration of the RCM with an intermediate resolution of 90 km covering the whole of Europe is first driven by the results of the GCM at its boundaries (not shown). The results of this experiment are subsequently applied as boundary conditions for the RCM with 30-km resolution. (Source: Gómez-Navarro et al. 2011.)

fields derived from observations. Consequently, the biases of the global model at large scale (see Section 3.5.2.3) generally have a significant impact on the RCM results. Nevertheless, RCMs have a clear added value compared to GCMs because they are able to adequately represent some regional-scale processes that are only crudely simulated in global models or must even be parameterised. In particular, they include a much better description of surface characteristics, including the topography, and their effect on the atmospheric state. The improvement depends on the variable and the region investigated. It is generally larger over mountainous areas than over wide plains and for precipitation, which displays complex spatial variations at all spatial scales, than for temperature. This is illustrated in Figure 3.7 for the Iberian Peninsula, where the temperature is much lower over mountainous areas such as the Pyrenees in the RCM than in the coarse-resolution GCM because the surface elevation is several hundred metres higher in the RCM. For precipitation, the impact of a better representation of the topography is particularly clear on the western side of the domain. In the RCM, the mountains receive a large amount of precipitation associated with the inflow of moist air of Atlantic origin, as observed, whilst precipitation is much smoother in the GCM, which does not include this mountain range.

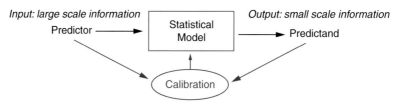

**Fig. 3.8**    Schematic of statistical downscaling illustrating how it is possible to estimate variables representing a small domain (predictand) from variables describing the climate on a large scale (predictor) thanks to a statistical relationship calibrated on the basis of the link between those variables over a known period. (Source: Modified from Benestad et al. 2008.)

## 3.2.5  Statistical Downscaling

Local/regional information also can be deduced from the results of a large-scale climate model using a statistical approach (Figure 3.8). It is generally referred to as 'statistical downscaling' (see, e.g., Benestad et al. 2008; Maraun et al. 2010) in contrast to the method described earlier based on RCM, which is termed 'dynamical downscaling'. The first step is to calibrate a statistical model relating one or a few variables characterising the climate on a large scale (referred to as the 'predictor') and the variable of interest on a smaller scale (referred to as the 'predictand') over periods during which information on both scales is available. The corresponding time series generally are taken from the results of GCM simulations covering the last few decades and recent observations at a particular weather station. The statistical model itself can display various levels of complexity. The lowest levels simply assume a linear relationship between the predictand and the predictor. For instance, the temperature at a particular location $T_{loc}$ can be related to the temperature in the nearest model point $T_{GCM}$ by a factor $\alpha_{sd}$ and an offset term $\beta_{sd}$. The relationship would take into account the difference in mean temperature due to a different elevation or different surface characteristics, as well as the different magnitude of climate fluctuations on the local and large scales:

$$T_{loc} = \alpha_{sd} T_{GCM} + \beta_{sd} \tag{3.4}$$

More sophisticated approaches are also applied, in which the predictand is a non-linear function of several predictors:

$$x_{loc} = F_{loc}\left[\mathbf{x}_{mod}\right] \tag{3.5}$$

where $x_{loc}$ is a chosen variable at a particular location. The function $F_{loc}$ depends on several model variables grouped in $\mathbf{x}_{mod}$. This approach is often referred to as 'model output statistics' (MOS), in particular, in weather forecast (see, e.g., Wilks 2011), as the local estimate is based on information provided by the model outputs using statistical methods to derive a link between those outputs and the real state of the system.

When the statistical model is fully tested using independent data not used for the calibration, it can be applied to estimate the climate change on a local scale in situations for which only large-scale model results are available. The approach assumes the stationarity for various climatic conditions of the relationship between the predictor and the predictand deduced from the calibration period. It is thus very important to adequately select the variables representing the large-scale climate which will be used as input to the statistical model to ensure a robust link with the local climate. This assumption about the stationarity is often considered to be a limitation of statistical downscaling compared to dynamical downscaling, which is based on physical equations that should be valid over a wider range of climates. Nevertheless, statistical downscaling is much faster, so a large number of tests can be performed, leading to a better estimate of the uncertainties of the outputs on a regional or local scale. Besides, RCMs also have limitations in their representation of the dynamics of the system, and statistical and dynamical downscaling thus should be viewed as complementary. As RCMs have a grid that is still too coarse for some applications, statistical downscaling also can be applied to the results of RCMs to infer information about local climate. Furthermore, the approach described in Figure 3.8 has strong similarities with the techniques applied to correct model biases [see Section 3.6.1; compare Eq. (3.4) with Eq. (3.43) and Eq. (3.5) with Eq. (3.45)], and statistical downscaling can be readily applied in conjunction with bias corrections.

## 3.3 Components of a Climate Model

### 3.3.1 Atmosphere

The basic laws that govern the dynamics of the atmosphere can be formulated as a set of seven equations with seven unknowns: the three components of the velocity $\vec{v}$ (components $u$, $v$ and $w$), the pressure $p$, the temperature $T$, the specific humidity $q$ and the density. Those seven equations are listed below (see, e.g., Marshall and Plumb 2008; Cushman-Roisin and Beckers 2011).

(1–3) Newton's second law, also referred to as the 'momentum balance', states that the forces applied to a parcel equal its mass times its acceleration (i.e., $\vec{F} = m\vec{a}$). This leads to three equations, one for each component of the velocity, written in vector form for the atmosphere as

$$\frac{d\vec{v}}{dt} = -\frac{1}{\rho}\vec{\nabla}p + \vec{g} + \vec{F}_{\text{fric}} - 2\vec{\Omega} \times \vec{v} \tag{3.6}$$

The right-hand side of Eq. (3.6) includes the forces divided by the mass. The four terms correspond to the force resulting from the pressure gradient, the gravitational force (where $\vec{g}$ is the apparent gravity vector, i.e., taking centrifugal force into account), the friction force $\vec{F}_{\text{fric}}$ and the **Coriolis force**, respectively. The Coriolis force is present because the velocities are computed relative to the Earth, so a term has to be included to account for the Earth rotation. $\vec{\Omega}$ is the angular velocity vector of the Earth, which is aligned with the Earth's axis of rotation and has a magnitude equal to angular speed ($2\pi\,\text{day}^{-1}$).

In Eq. (3.6), $d/dt$ is the 'material derivative', which measures the rate of change of a variable characterizing the state of an air parcel as a function of time. $d\vec{v}/dt$ is thus the acceleration of the parcel as required by classical mechanics. This formalism, following individual parcels, is referred to as the 'Lagrangian description' of the flow. Nevertheless, it is sometimes more convenient to estimate the rate of change of the variable at one fixed location than for different parcels as they transit through this location. This is referred to as the 'Eulerian description' of the flow. In other words, the Lagragian formalism is focussed on the velocity of an air parcel, whilst, in the Eulerian formalism, the key variable is the velocity at one point, as measured, for instance, by an anemometer. The local changes are given by the local derivate $\partial/\partial t$. They are related to the material derivative by a term associated with the transport:

$$\frac{d}{dt} = \frac{\partial}{\partial t} + \vec{v}\cdot\vec{\nabla}$$
$$= \frac{\partial}{\partial t} + u\frac{\partial}{\partial x} + v\frac{\partial}{\partial y} + w\frac{\partial}{\partial z} \tag{3.7}$$

(4) The conservation of mass, also termed the 'continuity equation', indicates that any local change in the density at one point must be balanced by an exchange with surrounding areas:

$$\frac{\partial\rho}{\partial t} = -\vec{\nabla}\cdot(\rho\vec{v}) \tag{3.8}$$

(5) The conservation of the mass of water vapour is based on the same principle as the conservation of mass but accounts for the local source and sink due to evaporation $E$ and condensation $C$, respectively:

$$\frac{\partial\rho q}{\partial t} = -\vec{\nabla}\cdot(\rho\vec{v}q) + \rho(E-C) \tag{3.9}$$

(6) The first law of thermodynamics (the conservation of energy) states that the variation in internal energy of a parcel, which can be related to temperature for an ideal gas, varies because of the heat input and the work associated with expansion or compression of the parcel. For the atmosphere, it can be written as

$$c_p\frac{dT}{dt} = Q + \frac{1}{\rho}\frac{dp}{dt} \tag{3.10}$$

where $Q$ is heating rate per unit mass, and $c_p$ is the specific heat at constant pressure.

(7) The equation of state [see Eq. (1.1)] is

$$p = \rho R_g T \tag{3.11}$$

Before these equations are used in models, some classical approximations are performed. For instance, assuming **hydrostatic equilibrium**, which is a good approximation at the scale of GCMs, provides a considerable simplification of the equation of motion along the vertical. In addition, the quasi-Boussinesq approximation states that the time variation of the density could be neglected compared to the other terms of the continuity equation, filtering the sound waves (see, e.g., Kalnay 2003).

The standard physical principles leading to these seven equations form the core of any atmospheric model, but they are not sufficient to reproduce the time development of the atmospheric state. Firstly, some terms present in the equations, such as frictional force [Eq. (3.6)], condensation and evaporation [Eq. (3.9)] and heating rate [Eq. (3.10)], need an explicit formulation based on additional physical laws. This part of the model is commonly referred to as the model 'physics', whilst calculation of the transport is called the 'dynamics'. The condensation/evaporation and heating rates are actually related through the **latent heat** exchanges associated with phase transition, whilst the heat production associated with the friction due to molecular viscosity is often neglected in climate models. Computing the heating rate also requires an estimate of the **longwave** and **shortwave** fluxes that govern the radiative budget of the atmosphere (see Section 2.1.6). This is achieved by applying the radiative transfer equations in each column of the atmospheric model (see, e.g., Wallace and Hobs 2006). The computation is separated into contributions from several frequency bands to account for the different radiative properties of the atmospheric compounds in the various part of the **electromagnetic spectrum** (see, e.g., Clough et al. 2005; Boucher 2012).

Secondly, Eqs. (3.6)–(3.10) are valid when all the temporal and spatial scales are adequately resolved, but this is not the case for a climate model because of the limited resolution of the **grid** (see Section 3.1.1). All the variables in those equations then can be expressed as the sum of an average over the grid during a model time step (denoted by the over-bar) plus a fluctuation that could not be simulated at the grid scale (denoted by a prime). This is referred to as the 'Reynolds decomposition' [for more details, see Kalnay (2003) and Cushman-Roisin and Beckers (2011)]. Applied, for instance, to the temperature and one component of the velocity, this gives

$$
\begin{aligned}
T &= \bar{T} + T' \\
u &= \bar{u} + u'
\end{aligned}
\tag{3.12}
$$

The goal is to derive equations for the mean quantities $\bar{T}, \bar{u}$ and so on, which are the values of interest at the model scale. This can be achieved by averaging Eqs. (3.6)–(3.10). As the mean of the fluctuations are by definition equal to zero, all the linear terms such as the local derivative have the same form for $T$ and $\bar{T}$ (i.e., the average of $\partial T/\partial t$ is simply $\partial \bar{T}/\partial t$). Nevertheless, as the mean of a product is different from the product of the means, the average of the velocity times the spatial derivative in Eq. (3.7) leads to new terms. For instance, the average of the term including the vertical velocity in Eq. (3.10) is given by

$$
\frac{\partial \overline{(wT)}}{\partial z} = \frac{\partial \left( \bar{w}\bar{T} \right)}{\partial z} + \frac{\partial \overline{(w'T')}}{\partial z}
\tag{3.13}
$$

In order to close the system of equations and find a solution, the product of fluctuations [the last term in Eq. (3.13)] must be expressed as a function of the mean quantities that can be resolved on the grid scale. Sub-grid-scale fluctuations generally are associated with turbulent motions, whose primary effect is to generate mixing by stimulating the contacts and thus the exchanges between adjacent fluid parcels. This is a large-scale analogue of the erratic motions at the molecular scale that are responsible for diffusion and heat conduction. Consequently, the large-scale influence of sub-grid-scale motion is often **parameterised** as a term proportional to the **gradient** in the mean quantity, similar to the one mentioned in Section 2.1.6 for heat conduction in soils (i.e., the **Fourier's law**):

$$\overline{(w'T')} = -K_T \frac{\partial \bar{T}}{\partial z} \tag{3.14}$$

where $K_T$ is termed the 'turbulent' or 'eddy diffusivity' because Eq. (3.14) has the same form as the mass flux due to molecular diffusion. $K_T$ depends strongly on the state of the atmosphere, in particular, its stability. For instance, $K_T$ is generally several orders of magnitude larger in the **planetary boundary layer**, where turbulence is generated by exchanges with the surface, than above it. This contrasts with the molecular diffusion coefficient or heat conductivity in a solid, which displays weaker variations in most cases relevant to climate. Besides, the value of the turbulent diffusivity is in most applications similar for temperature and humidity or for various gases. Indeed, the efficiency of the mixing is due to the flow itself rather than to intrinsic properties of each quantity, as for the exchanges at a molecular scale. The turbulence also affects the interactions between the atmosphere and the surface expressed by the boundary conditions of Eqs. (3.6)–(3.10) (see Section 3.3.8).

Equation (3.14) is well adapted to account for turbulent motions that are relatively isotropic in the horizontal and vertical directions. When the atmosphere becomes unstable (see Section 1.2.1), an additional **parameterisation** for the **convection** is required. The simplest one, referred to as 'convective adjustment', induces a rapid mixing of the grid cells along the vertical until neutral stability of the atmospheric column is restored. Current models use more advanced parameterisations, the so-called mass-flux schemes. They estimate explicitly the vertical mass fluxes associated with convection at different vertical levels on the basis of a simplified representation of convection dynamics (see, e.g., Tiedtke 1989; Emanuel 1991; Hourdin et al. 2013).

Furthermore, Eq. (3.12) assumes that a clear separation exists between large-scale variations, which can be represented by the variable averaged over the grid cell, and small-scale variations, which are accounted for by the fluctuation. This is not the case for atmospheric flows which vary on a very wide range of scales which interact with each other. For instance, some processes have a scale on the same order as or slightly larger than the grid spacing and can only be marginally resolved on the grid. This introduces additional difficulties in developing a parameterisation that ideally should correct for the biases introduced in the solution because of the inadequate representation of those processes and account for the effect of the scales not resolved at all.

Individual clouds are also too small to be represented explicitly on the scale used routinely by climate models. A large number of processes, often not well known, are involved in cloud formation, ranging from **cloud microphysics** to the role of large-scale motion. Consequently, the representation of clouds and their influence on the radiative fluxes is usually a source of considerable uncertainty in models. In some EMICS, the cloud amount is diagnosed based on the relative humidity. In this framework, it is assumed that clouds must be present at some locations if the mean relative humidity averaged over a grid element is above a threshold, chosen to account for the sub-grid-scale heterogeneity. This threshold is generally on the order 80–90%. Cloud formation also must be linked with the vertical velocity and coupled with the parameterisation of turbulence and convection because the vertical motion may lead to condensation of water vapour (see, e.g., Hourdin et al. 2013). In current models, this is often accounted for through supplementary prognostic equations for the liquid or solid (ice) water content of atmospheric parcels or other variables related to clouds. They are added to the basic set of Eqs. (3.6)–(3.11) in order to have a better representation of cloud processes as well as of precipitation (see, e.g., Morrison and Gettelman 2008). Additionally, some large-scale models are able to explicitly simulate some clouds or have cloud-resolving models included in each of their grid cells to replace cloud parameterisation (Grabowski and Smolarkiewicz 1999; Tao et al. 2009), but such configurations are not yet standard in climate modelling.

## 3.3.2 Ocean

The major equations that govern ocean dynamics are based on the same principles as the equations described in Section 3.3.1 for the atmosphere. The major difference compared to Eqs. (3.6)–(3.11) is that the equation for the specific humidity is not required for the ocean, whilst a new equation for the salinity needs to be introduced. The equation of state is also fundamentally different. Unlike the atmosphere, there is no simple law for the ocean, and the standard equation of state is expressed as a function of the pressure, the temperature and the salinity as a long polynomial series (see Section 1.3.1).

It is much easier to compute the heating rate in the ocean than in the atmosphere. In addition to heat exchanges at the surface, the only significant heat source in the ocean is the absorption of solar radiation. This is taken into account in the model via an exponential decay of the solar irradiance. The situation for salinity is even more straightforward, as there is no source or sink of salinity inside the ocean. The equations governing these two variables thus are relatively simple:

$$\frac{dT}{dt} = F_{\text{sol}} + F_{\text{diff}} \tag{3.15}$$

$$\frac{dS}{dt} = F_{\text{diff}} \tag{3.16}$$

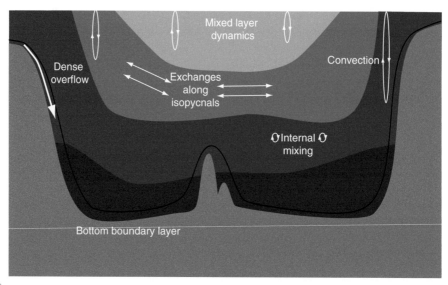

Schematic representation of some small-scale processes which have to be parameterised in global ocean models. (Source: Modified from http://www.gfdl.noaa.gov/ocean-models-at-gfdl.)

where $F_{sol}$ is the absorption of solar radiation in the ocean. Equation (3.15) generally does not apply to the in situ temperature but rather to the **potential temperature** in order to account more easily for the effect of the compressibility of sea water. The difference between those two temperatures is relatively low in the upper ocean, but it can reach several tenths of a degree in the deeper layers, an important difference in areas where the gradients are relatively small (see Section 1.3.3.2).

In Eqs. (3.15) and (3.16), we have explicitly added a term $F_{diff}$ to the right-hand side, representing the influence of the processes at scales which cannot be included in the model. They are often modelled as a diffusion term (see Section 3.3.1). In the simplest formulation, a **Laplacian** diffusion is retained by directly combining the equivalent for the ocean of Eqs. (3.13) and (3.14):

$$F_{diff} = \frac{\partial}{\partial x}\left(K_H \frac{\partial T}{\partial x}\right) + \frac{\partial}{\partial y}\left(K_H \frac{\partial T}{\partial y}\right) + \frac{\partial}{\partial z}\left(K_V \frac{\partial T}{\partial z}\right) \qquad (3.17)$$

where $K_H$ and $K_v$ are the diffusion coefficient in the horizontal and vertical, respectively. The diffusion coefficients $K_H$ and $K_v$ differ by several orders of magnitude as the sub-grid-scale processes in these two directions have different characteristics. This is due to the very different scales of ocean model grids on the vertical (a few hundred metres) and the horizontal (tens to hundreds of kilometres). For the friction force, a term similar to Eq. (3.17) is also often applied.

Actually, it appears that rather than separating horizontal and vertical directions, it is better to use a referential in the ocean interior which is aligned with the density surfaces because the density gradients damp the mixing. To this end, 'isopycnal' (along surfaces of equal density) and '**diapycnal**' (normal to surfaces of constant density) diffusion coefficients are calculated (Redi 1982; Cox 1987).

These coefficients can simply be chosen, or they can be computed using sophisticated modules (including turbulence models) which take into account the stirring of the winds, the influence of density gradients and the breaking of surface and internal waves (see Section 1.3.3.1). Furthermore, the **convection** occurring at high latitudes is generally represented by a convective adjustment scheme (see Section 3.3.1). Note that such a parameterisation only induces mixing of unstable water columns but no large-scale velocity. Large-scale vertical movements, associated, for instance, with the **thermohaline circulation**, should be explicitly simulated by the model.

Nevertheless, all small-scale processes cannot be represented by mixing. In particular, a standard parameterisation of the ocean **eddies** introduces a velocity that can be added to the large-scale velocity directly deduced from the equations at the grid scale (see, e.g., Gent et al. 1995). Besides, dense water formed at high latitudes can flow down the slope in narrow boundary currents called 'overflows'. They have a strong influence on the water mass properties but cannot be represented on the model grid scale. In such a case, a parameterisation of their effects as a transport process rather than a diffusion term appears more appropriate.

### 3.3.3  Sea Ice

The physical processes governing the development of sea ice can be divided conceptually into two parts (Figure 3.10). The first part covers the thermodynamic growth or decay of the ice, and the second part addresses ice dynamics and

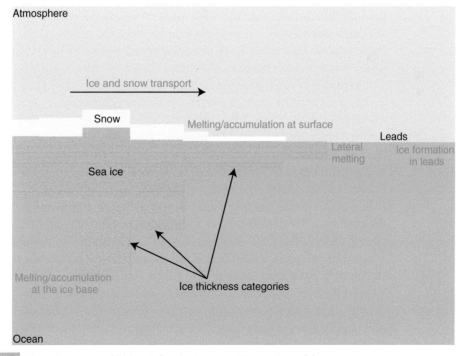

The main processes which have to be taken into account in a sea-ice model.

transport. For the thermodynamic processes, which depend on the exchanges with the atmosphere and the ocean, the horizontal heat conduction through ice can be safely neglected because the horizontal scale is much larger than the vertical scale. The thermodynamic code in a sea-ice model thus is applied separately for each grid cell to compute the 1-D energy balance over the vertical (see, e.g., Maykut and Untersteiner 1971). The heat conduction inside ice and snow then is ruled by

$$\rho_c c_c \frac{\partial T_c}{\partial t} = k_c \frac{\partial^2 T_c}{\partial z^2} \tag{3.18}$$

where $\rho_c$, $c_c$ and $k_c$ are the density, specific heat and thermal conductivity, and $T_c$ is the temperature. The subscript $c$ stands for either ice ($i$) or snow ($s$). Equation (3.18) is actually the equivalent of Eq. (3.10) for a solid. The left-hand side represents the changes in heat content, whilst the right-hand side can be related to the net local heating rate resulting from vertical conduction fluxes through the ice or snow. The left-hand side thus has a form very similar to Eq. (3.17), keeping only the vertical dimension.

The energy balance at the surface, which can be modelled similar to Eq. (2.37), allows computation of the surface temperature and of snow or ice melting. At the bottom of the sea ice, the local energy balance between the conduction through the ice and the flux from the ocean provides an estimate of ice melting or formation at the ice-ocean interface, the temperature there being considered equal to the freezing-point temperature.

The vertical heat flux resulting from conduction through the ice depends strongly on the thickness of the ice (Maykut 1982). In winter, the sea ice insulates the ocean from the cold atmosphere. The thicker the ice, the smaller are the conduction fluxes, reducing the heat losses of the ocean. As the majority of those oceanic heat losses must be compensated by sea-ice formation (and thus latent heat release) at the base of the floes, the ice growth is smaller for thicker ice. Consequently, a heterogeneous ice pack with areas covered by thick ice and others by thin ice will not have the same growth rate as a homogeneous ice layer with the same mean thickness. To represent this effect, current sea-ice models include a sub-grid-scale distribution of ice thickness (Thorndike et al. 1975). This distribution can be given as a simple function of the ice thickness but is much more often computed on the basis of additional equations for a number, generally between five and twenty, of different ice-thickness categories.

In addition to this thickness distribution of the ice, it is necessary to simulate the evolution of the fraction of each grid cell covered by ice, referred to as the 'ice concentration'. The remaining part corresponds to '**leads**' (or '**polynyas**'), defined in models as the open-water fraction present in each grid cell. This is achieved first by computing the energy budget for leads so as to determine whether new ice will form in the open ocean areas or whether lateral melting will occur. Second, ice divergence and convergence computed from the ice velocity (see below) will determine whether new areas of open water will be created because of ice export out of the grid cell or whether the leads area is reduced because of a net sea-ice inflow into the grid cell.

Large-scale dynamics may be computed assuming that sea ice is a 2-D continuum (see, e.g., Hibler 1979). This means that only one velocity is computed at the grid scale, whilst, in reality, the individual ice floes can move at different speeds. The hypothesis is adequate if, in a model grid box, a large number of ice floes of different sizes and thicknesses are present. In this case, estimating the velocity of each individual floes is not critical, and most of the effects of ice dynamics on large-scale variables can be represented by a mean transport at the grid scale. Newton's second law then gives

$$m\frac{d\vec{u}_i}{dt} = \vec{\tau}_{ai} + \vec{\tau}_{wi} - mf\vec{e}_z \times \vec{u}_i - mg\vec{\nabla}\eta + \vec{F}_{\text{int}} \tag{3.19}$$

where $m$ is the mass of snow and ice per unit area, $\vec{u}_i$ is the ice velocity and $f$, $\vec{e}_z$, $g$ and $\eta$ are, respectively, the Coriolis parameter, a unit vector pointing upward, the gravitational acceleration and the sea-surface elevation. $\vec{\tau}_{ai}$ and $\vec{\tau}_{wi}$ are the drag forces per unit area from the air and water, representing the interactions with the atmosphere and the ocean. The third term is the **Coriolis force**. The fourth is the force due to the oceanic tilt, since the ocean is not perfectly flat, as discussed in the framework of the **geostrophic equilibrium**. The internal forces $\vec{F}_{\text{int}}$ are due to interactions between floes, such as collisions, and the resistance of the ice before it can be broken to form leads. $\vec{F}_{\text{int}}$ is a function of ice thickness and concentration.

The velocity obtained from Eq. (3.19) is used in computation of the horizontal transport of the model state variables such as the ice thickness, concentration of each ice-thickness category and internal sea-ice temperature and salinity. The time variations of these variables thus are ruled by both the horizontal transport and the vertical exchanges shown, for instance, in Eq. (3.18). This illustrates that the classical separation between thermodynamics and dynamics in sea-ice models is only conceptual.

## 3.3.4 Land Surface

As with sea ice, horizontal heat conduction in soil can be safely neglected. Therefore, thermodynamic processes are only computed along the vertical for each grid point [in a similar way to Eq. (3.18)]. In the first generation of land-surface models, only one soil layer was considered. Using the same notations as in Eq. (2.37), soil temperature can be computed from the energy balance at the surface:

$$\rho c_{\text{ground}} h_{su} \frac{\partial T_s}{\partial t} = (1 - \alpha) F_{\text{sol}} + F_{IR\downarrow} + F_{IR\uparrow} + F_{LH} + L_v E + F_{\text{cond}} \tag{3.20}$$

When a snow layer is present, computation of the development of snow depth, density and concentration is part of the surface energy balance. This is very important for the **albedo**, whose **parameterisation** as a function of soil characteristics (i.e., snow depth, snow age, vegetation type, etc.) is a crucial element of surface models.

The latent heat flux $F_{LH}$ in Eq. (2.37) has been replaced in Eq. (3.20) by the latent heat of fusion times the evaporation rate ($L_v E$), as is classically done in

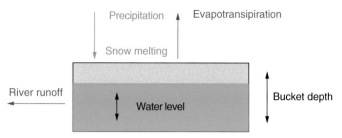

**Fig. 3.11** The schematic representation of the hydrological cycle over a land surface in a simple bucket model.

land-surface models. This evaporation rate depends on the characteristics of the soil and vegetation cover, as well as on water availability. It can be expressed with the help of the 'moisture availability function' $\beta$ ($0 < \beta < 1$), defined as the ratio between the **evapotranspiration** rate of the surface $E$ and the potential evapotranspiration $E_p$, that is, the evaporation that would occur on a homogeneous wet surface:

$$\beta = E/E_p \qquad (3.21)$$

A land-surface model also simulates the water content of the soil. In the simple early models, this was represented by a so-called bucket model (Figure 3.11). The water budget of the 'bucket' is computed following an equation similar to Eq. (2.40). If, after taking into account precipitation and surface evaporation, the amount of water in the soil exceeds a threshold equivalent of 15–30 cm $H_2O$ averaged over the grid cell, the excess water is immediately transferred to a prescribed oceanic grid point through river runoff. The water content of the soil then is used to compute the moisture availability function $\beta$. This value is equal to 0 for a dry soil and 1 at or above a prescribed value of the soil moisture content, which is often chosen equal to the maximum capacity of the bucket model. In this framework, the spatial distribution of the parameters characterising the surface, such as the albedo, the surface roughness or the bucket depth, are prescribed on the basis of observations.

This bucket model is obviously an overly strong simplification of a complex system. More sophisticated representations of the surface processes are required to have an accurate estimate of the energy and water exchanged between the atmosphere and the land. This has led to the development of components classified as second-generation models (Sellers et al. 1997) which explicitly account for the biophysical control of evaporation through the **stomatas** and for the interactions among the **canopy**, the soil and the roots (see Section 2.2.2). Additionally, the so-called third-generation models include a representation of plant physiology, in particular, of photosynthesis and the role of nutrients, coupled to the other components of the model. As part of these developments, multi-layer soil models are now standard in state-of-the-art GCMs in order to have a better estimate of the heat and water storage and the different time scales associated with the response of the surface and deeper layers (Figure 3.12). Horizontally, instead of

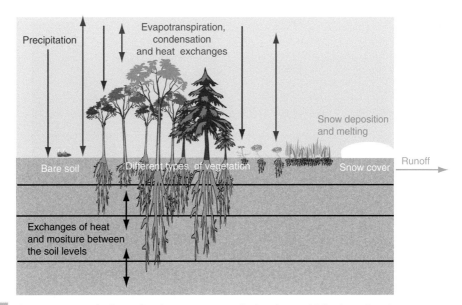

**Fig. 3.12**    The main processes that have to be taken into account in a land-surface model. For clarity, the carbon storage in plants and in soils, as well as the exchanges between these reservoirs and with the atmosphere, are not shown.

a homogeneous description as proposed in Eq. (3.20), the heterogeneous nature of the land surface, covered by different types of vegetation, as well as soils, lakes and cities, is generally modelled by sharing the individual grid cells in different regions with different characteristics and thus different surface temperatures. Furthermore, current models have a sophisticated river-routing scheme which accounts for the duration of water transport as well as for evaporation during the journey to the ocean or the interior sea. These improvements are also essential in an adequate representation of the carbon cycle on land (see Section 2.3.3). At present, only a few climate models include a representation of **permafrost**, but this is likely to improve because of the large modifications in the extent of permafrost that are expected during the twenty-first century.

Some models take the community composition and vegetation structure as a boundary condition or forcing (if land-use changes are specified, for instance). They then use this information to determine the physical characteristics of the surface and the soil, as well as the carbon storage over land. By contrast, dynamic global vegetation models (DGVMs) explicitly compute the transient dynamics of the vegetation cover in response to climate changes and to disturbances such as fires. DGVMs also can provide the distribution of **biomes** which are in equilibrium with climate conditions (Figure 3.13). Of course, it is impossible to represent the fraction covered by each of thousands of different plant species in DGVMs. The plants thus are grouped into '**plant functional types**' (PFTs) which share common characteristics. Very simple models only use two PFTs (e.g., trees and grass, the area not covered by trees or grass corresponding to the fraction of desert in a **grid** element), whilst more sophisticated models use more than ten different PFTs.

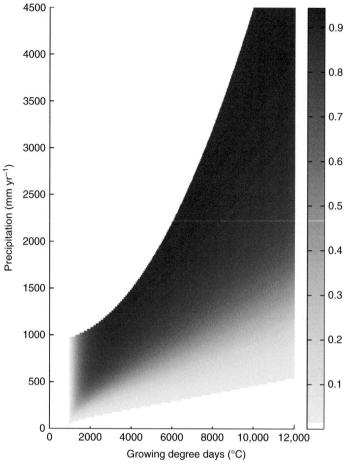

**Fig. 3.13**  The equilibrium fraction of trees in a model that includes two plant functional types and whose community composition is influenced only by precipitation and the growing degree days (GDDs). 'Growing degree days' are defined as the sum of the daily mean near-surface temperatures for the days of a year with a temperature higher than 0°C. (Source: Courtesy of V. Zunz, using the parameterisation proposed in Brovkin et al. 1997. Reproduced with permission.)

### 3.3.5  Marine Biogeochemistry

Models of biogeochemical cycles in the oceans are based on a set of equations whose formulation is very close to that of Eqs. (3.15) and (3.16) for the ocean temperature and salinity:

$$\frac{dTrac}{dt} = F_{\mathrm{diff}} + F_{\mathrm{sources}} - F_{\mathrm{sinks}} \qquad (3.22)$$

where *Trac* is a biogeochemical variable. Those variables are often called 'tracers' because they are transported and diffused by the oceanic flow (the left-hand side of the equation and the term $F_{\mathrm{diff}}$).

*Trac* can represent DIC, alkalinity (see Section 2.3.2), the concentration of various chemical species (including the nutrients necessary for phytoplankton growth), detritus and the biomass of different groups of phytoplankton, zooplankton and (more rarely) higher trophic levels. Simple carbon-cycle models include a few **state variables**. Some are referred to as nutrient-phytoplankton-zooplankton-detritus (NPZD)–type models, in reference to the minimum set of four variables (nutrients, phytoplankton, zooplankton, detritus) represented (see, e.g., Oschlies and Garçon 1999). Nevertheless, various levels of complexity exist among marine biogeochemistry models, and sometimes NPZD models form the basis of more complex models. Presently, the most sophisticated biogeochemical components used in ESMs can have more than twenty variables, taking into account the potential limitation of production by various nutrients (e.g., phosphate, nitrate, silicates, dissolved iron), the biogeochemical cycles of several species of nitrogen and other chemical elements, and many planktonic groups (see, e.g., Ilyina et al. 2013; Cabré et al. 2015).

The $F_{sources}$ and $F_{sinks}$ terms account for the increase or decrease in the tracer concentration in response to biogeochemical processes, including a representation of the processes described in Section 2.3 (Figure 3.14). For instance, for a particular phytoplankton group, the 'sources' term could be related to growth of the biomass by photosynthesis, whilst the 'sinks' could be the consumption of phytoplankton by zooplankton as well as the mortality of the cells. In simple models, these terms are parameterised as a function of a few variables, whilst others base the estimation of $F_{sources}$ and $F_{sinks}$ terms on a more comprehensive representation of ecosystem dynamics. In addition to the processes taking place in the

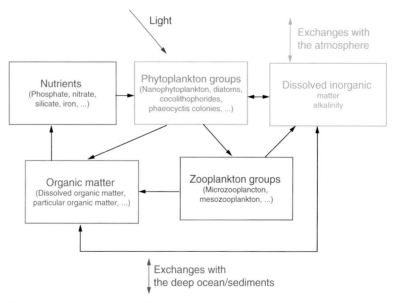

**Fig. 3.14** A simplified scheme representing some of the variables of a biogeochemical model. The interactions between the groups are complex as the different types of phytoplankton need different nutrients and are grazed by different types of zooplankton.

water column, some models include a comprehensive ocean sediment component so as to be able to study the long-term changes in the carbon cycle.

## 3.3.6 Ice Sheets

As discussed for sea ice, ice-sheet models can be decomposed into two major components: a dynamic core which computes the flow of ice and a thermodynamic part which estimates the changes in ice temperature, snow accumulation, and melting (see, e.g., Winkelmann et al. 2011). The ice velocity can be computed using the full 3-D equation. This is affordable for regional models focussing on a relatively small area, but approximations are often necessary for climate models which compute the development of whole ice sheets on long time scales.

Using a similar approach as for Eq. (3.9), the conservation of ice volume can be written as

$$\frac{\partial H}{\partial t} = -\vec{\nabla} \cdot (\vec{u}_m H) + M_b \qquad (3.23)$$

where $H$ is the ice-sheet thickness (i.e., the volume per unit surface), $\vec{u}_m$ is the depth-averaged horizontal velocity field and $M_b$ is the mass balance accounting for snow accumulation as well as basal and surface melting. Surface melting can be deduced from the energy budget at the surface [similar to Eq. (2.37); see also Section 3.3.3]. Simpler formulations of surface melting are based on positive-degree-day methods, which relate the melting to the temperature during the days with temperatures above 0°C (see Figure 3.13 for another application of the degree-day method). An important element in the mass balance at the surface of the ice sheets is the position of the equilibrium line between the accumulation and ablation regions (Figure 3.15), which correspond to areas characterised by a net snow gain and a net snow and ice melt at the surface on a yearly average, respectively. On the Greenland ice sheet, in present-day conditions, ablation occurs in many areas, whilst on the colder Antarctic ice sheet, it is restricted to a few regions only.

The melting at the ice base is deduced from the balance between the heat conduction in the ice and in the ground, taking into account the geothermal heat flux. Conditions at the ice base, and in particular, the presence of water or ice close to the melting point at the corresponding pressure, have a large impact on the ice velocity. Indeed, such a condition reduces the stresses greatly compared to the situation where the ice is well below the freezing point.

An additional element in ice-sheet models is representation of interactions between the ocean and **ice shelves**, whose extent is conditioned by the ocean temperature and the global mean sea level. As the later, in turn, depends on the volume of land ice, this leads to potential **feedbacks**. Because of local melting and iceberg calving, ice shelves can make a large contribution to the mass balance of the ice sheets, as is currently the case for Antarctica. Furthermore, they generate stresses that tend to slow down the ice flow on land. Indeed, observations

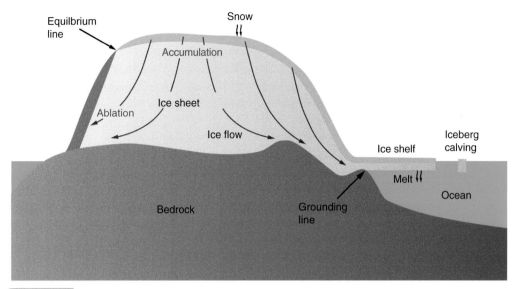

**Fig. 3.15**    The main processes which have to be taken into account in an ice-sheet model.

have shown that the recent breakdown of some ice shelves has produced, in some regions, an acceleration of the land ice (see also Section 6.3.3).

Finally, ice-sheet models also should take into account the interactions with the underlying bedrock. In particular, as the load of the ice sheet tends to depress the bedrock, a bedrock adjustment model is needed to compute the position of the ground as a function of the ice mass. Knowing the ice thickness, this yields the elevation of the ice sheet.

### 3.3.7  Aerosols and Atmospheric Chemistry

The equations governing the development of **dry air** and atmospheric water vapour were described in Section 3.3.1. Nevertheless, the atmosphere also includes minor gases and '**aerosols**', defined here as relatively small solid or liquid particles in suspension in a gas, which play an important role in the radiative balance of the Earth (see Sections 2.1.2 and 4.1.2). Although the atmospheric composition and the distribution of various aerosols can be imposed as a forcing, many models now compute their development interactively.

As the radiative properties of aerosols depend on their size, it is necessary to determine the spatial and temporal variations in their size distribution. This distribution is generally approximated through a few size classes, from finest to coarsest (see, e.g., Stier et al. 2005; Boucher 2012; Mann et al. 2014). The greenhouse gases such as methane, nitrous oxide and ozone are involved in many chemical reactions with other atmospheric components. Consequently, chemical models have to include a large number of species that ranges from around ten to more than one-hundred in the most complex cases (see, e.g., Lamarque et al. 2014).

In addition to representation of interactions between particles or chemical reactions in the atmosphere, determining the surface sources of the various chemical species, of the aerosols and of their precursors (which would lead to the formation of some aerosols after transformation), is a key element in chemistry and aerosol modelling. Besides those sources, sinks by dry deposition (sedimentation and then direct contact with the surface) or by wet deposition (precipitation of rain drops and ice crystals including some chemical species or aerosols) also have to be taken into account.

### 3.3.8  Earth System Models: Coupling between the Components

The interactions between the various components of the climate system play a crucial role in the dynamic of climate. Wind stress, heat and freshwater fluxes at the ocean surface are the main drivers of the ocean circulation (see Section 1.3.2). The evaporation at the ocean surface is the largest source of water vapour for the atmosphere, which influences the radiative properties of the air (see Section 2.1.2) and atmospheric heat transport (see Section 2.1.5). Snow falling on ice sheets is an essential element of their mass balance. Many other examples can be cited.

Some of these interactions are quite straightforward to compute from the models' state variables, whilst more sophisticated **parameterisations** are required for others. For instance, the parameterisation of wind stress and heat flux at the atmospheric base [e.g., Eqs. (2.34) and (2.35)] can be derived from theories about the **atmospheric boundary layer**. However, this computation still requires empirical parameters which depend on the characteristics of the surface, introducing some uncertainties into estimation of the flux.

The technical coupling of the various components to obtain a climate or Earth system model brings additional difficulties. The numerical codes generally have been developed independently by different groups using different coding standards and different **numerical grids**. It is thus necessary to design an interface and use a 'coupler', that is, a code specifically adapted to represent the exchanges between the various components. Additionally, the critical scale for one component may not be represented adequately in another component, leading to the need for some downscaling. For instance, this causes trouble for the coupling between ice-sheet and atmospheric models as the ablation zones responsible for a large fraction of the summer melt are often too small to be sufficiently well simulated on the scale of the AGCM.

The preceding sections dealt with some important aspects of current climate or ESMs, but new elements are regularly included to obtain an increasingly comprehensive representation of the complex interactions within the system. Nevertheless, because of the cost in computer time required by each component, many simulations are still performed with only the ocean, the atmosphere and the sea-ice modules activated. This also implies a trade-off between the grid resolution, the duration of the simulation and the number of components that could be interactively modelled with available computers. Besides, activating only selected elements also can be a choice in order to focus on certain specific processes (see Section 3.1.2).

## 3.4 Numerical Resolution of the Equations

To be handled by a computer, the mathematical models developed for each component of the climate system (see Section 3.3) have to be expressed as numerical models. In particular, the derivatives present in the **partial differential equations** (PDEs) described earlier have to be replaced by simpler operations such as sums and products which can be performed readily by the computer. The principles used to develop such a numerical model and obtain a numerical solution of the equations are explained in the following sub-sections. The goal here is to briefly discuss the interest and limitations of the methods in order to be able to critically assess the performance of climate models. Readers interested in a more comprehensive description of the numerical analysis applied in climate science are referred, for instance, to Cushman-Roisin and Beckers (2011).

Nevertheless, before solving those equations, it is first necessary to ensure that the problem is well posed mathematically, that is, that it has a unique solution that depends on the initial and boundary conditions. This requires that the initial and boundary conditions be specified properly. For instance, to solve the equation for temperature in the ocean knowing the velocity field [Eqs. (3.15) and (3.17)], we must specify the initial temperature over the whole domain at a time $t_0$ as well as one boundary condition over all the points of the spatial boundaries of the domain, which can be the value of the heat flux or the temperature there. In the following sub-sections we will consider that all the problems investigated are well posed.

### 3.4.1 A Simple Example Using the Finite-Difference Method

The numerical method which allows us to solve differential equations that is probably the easiest to understand consists of approximating the derivatives by finite differences. This is called the 'finite-difference method', and the process is sometimes referred to as 'discretisation' of the equations. The solution is no longer a continuous function (as for **partial differential equations**) but a discrete one, defined only for specific times separated by the time step $\Delta t$ and specific locations separated by the spatial step $\Delta x$ (plus $\Delta y$ and $\Delta z$ for a problem with three spatial dimensions).

Consider, for instance, the ordinary differential equation

$$\frac{du}{dt} = F(u,t) \tag{3.24}$$

where $t$ is the time, $u$ is a state variable (e.g., velocity) which depends here only on time and $F$ is a function of $u$ and $t$. Solving Eq. (3.24) knowing the initial condition $u(t_0)$ can be interpreted as finding $u(t)$, the integral of $F$ between $t_0$ and $t$. This is the reason why a model simulation is sometimes referred to as an 'integration'.

The integration of Eq. (3.24) is not straightforward as $F$ depends on the unknown $u$. We will thus consider a simpler case where $F$ is function of $t$ only:

$$\frac{du}{dt} = A\cos(t) \tag{3.25}$$

where $A$ is a constant.

The mathematical definition of the derivative according to time of a function $u(t)$ is

$$\frac{du}{dt} = \lim_{\Delta t \to 0} \frac{u(t+\Delta t)-u(t)}{\Delta t} \tag{3.26}$$

Numerically, we cannot have an exact representation of this limit with a time step $\Delta t$ which may be small but always must be larger than zero. The derivative is then approximated by a finite difference. If we assume that $\Delta t$ is constant and that the initial time is $t = 0$, we can write that $t = n\Delta t$, where $n$ is the number of time steps since the beginning of the simulation. Using the standard notation $u(t+\Delta t) = u\big[(n+1)\Delta t\big] = u^{n+1}$, this leads to

$$\frac{du}{dt} \simeq \frac{u(t+\Delta t)-u(t)}{\Delta t}$$
$$\simeq \frac{u^{n+1}-u^n}{\Delta t} \tag{3.27}$$

If we apply this approximation in Eq. (3.25), we have

$$\frac{u^{n+1}-u^n}{\Delta t} = A\cos(n\Delta t) \tag{3.28}$$

and then

$$u^{n+1} = u^n + A\Delta t \cos(n\Delta t) \tag{3.29}$$

Knowing the initial condition $u(t_0) = u^0$, we can compute $u^1$, then $u^2$ and so on for any time $n\Delta t$ (Figure 3.16). Such problems are called 'initial-value problems' because once the initial value is specified, the unknown can be estimated for any time by advancing or 'marching' in time.

## 3.4.2 Consistence, Convergence, Stability and Accuracy

The preceding example has illustrated the procedure to solve numerically a differential equation, but more formally, two fundamental properties must hold to consider that a numerical method is adequate. First, the finite-difference equation must be *consistent* with the PDE. This means that as $\Delta t \to 0$ (and $\Delta x \to 0$, $\Delta y \to 0$ and $\Delta z \to 0$ if the problem has one or more spatial dimensions), the finite differential equation coincides with the **partial differential equation**. This is essential to ensure that the equation which has been solved numerically is a reasonable approximation of the mathematical model. It can be

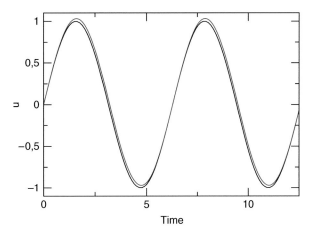

**Fig. 3.16** The analytical solution of Eq. (3.25) using $A=1$ and $u(t=0)=0$ (black) and the numerical solution using Eq. (3.28) with a time step $\Delta t = \pi/50$ (red). The discrete solutions $u^n$ at times $n\Delta t$ have been joined by straight lines.

checked by using **Taylor series** expansions. For the left-hand side of Eq. (3.28), this gives

$$u^{n+1} = u^n + \frac{du}{dt}\Delta t + \frac{1}{2}\frac{d^2u}{dt^2}\Delta t^2 + \text{higher-order terms} \qquad (3.30)$$

and thus

$$\frac{u^{n+1}-u^n}{\Delta t} = \frac{du}{dt} + \frac{1}{2}\frac{d^2u}{dt^2}\Delta t + \text{higher-order terms} \qquad (3.31)$$

As expected, the approximation of the derivative applied in Eq. (3.28) effectively tends to $du/dt$ as $\Delta t$ tends to 0. For the right-hand side of Eq. (3.28), the demonstration is straightforward. This shows that the numerical scheme proposed earlier is consistent.

The introduction of Taylor series expansion allows estimation of the 'truncation error', which can be defined as the error made by replacing the exact derivative by a finite difference. For small values of $\Delta t$, the dominant term of this error in Eq. (3.31) is $0.5\big(d^2u/dt^2\big)\Delta t$. Consequently, the error is said to be of the first order in time as the power of $\Delta t$ in this term is 1. If for a different numerical scheme the dominant term in the truncation error were proportional to the square of $\Delta t$, the scheme would have been said to be of the second order, and so on. The higher the order of the numerical scheme, the faster the truncation error decreases for values of $\Delta t$ tending to zero. Nevertheless, the complexity of the implementation and the number of operations required (and thus the cost in computer time) also increase for those schemes.

It can be shown that the main effect of the truncation error is often equivalent to the addition of an artificial term in the differential equation with the same mathematical form as diffusion (see, e.g., Cushman-Roisin and Beckers 2011). This implies that the truncation error will introduce a spurious smoothing of the numerical solution compared to the exact solution, similar to what would occur with a real physical diffusion term in the mathematical model [e.g., Eq. (3.17)]. This type of error resulting from numerical implementation of the mathematical model thus is referred to as 'numerical diffusion'.

The second property that has to be verified by the numerical scheme is that the solution of the finite-difference equation *converges* to the solution of the PDE as the time step (and grid spacing) tends to zero. Whilst the consistency deals with the equation itself, the convergence focusses on the numerical solution and ensures that it tends to the exact solution as required. In the example in Section 3.4.1, this means that

$$u^n \to u(t) \quad \text{when} \quad \Delta t \to 0 \qquad (3.32)$$

Another characteristic of a numerical scheme is its *stability*. Precise and general definitions are available, but we will simply consider here that a numerical scheme is computationally stable over a fixed interval $T$ if the solution of the finite-difference equation remains bounded for all time $t = n\Delta t < T$. By contrast, for an unstable scheme, the magnitude of some variables may increase without limits, leading to a stop of the execution of the numerical code, for instance, because some numbers become too big to be handled by the computer. In a more colourful language, it can be said that in this case the numerical model 'explodes' or 'blows up'.

The convergence and computational stability are related, thanks to the Lax-Richtmyer theorem, which states that for a well-posed initial-value problem, a consistent finite-difference numerical scheme for a linear PDE is convergent if and only if it is stable. As stability is easier to study than convergence, the practical methods used to test the convergence of a numerical scheme are based on an analysis of its stability. In some cases, it is possible to explicitly demonstrate that the solution is bounded, a property generally conditioned by a criterion that governs $\Delta t$ and $\Delta x$ (see examples in Section 3.4.3). A more general way to determine the largest time and spatial step allowed is the von Neumann method, in which the stability of the finite-difference equation is analysed by expressing the solution as an expansion of an appropriate set of basis functions, generally **Fourier series** (see, e.g., Kalnay 2003; Cushman-Roisin and Beckers 2011).

The consistency and stability must be investigated, but determining the behaviour of a numerical scheme as the time step and/or grid spacing tends to zero is not enough as, in practical implementations, this time step and grid spacing always have a finite value. A point that is particularly important in climatology is to acknowledge that the differential equation satisfies some important properties that are not necessarily verified in practice by the numerical scheme. Consider, for instance, the mass of salt in an ocean model, which is conserved by the PDE in Eq. (3.16). The numerical scheme also should conserves mass when $\Delta t$ and $\Delta x$ tend to zero but may lose or artificially create mass at each time step when $\Delta t$ has the finite value selected to solve the problem. After many time steps, the numerical errors may accumulate to a point that a significant amount of salt in the ocean has disappeared, leading to a totally unphysical solution.

Consequently, in this case, it is necessary to choose a numerical scheme that conserves the mass of salt, whatever the value of the time step and not only at the limit for the time step going to zero. Similar arguments can be given for the conservation of energy, the mass of the atmosphere and so on.

In addition, the consistency and stability of the scheme do not ensure that the processes of interest are represented with a sufficient accuracy. The physical and numerical aspects are often related, some numerical instabilities being, for instance, generated by the inadequate representation of a mechanism which leads to an unphysical behaviour and then the blow-up of the code. Nevertheless, large errors can already be present before such explosion. By contrast, very stable schemes may strongly damp the temporal and spatial variations of the fields, leading to a solution which is bounded but far away from the true values. A natural criterion is to choose a grid spacing and time step which are smaller than the characteristic scales of the process studied. For instance, to represent the daily cycle of the sea surface temperature, a minimum time step of 1–2 hours seems reasonable even if the numerical scheme may be stable for larger values. Furthermore, the properties and accuracy of the various schemes should be analysed in idealised test cases to determine the one which is the most adapted to the problem.

The resolution and numerical methods selected in a particular model finally result from a balance between the various constrains related to the characteristics of the problem analysed and the resources in computer time. The architecture of the computer itself may have consequences. GCM simulations are now performed on machines including thousands of processors which realise many operations simultaneously in parallel. After having completed the elementary computations, the processors interact with each other before going to the next step. As this communication between processors takes time, the numerical code has to be optimised to minimise the exchanges, and some numerical methods appears more adapted to this goal than others.

This brief discussion underlines the need to keep in mind that the numerical resolution introduces errors which should be smaller than the uncertainties of the mathematical model itself to have a solution that fairly represents the behaviour of the modelled climate. The estimate of the magnitude of the error associated with the numerical resolution also should be a guide for model development in order to set up priorities. For instance, it would be a nonsense to improve the representation of the physical diffusion present in the equation of the mathematical model if the expected changes are much smaller than the numerical diffusion that is present because of truncation errors.

### 3.4.3 Time and Space Discretisations Using Finite Differences

Many options are available for discretising an equation, and the choice depends on the properties required for the numerical scheme. We presented a first example of time discretisation in Section 3.4.1 which is called the 'forward Euler method':

$$\frac{u^{n+1} - u^n}{\Delta t} = F\left(u^n\right) \tag{3.33}$$

for a right-hand side represented in a general way as the function $F(u^n)$. This simple scheme is still applied in some practical implementations but may be unstable for some types of equations.

An alternative scheme is a centred difference (leapfrog scheme):

$$\frac{u^{n+1} - u^{n-1}}{2\Delta t} = F(u^n) \tag{3.34}$$

which has a second-order truncation error and thus is, in principle, more precise than the first-order schemes. However, this scheme allows the presence and growth of unphysical modes and thus is generally stabilised by associating it with a time filter.

Equations (3.33) and (3.34) correspond to so-called explicit schemes. In implicit schemes, the right-hand side is a function of both $u^n$ and $u^{n+1}$. If $F$ is a function of $u^{n+1}$ only, the scheme is called 'fully implicit' or 'backward'. In explicit schemes, expressing $u^{n+1}$ as a function of the value at the preceding time step is immediate, as shown in Eq. (3.29). By contrast, implicit schemes require an equation or a system of equations to be solved to obtain $u^{n+1}$, equations which could be non-linear. Implicit schemes thus can be relatively expensive in terms of computer time. However, implicit schemes are stable for longer time steps, which in some circumstances is a clear advantage.

The same variety of numerical schemes is available for space discretisation. Consider the diffusion equation [like Eq. (3.18)]

$$\frac{\partial u}{\partial t} = k\frac{\partial^2 u}{\partial x^2} \tag{3.35}$$

where $k$ is a constant. This can be discretised as

$$\frac{u_j^{n+1} - u_j^n}{\Delta t} = k\frac{u_{j+1}^n - 2u_j^n + u_{j-1}^n}{\Delta x^2} \tag{3.36}$$

The index $j$ refers to point number $j$ of the spatial grid, which is at a distance $(j-1)\Delta x$ from the first grid point if the grid spacing $\Delta x$ is constant. It can be easily shown (see, e.g., Cushman-Roisin and Beckers 2011) that this scheme is consistent and that the truncation error is first order in time and second order in space. It is stable if

$$k\frac{\Delta t}{\Delta x^2} \leq \frac{1}{2} \tag{3.37}$$

illustrating the link usually present between $\Delta t$ and $\Delta x$. Using this scheme, the solution at point $j$ is updated at each time step $n + 1$ from the values computed at time step $n$ for points $j - 1$, $j$ and $j + 1$ (Figure 3.17).

Another standard case is the advection equation

$$\frac{\partial u}{\partial t} + c\frac{\partial u}{\partial x} = 0 \tag{3.38}$$

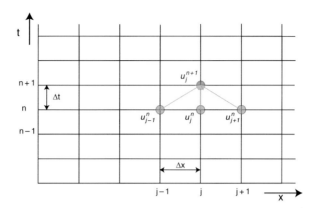

Schematic representation of the grid structure in space and time with one spatial dimension for the numerical scheme proposed in Eq. (3.36) showing that $u_j^{n+1}$ depends on $u_{j-1}^n$, $u_j^n$ and $u_{j+1}^n$.

which corresponds, for instance, to Eq. (3.16) if the transport in one direction with a velocity $c$ ($c > 0$) is the dominant term and if diffusion can be neglected. For this equation, the so-called upwind or upstream scheme gives

$$\frac{u_j^{n+1} - u_j^n}{\Delta t} + c\frac{u_j^n - u_{j-1}^n}{\Delta x} = 0 \tag{3.39}$$

It is associated with large numerical diffusion which reduces the magnitude of the gradients compared to the true solution of the equations. Consequently, more sophisticated schemes are generally applied in current models. The upwind scheme is stable if the Courant number $C$ is smaller than 1:

$$C = c\frac{\Delta t}{\Delta x} \leq 1 \tag{3.40}$$

This stability criterion is called the 'Courant-Frederichs-Lewy (CFL) condition'. An interpretation of this condition is to state that the distance travelled by a fluid parcel during a time step (equal to $c\Delta t$) should be smaller than the grid spacing $\Delta x$. Indeed, if the Courant number is higher than 1, the solution for $u_j^{n+1}$ should be influenced after one time step by the conditions at $u_{j-2}^n$, but this is not allowed by the numerical scheme which computes $u_j^{n+1}$ only from $u_{j-1}^n$ and $u_j^n$. In other words, the physical propagation of the signal is faster than allowed by the numerical scheme, leading to an instability.

Like Eq. (3.37), Eq. (3.40) also illustrates the link between the time step and the grid spacing which has to be respected for numerical stability. Because transport is generally the process which limits the stability of a numerical model, an increase in the resolution by a factor of 2 implies a reduction of the time step by a factor of 2. Furthermore, the equations show that the time step is imposed by the smallest grid spacing present in the whole grid. Since meridians converge to one point at the North Pole, grids based on classical latitude/longitudes have smaller grid spacing in longitude as latitude increases. This finally leads to a singularity at the poles where the grid points corresponding to all longitudes collapse to a single

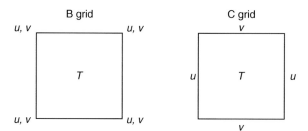

**Fig. 3.18** The location of some variables on the staggered B and C grids according to Arakawa and Lamb's (1977) classification.

point. To avoid using a too small time steps because of the smaller grid spacing and the troubles associated with the singularity, some spatial filters can be applied to stabilise the solution at high latitudes, or grids which are more sophisticated than latitude/longitude grids can be chosen (see Figure 3.2). Alternative numerical methods, not constrained by a criterion similar to Eq. (3.40), are also available.

In the preceding three examples [Eqs. (3.25), (3.35) and (3.38)], only one equation for one variable was solved. However, for many components of the climate system, equations for several variables must be solved simultaneously. For numerical reasons, those variables are not necessarily located at the same place on the grid, leading to what are called 'staggered grids'. Arakawa and Lamb (1977) proposed a classification of these grids. Two popular ones are the B and the C grids. If we consider an elementary square of the grid for an ocean model, for instance, for the B grid, the temperature $T$ (as well as the salinity, pressure and density) is computed at the centre of the cells, whilst the velocity components $u$ and $v$ are obtained at the corners of the grid elements (Figure 3.18). For the C grid, the velocities are computed on the sides of the elements. Staggered grids are also widely used for vertical elements, with the vertical velocity usually computed at the boundary between the layers, whilst the temperature is defined at the centre of the layers.

### 3.4.4 Spectral Representation, Finite-Volume and Finite-Element Methods

In addition to finite differences, several other methods can be used to discretise equations. One of them is to integrate the basic equations over a finite volume before the discretisation. This finite-volume method has the advantage that it explicitly and easily ensures the conservation of some important properties. If particular hypotheses are made about the changes in the variables inside the volumes, numerical schemes similar to those described using finite-difference methods can be obtained.

Alternatively, the numerical solution can be discretised in space as a sum of $K$ basis functions $\phi_k(x)$ multiplied by coefficients $A_k$ which depend on the time:

$$u(x,t) = \sum_{k=1}^{K} A_k(t)\varphi_k(x) \tag{3.41}$$

with a trivial extension to more than one spatial dimension. The unknown of the problem thus is the coefficient $A_k$ and not the value of the variable at some locations, as in the finite-difference method. These coefficient $A_k$ are chosen in order to obtain a function which approximates as accurately as possible the true solution.

To obtain good results with a reasonable number of operations, and thus a reasonable cost, the choice of the basis function must be adequate. In the so-called spectral methods, **Fourier series** are chosen for 1-D problems, and the solution is represented as the sum of sine waves [for more detail on spectral methods, see, e.g., Kalnay (2003)]. For a situation with spherical geometry such as the Earth, spherical harmonics are preferred. They are the product of Fourier series in longitude and associated **Legendre polynomials** in latitude. The big advantage of the spectral method is that the space derivation of $u(x, t)$ can be analytically computed from $d\phi_k(x)/dx$ without any additional approximation. This is one of the reasons why most atmospheric models employ this approach. It also avoids most of the problems associated with the singularity at the poles (see Section 3.4.3).

The larger the number of basis functions retained, the more precise is the spatial representation of the solution. However, for practical reasons, a limit has to be set. For a 1-D case, this is easily measured by the index $K$ in Eq. (3.41). A wider range of options is available for 2-D problems. A classical choice is to use a triangular truncation, meaning that the basis functions are chosen so that the largest Fourier wave number and the highest degree of the associated Legendre polynomial are equal [for a more precise definition, see McGuffie and Henderson-Sellers (2014)]. This type of truncation is referred to by the letter 'T', with, for instance, a spectral model with a truncation 'T106' having 106 sinusoidal basis functions in longitude.

The spectral and grid-point (also referred to as 'physical space') representations are complementary. The first provides an accurate computation of horizontal derivatives, whilst the second forms a convenient framework to evaluate the physics of the model (e.g., computation of the radiative transfer). When the two approaches are combined, some transfers of the variables from the physical to the spectral space are required. There must be a correspondence between the number of basis functions and the horizontal resolution of the corresponding grid to avoid numerical problems during this transformation. This is the reason why an equivalent in grid spacing is generally given with the spectral resolution. The outputs of spectral models are also provided on a **grid** as for models based on finite differences. The estimates of physical variables at some locations are indeed more explicit and easier to handle than the values of the coefficients of the basis functions. Nevertheless, a grid spacing which is slightly smaller than the scale actually resolved by the spectral representation generally is used to reduce the inaccuracies associated with the transformation from the physical to the spectral space. This leads to an apparently higher resolution in the physical space (see, e.g., Neelin 2011).

The finite-element method is based on a similar approximation to Eq. (3.41), but instead of functions $\phi_k(x)$ which cover the whole domain as in the spectral method, local basis functions are used. For example, $\phi_k(x)$ can be a piecewise linear function equal to 1 at a grid point and 0 at all other points. The finite-element method is widely applied in many fields but is not yet common in climate modelling, although some ocean models now use this approach (see, e.g., Hanert et al. 2005; Wang et al. 2014).

## 3.5   Model Evaluation

### 3.5.1 Testing, Verification and Validation

Despite a very careful design, there is no guarantee that a numerical model will be adequate for its intended use: some processes treated as negligible can turn out to be more important than initially thought; a **parameterisation** may not be valid in the particular conditions of interest or may be incompatible with other hypotheses employed; the selection of parameters within their range of uncertainty can be far from optimal; and so on. As a consequence, climate models have to be tested to assess their quality and evaluate their performance (e.g., Schmidt and Sherwood 2014). In this framework, it is always necessary to keep in mind the scientific objectives of the study (or studies) which will be conducted using a particular model. Although the principles remain the same, the tests performed with a model created to study the development of the global carbon cycle over the last million years (see Section 5.4.2) are clearly different from those for a model providing projections of future climate changes at the highest possible **resolution** (see Chapter 6).

A first step is to ensure that the numerical model solves the equations of the physical model adequately. This procedure, often referred to as 'verification' (Figure 3.19), deals only with numerical resolution of the equations in the model, not with agreement between the model results and reality (see, e.g., Oberkampf et al. 2002). It checks that no coding errors have been introduced into the program. It is also possible to formally state that some parts of the code are correct, for instance, the part that solves large systems of $n$ linear algebraic equations with $n$ unknowns (which are often produced as part of the numerical resolution of the **partial differential equations** on the model **grid**). The accuracy of the numerical methods used to solve the model equations also must be tested. Different methods are available. A standard approach is to compare the numerical solution with the analytical solution for highly idealised test cases for which an exact solution is available.

The next step is the validation process, that is, determining whether the model results are sufficiently close to reality. To do this, the results of some simulations have to be compared with observations obtained in similar conditions. In particular, this implies that the boundary conditions and forcings must be correctly specified to represent the observed situation. Validation first must be performed on the representation of individual physical processes, such as, for instance, the formulation of the changes in the snow **albedo** in response to surface melting and temperature variations. This is generally achieved for particular locations during field

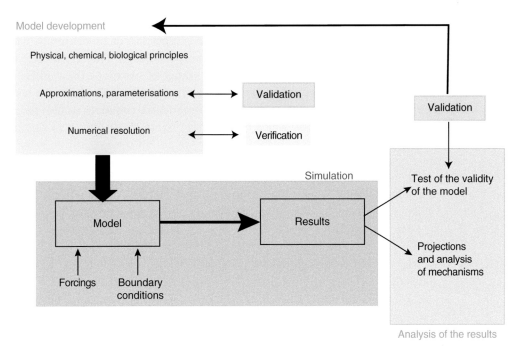

**Fig. 3.19**   A modified version of Figure 3.1 illustrating the verification and validation processes. An additional arrow from the analysis of the results towards model development has been added to show that validation is a continuous process.

campaigns specifically designed to study the process. They provide a much larger amount of very specific data than global databases, allowing a detailed evaluation of the performance of the model on this topic. On a larger scale, the different components of the model (i.e., atmosphere, ocean, sea ice, etc.; see Section 3.3) have to be tested independently. The goal is to ensure that their behaviour is reasonable when the boundary conditions at the interface with the other components are specified from observations. Finally, the results of the whole coupled model have to be compared with observations. All these steps are necessary because unexpected problems are always possible after the different elements are coupled together because of non-linear interactions between the components. Some problems with the model also can be masked by the formulation of the boundary conditions when components are run individually. Besides, having a coupled model providing results in agreement with a few key observations is clearly not enough. In order to test whether those results occur for the correct reasons, it is necessary to check as much as possible that all the elements of the model are working properly and that the satisfactory overall behaviour of the model is not due to several errors in its various elements cancelling each other out.

When discussing verification and validation, we must always recognise that both of them can only be partial for a climate model, except maybe in some trivial cases (see, e.g., Oreskes et al. 1994). The accuracy of the numerical solution can only be estimated for small elements of the code or in very special (simplified) conditions. Indeed, if it were possible to obtain a very accurate solution to compare with the numerical model results in all possible cases, there would be no point in developing a numerical model. The comparison of model results with observations is also limited to certain particular conditions. They are selected as a balance among the wish to cover the widest range of situations, the availability of data and the time (human and computer) which can be allocated to the task. Completely validating a climate model in all its potential situations would require an infinite number of tests, which is obviously impossible. *A climate model thus can never be considered as formally verified or validated.* Consequently, there is no way to guarantee that the results of the model will be totally adequate for certain conditions which differ from those used in the validation process, in particular, for a very complex system such as the climate. A model is sometimes said to be validated if it has successfully passed a reasonable number of tests, in particular, some classical benchmarks (see Section 3.5.2). In such a case, the credibility of projections performed with such a model can be high, but assessing precisely the uncertainties in the projections associated with model formulation remains a difficult issue.

The term 'a validated model' and phrases such as 'the model has been validated' therefore should be avoided. Rather, verification and validation should be considered to be processes which never lead to a final, definitive product. The model should be continuously re-tested as new data or experimental results become available. The building of a model can be viewed in the same way the creation of a scientific theory. Hypotheses are formulated, and a first version of the model is developed. The results of the model then are compared to observations. If the model results are in good agreement with the data, the model can be said to be confirmed for those conditions, thus increasing its credibility. Nevertheless, this does not mean that the model is validated for all possible cases. If the model results do not compare well with observations (taking into account their uncertainties), the model should be improved. This could lead to additional terms in the governing equations or to the inclusion of new processes by new equations and new parameterisations.

Alternatively, a disagreement between the model and observations can be related to inadequate values for certain parameters which are not precisely known [e.g., the exchange coefficients in Eqs. (2.34) and (2.35)]. Adjusting those parameters is part of **calibration** of the model, also referred to as 'tuning'. Ideally, the decision between such a modification of the parameters and an improvement in the representation of some processes should be based on our physical understanding of the system. In practice, the choice is generally not straightforward. Nevertheless, if some robust biases are found for a wide range of parameters, this suggests that the source of the error is not in the selection of their precise values.

The calibration of physical parameters is required and is perfectly justified because there is no a priori reason to select one particular value in their observed range. Furthermore, some parameters may not have a direct observed equivalent as they represent an aggregate of certain properties at the scale of the model grid (see, e.g., Section 3.3.1). It is also valid to calibrate the numerical parameters to obtain the most accurate numerical solution of the equations. However, care has to be taken to ensure that the calibration is not a way to compensate for the model errors and to artificially mask certain deficiencies. If this does occur, there is a high probability that the selected parameters will not provide satisfactory results for other conditions than the ones used for the tuning. Performing many tests for widely different situations and for various elements of the model should limit the risk, but the number of observations is often too small to ensure that the problem has been completely avoided.

An additional problem with constant improvement of the model and its calibration as soon as new data becomes available is the absence of independent data to really test the performance of the model. Some of the available information must be used for model development and calibration, and some should be kept to assess its accuracy. Another good model practise is to choose or design model components for which the selection of one particular value of the lesser known parameters has only a small impact on model results, so reducing the importance of the calibration.

Each modelling centre follows a similar overall methodology for calibration, but each has its specificity. A set of priorities has to be determined first to decide which variable should be in agreement with observations to consider that model calibration is satisfactory. This may lead to a kind of cost function that summarises the importance of the relative errors on all those variables. This cost function can be provided in an explicit mathematical form or more implicitly, just loosely guiding the choice of model developers (see also Section 3.5.2.1). The list of parameters which can be modified also should be selected. For instance, Mauritsen et al. (2012) have chosen to adjust a small number of parameters to keep track as much as possible of the effect of each of them and of the differences between model versions. Mauritsen et al. (2012) also propose a small number of variables as targets for the calibration: the global mean temperature for present-day conditions, the radiative balance at the top of the atmosphere, the mean sea-ice thickness in the Arctic, and so on. Other characteristics of model behaviour, such as global surface warming over the twentieth century, are not tuned but provide a test for the model.

Model developers and users also may decide that if the model cannot properly reproduce the observations in certain special cases, this indicates that it is not valid for such conditions, although it can still be used in other situations where the tests indicate better behaviour. As an extreme case, we can imagine a climate model that cannot simulate the climate of Mars correctly without important modifications. Nevertheless, this does not invalidate it for modelling conditions on Earth. However, if it works well for both Mars and Earth with only changes in a few parameters, this is a good test of its robustness. More generally, no single model could be considered to be best for all the regions and variables analysed at all the time scales. The way the processes are represented and the values selected for the parameters result from trade-offs as alternative choices would improve the results in some regions for some variables and worsen them in others. Consequently, a model should be adapted to the scientific goal of the study, and the results of different models should be compared to test the validity of the conclusions.

## 3.5.2  Evaluating Model Performance

Section 3.5.1 has stressed the absolute necessity of testing the quality of model results as thoroughly as possible. This requires a method to measure the agreement between model results and observations (Section 3.5.2.1) as well as a plan for simulations which should be performed for a reasonable benchmark of model behaviour. Some standard simulations will be presented in Section 3.5.2.2, but we will not discuss the tests specifically designed to analyse the accuracy of numerical methods or of a particular parameterisation.

The model-data comparison should take into account the uncertainties in both the model results and the observations. Errors in the observations can be directly related to the precision of the instruments or in the indirect method used to retrieve the climate signals from the observed fields (see, e.g., Section 5.3). Additionally, because of small-scale variability, a few observations in a grid cell often are not enough to estimate precisely the mean at the grid scale as given by the model. The associated error is called the 'error of representativeness'. Furthermore, some uncertainties can be caused by the **internal variability** of the system (see Section 5.1.1) because observations and model results covering a relatively short period are not necessarily characteristic of the mean behaviour of the system.

### 3.5.2.1  Performance Metrics

The model-data agreement can be estimated more or less intuitively by visually comparing maps or plots describing both the simulation results and the observations. This is a necessary first step which may lead to the definition of appropriate metrics characterising the behaviour of the model (see, e.g., Gleckler et al. 2008). These metrics provide a quantitative and objective assessment of the performance of the model, although selection of the metrics themselves among all possibilities is somehow subjective. The metrics also allow quantification of model improvements from one version to the next.

The simplest metric is probably the difference between observations and model results for a specific quantity, such as the annual mean surface temperature $T_s$. If the climatological mean is analysed, this difference is often referred to as the 'model bias', but the term can be applied more generally to measure the disagreement between model and data over a specific period. The absolute value of this difference is called the 'absolute error'.

The bias can be computed for the local temperature or for the mean over a region, but the information given at local scale also can be synthesised in one single quantity, for instance, the root-mean-square error (RMSE):

$$\text{RMSE} = \sqrt{\frac{1}{n}\sum_{k=1}^{n}\left(T_{s,\text{mod}}^{k} - T_{s,\text{obs}}^{k}\right)^{2}} \qquad (3.42)$$

where $n$ is the number of grid points for which observations are available, $T_{s,\text{mod}}^{k}$ is the model surface air temperature at point $k$ and $T_{s,\text{obs}}^{k}$ is the observed surface air temperature at the same point. This estimate could be improved by taking into account the area of each grid point or by giving greater weight to the regions of greater interest. If many variables have to be included in the metric, the RMSE of the different variables can be combined in various ways.

Many other diagnostics can be performed, and the choice should be guided by the precise objective of the model evaluation. For instance, the variance of the variable or the trend over a particular period in the model can be compared to observations both at the grid scale or focussing on some modes of variability (see Sections 5.2). The correspondence between a spatial pattern simulated by the model and the observed pattern can be measured by spatial correlations. In the following sub-sections and chapters, we will use only relatively simple diagnostics. More complex methods may be required for certain specific purposes, in particular, to characterise precisely the disagreement between model results and observations and determine its origin.

### 3.5.2.2  Standard Tests and Model Inter-Comparison Projects

The first requirement is that the model is able to simulate reasonably well the climate in recent decades for which we have a large number of precise observations for many variables (Figure 3.20). This implies performing simulations including the evolution of both natural and anthropogenic forcings (see Section 5.7) over that period. In these simulations, the long-term average of various variables, in all the model components, is compared with observations, generally interpolated on a common **grid**. Furthermore, the ability of the model to reproduce the observed climate variability on all time scales must be checked. This ranges from temperature extremes such as heat waves to the most important modes of large-scale variability such as the El Niño–Southern Oscillation and the North Atlantic Oscillation (see Section 5.2). Finally, when driven by an adequate forcing, the climate models must be able to reproduce the observed warming at the Earth's surface over the last 150 years as well as other recent climate changes.

Recent decades cover only a small fraction of the climate variations known since the Earth's formation (see Chapter 5) and expected in the future (see Chapter 6). To test the ability of models to simulate different climates, it is thus necessary to try to reproduce some past conditions. The quality of the available

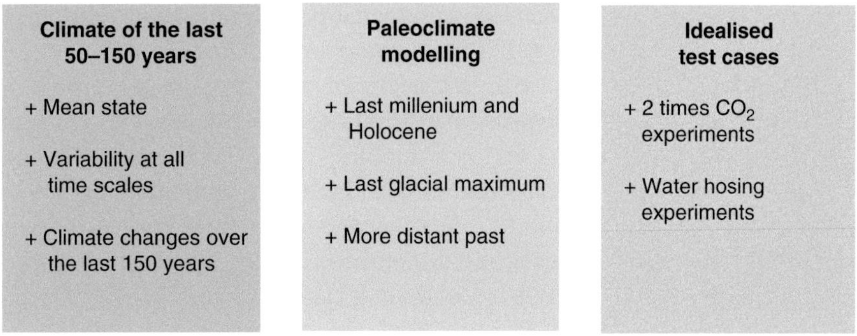

| Climate of the last 50–150 years | Paleoclimate modelling | Idealised test cases |
|---|---|---|
| + Mean state | + Last millenium and Holocene | + 2 times $CO_2$ experiments |
| + Variability at all time scales | + Last glacial maximum | + Water hosing experiments |
| + Climate changes over the last 150 years | + More distant past | |

**Fig. 3.20**  Classical tests performed on climate models.

observational data is (much) lower than that for recent decades, and it may be hard to draw strong quantitative conclusions from model-data comparisons for certain past periods. Nevertheless, this is the only sample of possible states of the Earth's climate available to us. Consequently, a second natural test period (see Figure 3.20) is the Holocene and the last millennium, for which we have a reasonably good knowledge of climate variations (see Section 5.6). Although significant uncertainties are present, the forcing is much better known than for earlier periods. Furthermore, the boundary conditions (e.g., the topography or ocean bathymetry; see Section 1.5) are similar to those of the present.

The last glacial maximum (LGM, around 21,000 years ago) is also a key period because it represents a relatively recent climate clearly different from that of recent decades (see Section 5.5). To perform simulations for this period, it is necessary to specify from available reconstructions the position and shape of the large ice sheets present on continents, the changes in the land-sea boundaries and ocean depth due to the lower sea levels and the modifications in vegetation cover and the radiative properties of the atmosphere (in particular due to the higher dust content), unless some of these variables are computed interactively. All these elements can be sources of uncertainty for the climate simulation. Pre-quaternary climates (see Section 5.3) offer an even wider range of variations, but the uncertainties in the forcing, boundary conditions and climate reconstructions are larger. They thus provide a very interesting but challenging test bed for modellers. In Section 3.5.2.3, we will briefly discuss the ability of current GCMs to reproduce the mean climate of the last decades. The **skill** of various models in the simulation of past changes, including the last century, will be addressed, when appropriate, in Chapter 5.

In addition to the simulations performed in realistic conditions which can be compared with observations, it is instructive to have numerical experiments in more idealised conditions to isolate some specific processes or the responses to individual forcings. Indeed, when comparing conditions during two different past periods, it is not possible to disentangle the contribution of each modification of the boundary conditions or the forcing to the observed and simulated climate changes.

In this framework, simulations with a constant forcing set at the mean for recent decades or for pre-industrial conditions (i.e., before any significant anthropogenic forcing from changes in greenhouse gas concentration, generally 1750 or 1850) are conducted to characterise a quasi-equilibrium behaviour of the model. The response to a forcing change then can be measured by comparison with this reference state. To document the model response to a simple, well-defined perturbation, two standard experiments classically are conducted. The first is a doubling of the atmospheric $CO_2$ concentration in the model, a test required to estimate the **equilibrium climate sensitivity** of the model (see Section 4.1.3). In second experiment, known as 'water hosing', large amounts of freshwater are poured into the North Atlantic to analyse the climate changes induced by the associated modification of the oceanic circulation (see Section 4.2.5).

When comparing the results of simulations performed with various models, it is important to identify the influence of the choices in model design from the influence coming from differences in the details of implementation of the boundary conditions. Consequently, some tests, in idealised and realistic settings, are performed using a controlled experimental design, identical for each model. If all the models agree with each other on a particular point, it then is possible to state that this is a robust behaviour across models and not just the characteristic of a specific model. If in addition this result is compatible with observations, this is clearly strengthens our confidence in simulated output and our understanding of the associated processes. Besides, when using identical setups, a difference between model results must be found in the models themselves and not to a slightly modified experimental design. This then indicates that the response of models is still uncertain in this aspect, and the origin of this uncertainty should then be investigated. Nevertheless, this is a complex task as many elements generally vary among the models.

The conditions of the recommended experiments are defined in what is referred to as 'model inter-comparison projects' (MIPs). The results of such inter-comparisons are archived in databases to ensure a wide access so that they can be analysed independently by large numbers of scientists. We will discuss in Chapters 5 and 6 mainly the Coupled Model Inter-Comparison Project Phase 5 (CMIP5; http://cmip-pcmdi.llnl.gov/cmip5/) and the Paleoclimate Modelling Inter-Comparison Project Phase 3 (PMIP3; https://pmip3.lsce.ipsl.fr/). CMIP5 is a huge co-ordinated international effort involving more than twenty modelling groups worldwide which have launched simulations with more than fifty different models (Taylor et al. 2012). It involves various type of experiments focussing on the last 150 years (the so-called historical period); on key past periods in collaboration with PMIP such as the Last Glacial Maximum, the mid-Holocene and the past millennium; as well as on future climate change. It also proposes idealised simulations. Some of the experiments are mandatory for participation in the exercise, in particular, certain simulations that are critical for evaluation of the models. For many other experiments, each modelling group has the choice to perform them or not depending of their research priorities and the required computer time. CMIP5 provides a massive number of model outputs, corresponding to several petabytes (1 PB = 1000 TB = $10^6$ GB), which can be freely analysed by the whole scientific community.

### 3.5.2.3   Ability of Models to Reproduce the Current Climate

The analysis presented in this sub-section is based on assessment of the numerical experiments performed in the framework of CMIP5 included in the 'Fifth Assessment Report of the Intergovernmental Panel on Climate Change' (IPCC AR5) (Flato et al. 2013). We will focus on three of the variables only (Figures 3.21 to 3.23) as they illustrate the main characteristics of the results of the **ensemble of simulations** performed with different models, which hereafter are referred to as 'multi-model ensembles'.

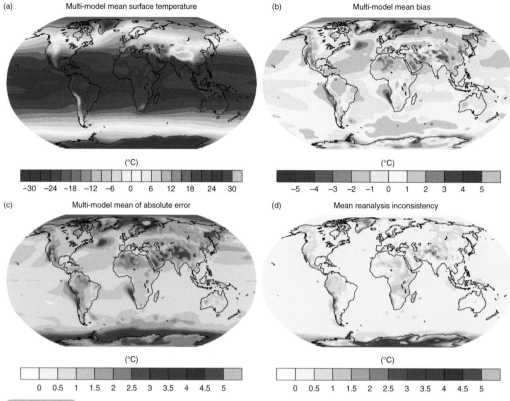

**Fig. 3.21** Annual mean surface (2 m) air temperature (°C) for the period 1980–2005. (a) Mean of atmosphere-ocean general circulation model (AOGCM) simulations performed in the CMIP5 historical experiment. (b) Difference between the multi-model mean displayed in (a) and the climatology from ECMWF reanalysis of the global atmosphere and surface conditions from ERA-Interim (Dee et al. 2011). (c) Mean of the absolute value of the error of individual AOGCMs with respect to the climatology from ERA-Interim. (d) Mean inconsistency among three different reanalyses: ERA-Interim, ERA 40-year reanalysis (ERA40) and Japanese 25-year reanalysis (JRA-25) measured by the mean of the absolute pairwise differences between those fields for their common period (1979–2001). (Source: Flato et al. 2013.)

All the AOGCMs reproduce very well the large-scale distribution of surface temperature. The spatial correlation with the observed pattern (see Figure 1.4) is above 0.95 for all the models. The model biases can be larger at regional scale, in particular, in mountain regions such as the Andes and the Himalaya, as well as over Greenland and Antarctica (Figure 3.21b). Over the ocean, the multi-model mean overestimates the temperature off the East Coast of the United States and underestimate it in the Barents Sea, north of Norway and Finland. This is due to the presence of oceanic currents in those zones which transport large amount of heat (see Section 1.3.2) and are not adequately represented in most current models. The simulated surface temperature is also too high on average in the southeast Atlantic and southeast Pacific off the coast of Africa and South America partly because of biases in the simulation of the effect of oceanic **upwelling** and low clouds there.

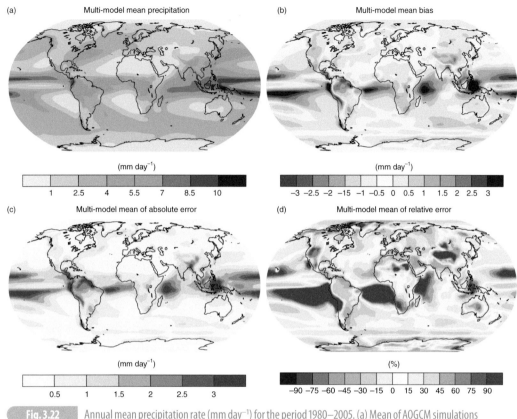

(a) Multi-model mean precipitation

(b) Multi-model mean bias

(mm day$^{-1}$)

1    2.5    4    5.5    7    8.5    10

(mm day$^{-1}$)

−3 −2.5 −2 −15 −1 −0.5  0  0.5  1  1.5  2  2.5  3

(c) Multi-model mean of absolute error

(d) Multi-model mean of relative error

(mm day$^{-1}$)

0.5    1    1.5    2    2.5    3

(%)

−90 −75 −60 −45 −30 −15  0  15  30  45  60  75  90

**Fig. 3.22**   Annual mean precipitation rate (mm day$^{-1}$) for the period 1980–2005. (a) Mean of AOGCM simulations performed in the CMIP5 historical experiment. (b) Difference between the multi-model mean displayed in (a) and precipitation analyses from the Global Precipitation Climatology Project (Adler et al. 2003). (c) Mean of the absolute value of the error of individual AOGCMs with respect to observations. (d) Multi-model mean error relative to the multi-model mean precipitation itself. (Source: Flato et al. 2013.)

The multi-model mean provides a good summary of the information included in the ensemble, but the averaging procedure may lead to a reduction in the error compared to individual simulations, thanks to a compensation between positive and negative biases in different numerical experiments. This is the reason why it is always instructive to compare the error of the multi-model mean to the one of each model. This has been done in Figure 3.21c, where the absolute error of each model has been computed first before performing the average over all the models. In this case, no compensation is possible. The similarity between Figure 3.21b and 3.21c indicates that in many regions the biases of the multi-model mean are well representative of a common behaviour of many models. Finally, some of the model-data disagreements may be due to uncertainties in the estimates derived from the observations themselves. A comparison of the temperature provided by various **reanalyses** indicates that this uncertainty is lower than model biases over most of the Earth but can be significant in some regions, for instance, at high latitudes or over Africa (Figure 3.21d).

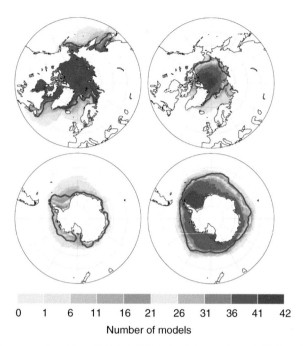

0    1    6    11    16    21    26    31    36    41    42

Number of models

**Fig. 3.23**    Sea-ice distribution averaged over the period 1986–2005 in simulations performed with forty-two AOGCMs in the framework of CMIP5. The upper panels correspond to the northern hemisphere and the lower panels to the southern hemisphere. Left column is for February, right one for September. The figure indicates the number of models that simulate at least 15% of the area covered by sea ice. Consequently, in the grey area, all the forty-two models simulate the presence of ice, whilst only one model has a significant amount of ice in the yellow zone. The observed 15% concentration limits, indicating roughly the mean position of the ice edge (red line), are based on the Hadley Centre Sea Ice and Sea Surface Temperature (HadISST) data set (Rayner et al. 2003) (see Figure 1.19). (Source: Flato et al. 2013, as adapted from Pavlova et al. 2011.)

The quality of the simulations is lower for precipitation than for surface temperature. The temperature distribution is mainly driven by the amount of radiation received in various regions as well as by large-scale heat transport and storage, mechanisms which are relatively well understood and modelled. The precipitation also depends on the large-scale circulation but is more strongly influenced by smaller-scale processes related, for instance, to topography, convection and vertical movement in the atmosphere. Many of these processes must be parameterised, introducing uncertainties in the model results. The observed large-scale patterns (see Figure 1.8) are still relatively well reproduced, with a spatial correlation between AOGCM results and observations generally larger than 0.70 for the models included in CMIP5 (Figure 3.22a). Besides, the biases are large locally (Figure 3.22b and 3.22c), the relative error reaching nearly 100% in some regions (Figure 3.22d).

Sea ice is a clear example in which the multi-model average introduces a strong compensation between the models that underestimate the sea-ice extent and those which overestimate it. This is illustrated in Figure 3.23, where the limit for 50% of the model displaying a significant amount of sea ice for a particular location is close to the observed ice edge for all seasons and in both hemispheres (this thus implies that about 50% of the models have a too extensive ice cover and 50% have a too low ice concentration in each region). As a consequence, the multi-model mean is very close to observations and provides a better representation of the sea-ice extent than any of the individual simulations. Besides, individual simulations may have very large errors, some models having, for instance, no ice remaining in summer in the Southern Ocean or nearly two times the observed extent in the summer in the Arctic. This underlines the fact that the multi-model mean does not represent a physically consistent state but rather an aggregate of very different behaviours which must be individually analysed to understand the processes responsible for the simulated results.

## 3.6 Combining Model Results and Observations

### 3.6.1 Correction of Model Biases

Section 3.5.2.3 has illustrated some typical biases of current climate models. A continuous activity of modelling groups is to propose new developments to improve the models and reduce those biases. Nevertheless, they cannot be totally removed as models are always imperfect representations of reality. Consequently, the discrepancies between model results and observations has to be taken into account when interpreting certain observed variations on the basis of model results and when using model outputs to make **predictions** or **projections** of future climate changes.

It is first possible to do this qualitatively, the analysis of the biases on some variables providing an estimate of the confidence on the model results. As discussed earlier, the confidence on conclusions derived from model outputs is larger if the model reproduces well the past conditions and if we have a good physical understanding of the processes involved. Besides, for many applications, a quantitative approach is required. This is the case, for instance, when the outputs of climate models are used to drive impact models which, for instance, estimate the changes in crop yields as a function of future precipitation and temperature developments. This quantitative approach is sometimes referred to as 'calibration' of model outputs, but to avoid confusion with the model calibration discussed in Section 3.5.1, we will employ here the terms 'correction' and 'adjustment'.

The simplest adjustment is based on the assumption that the bias is constant over the period studied. It is equivalent to suppose that the simulated changes between a reference period (generally the last decades) and a target period (e.g., the end of the twenty-first century) are well simulated despite the model biases (Figure 3.24a). Imagine that, for a particular region, the simulated temperature for the reference period 1986–2005 is 15°C and the projected temperature in a given **scenario** (see Section 6.1) for the end of the twenty-first century is 18°C, corresponding to a warming of 3°C. However, the model is 2°C too warm for

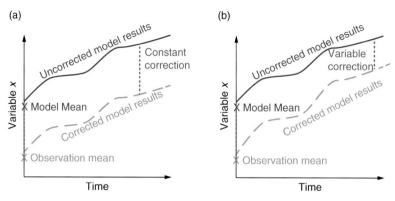

Schematic representation of possible adjustments of model results to take into account biases. (a) The model results are shifted to compensate for the bias on the mean state for the reference period, following Eq. (3.43). (b) In addition to correction of the mean state displayed in (a), the trend has been adjusted as the model underestimated this trend for previous periods over which it was tested. The blue and green crosses display the model and observation averages over the reference period, respectively.

the reference period compared to the observed temperature of 13°C. Thus 18°C appears likely as an overestimate of the absolute temperature for the end of the twenty-first century because the model already has a clear positive bias for present conditions. The hypothesis then is to assume that the warming of 3°C provided by the model is a reasonable estimate of future changes, the model having the right sensitivity to the perturbation. Consequently, this 3°C warming is added to the observed value of 13°C, giving an estimate for the end of the twenty-first century of 16°C instead of 18°C before correction. More generally, the corrected variable $x_{\mathrm{corr}}$ at a time $t$ can be given by

$$
\begin{aligned}
x_{\mathrm{corr}}\left(t\right) &= x_{\mathrm{mod}}\left(t\right) - \left(\bar{x}_{\mathrm{mod}} - \bar{x}_{\mathrm{obs}}\right) \\
&= \bar{x}_{\mathrm{obs}} + \left[x_{\mathrm{mod}}\left(t\right) - \bar{x}_{\mathrm{mod}}\right]
\end{aligned}
\tag{3.43}
$$

where $x_{\mathrm{mod}}$ is the raw variable simulated by the model. $\bar{x}_{\mathrm{mod}}$ and $\bar{x}_{\mathrm{obs}}$ are the mean of the simulated variable and the observations over the reference period, respectively. The quantity $\left(\bar{x}_{\mathrm{mod}} - \bar{x}_{\mathrm{obs}}\right)$ is the mean model bias over the reference period.

A similar kind of correction is implicit when displaying **anomalies**, that is, the difference $\left[x_{\mathrm{mod}}(t) - \bar{x}_{\mathrm{mod}}\right]$ between the target period and a reference period. By construction, any bias on the mean state is indeed removed. This is directly deduced from Eq. (3.43) as the anomaly $\left[x_{\mathrm{mod}}(t) - \bar{x}_{\mathrm{mod}}\right]$ is equal to $\left[x_{\mathrm{corr}}(t) - \bar{x}_{\mathrm{obs}}\right]$. The additional advantages of anomalies is that they underline clearly the changes between two periods which can be smaller than the geographical variations of the corresponding variables and thus can be hard to see if the full field was displayed. This is the reason why the climatic fields for past or future periods are often presented in this way (see Chapters 5 and 6).

Adding or removing a constant number from model outputs as in Eq. (3.43) can be problematic for some variables as it may lead to totally unphysical values such as negative precipitation or negative sea-ice concentration. Consequently, the simplest correction for precipitation is generally applied as a multiplying factor equal to the ratio between observed and simulated means for the reference period:

$$
\begin{aligned}
x_{\mathrm{corr}}\left(t\right) &= x_{\mathrm{mod}}\left(t\right)\frac{\overline{x}_{\mathrm{obs}}}{\overline{x}_{\mathrm{mod}}} \\
&= \overline{x}_{\mathrm{obs}}\,\frac{x_{\mathrm{mod}}\left(t\right)}{\overline{x}_{\mathrm{mod}}}
\end{aligned}
\tag{3.44}
$$

This implies that if the simulated precipitation rate decreases by 50% between the reference and target periods, the corrected value will be the observed precipitation for the reference period multiplied by 0.5. This allows avoiding negative precipitation in the case of a strong model bias for the reference state whilst keeping the signal of a significant decrease given by the model. The situation is more problematic for sea-ice concentration, and no simple solution is universally adapted. Consider, for instance, a model which simulates no ice at one location in summer, whilst observations give a mean concentration of 50% over the reference period. In case of warming, the state of sea ice at this point will remain stationary in the model because it still will simulate the absence of ice. Thus, it would not provide any direct information on the future pace of sea-ice retreat locally caused by the warming and, thus, no clue, for instance, on the date of disappearance of the ice in that region without relying on more sophisticated approaches for the adjustment of model results.

The basic hypothesis behind Eqs. (3.43) and (3.44) is that the correction is independent of the state considered and thus can be taken as a constant. This is likely valid for small biases and small climate changes but may be too strong an approximation in many cases. In particular, the model may overestimate or underestimate systematically the response to a perturbation. Additionally, the magnitude of this response may be related to the model mean state, or the model may display a spurious long-term drift away from observations. This introduces the need for non-stationary elements in the correction (see, e.g., Kharin et al. 2012; Kerkhoff et al. 2014).

In this framework, it is instructive to distinguish two cases. Firstly, a simulation can be started from a previous (quasi-)equilibrium state of the model or analysed only when such an equilibrium is achieved. The differences between the period studied and the reference period are then conditioned by the model representation of the dynamics of the system and by its response to the forcing. The model biases can be corrected following Eq. (3.43) (Figure 3.24a). Alternatively, a non-stationary correction can be based on previous simulations performed in similar conditions which have been compared with observations. For instance, in Figure 3.24b, if a model has been shown to underestimate systematically the warming trend due to a particular forcing over past periods, this bias can be corrected by an amplification of the long-term trend.

Secondly, a transient simulation can be initialised from a state far from the equilibrium of the model. This may have two origins. Firstly, the equilibrium of the climate system is reached only after a long time, several thousands of years

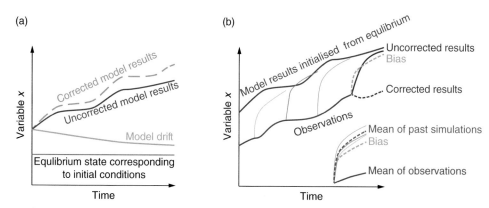

**Fig. 3.25** Illustration of methods for correcting model drift. (a) The simulation is initialised from conditions far away from a stationary state of the model whilst the system is supposed to be close to equilibrium at that time. If an experiment is performed from this initial state using constant forcing, the model drifts towards the model equilibrium corresponding to the boundary conditions imposed (orange curve). If the forcing varies, the model results are thus a combination of this drift and of the model response to forcing (dark-blue curve). The effect of the drift can be removed by subtracting the spurious drift represented by the orange curve to get the corrected model results (dashed green curve) which represent only the model response to forcing. (b) If the model is initialised directly from observations (brown curve), it will drift towards a state which would be obtained in a simulation started from an equilibrium state (potentially much before the period displayed in the figure) (dark-blue curve). If previous simulations are available (light-green, dark-green and light-blue curves), the mean bias of the model can be obtained (orange dashed curve) by subtracting from the mean of those previous simulations (dashed red curve) the average of observations (brown curve) over the same periods, as illustrated in the inset at the bottom right of the figure. The mean bias is then subtracted from the most recent simulation (violet curve) to have results adjusted from the drift (dotted violet curve).

for the deep ocean. This is not always affordable for complex models needing a lot of computer time for each year of simulation. If this is the case, the model continues to drift after the beginning of the simulation to its own equilibrium state (Figure 3.25a). For some variables, this drift is larger than the signal itself and therefore must be removed before any analysis. This can be done, for instance, by performing in parallel to the simulation of interest a similar experiment started from the same initial state but with constant forcing. The time development of this experiment then is only due to the model drift to its own equilibrium. If we make the hypothesis that the way the model reaches this equilibrium from a specific initial state is more or less independent of the conditions of the experiment, the time evolution of this parallel experiment can be removed from the simulation of interest. This then provides a reasonable correction for the drift due only to the experimental design. Secondly, it may be instructive to start a simulation from a state as close as possible to observations to analyse the influence of the actual conditions on the future evolution of the system (see Section 6.2.2). Such predictions can be adjusted for the drift towards the equilibrium state of the model by first performing the average over many equivalent predictions made for previous periods. The difference between this mean simulated change and the mean of observations is then computed, providing an estimate of the systematic drift that can be subtracted from new simulations (Figure 3.25b). Unfortunately, this method is only possible for short-term predictions as many tests have to be performed over the period for which enough data are available to evaluate the **skill** of the model forecasts.

Correcting the mean state of the system is a first step, but many other elements can or must be adjusted. For many applications, knowing the probability of occurrence of some extreme events such as droughts is more important than the mean itself. If the model does not reproduce the observed variance well, and more generally, the probability distribution of various types of events for the reference period, corrections also may be applied to reduce those model errors.

The different types of adjustments just mentioned may be combined, leading to an expression of the form

$$x_{\mathrm{corr}}\left(t\right) = F_{\mathrm{corr}}\left[\mathbf{x}_{\mathrm{mod}}\left(t\right), t\right] \qquad (3.45)$$

where $F_{\mathrm{corr}}$ is a function of variables simulated by the model for the period of interest and of time $t$ included in $\mathbf{x}_{\mathrm{mod}}$. The expression for $F_{\mathrm{corr}}$ is estimated from the comparison of model results and observations in various past conditions. This corresponds to the MOS approach applied to bias correction (see Section 3.2.5). The exact formulation of $F_{\mathrm{corr}}$ depends on the implicit or explicit hypotheses retained to correct model errors. These hypotheses lead to uncertainties that must be kept in mind as different results can be obtained as a function of the correction method selected (Ho et al. 2012). This is also the reason why, in many cases, raw results are presented, in particular, if the biases are relatively small.

## 3.6.2  Data Assimilation

Observations and model results are complementary: observations provide a description of the real state of the system, whilst models represent our knowledge of the climate dynamics, expressed through mathematical laws. On the one hand, preceding sub-sections have demonstrated that observations are essential to evaluate model performance and to propose corrections for model biases. On the other hand, models are essential tools to interpret the observed signal. The goal of data assimilation is to combine more tightly these two sources of information to estimate as accurately as possible the state of the system (see, e.g., Talagrand 1997). To underline this goal, data assimilation is also referred to as 'state' or 'field estimation'.

Several practical applications are standard. Data assimilation can provide the best possible description of present-day conditions. This 'analysis', combing data and model results, informs potential users of the current state of the system and gives the initial conditions for forecasts (see below and Section 6.2.2). Data assimilation also can cover past periods of various duration to reconstruct climate variations in a consistent way, leading to what is often called a '**reanalysis**'.

The optimal combination of different sources of information is a standard problem in many disciplines. To illustrate the principle without much mathematical complexity, we will start with a simple, classical example. Consider that we have two independent measurements of the same quantity, for instance, the temperature time series $T_1(t)$ and $T_2(t)$ given at one particular location by two different thermometers. We would like to estimate the 'true temperature' $T_t(t)$ from a linear combination of these two measurements given by the analysis $T_a(t)$:

$$T_a(t) = w_1 T_1(t) + w_2 T_2(t) \tag{3.46}$$

where $w_1$ and $w_2$ are the weights given to each time series. Assuming that none of the observation time series are biased [or that the bias has been removed before using it in Eq. (3.46)], the sum of the weights should be equal to 1 to avoid a bias in $T_a(t)$:

$$w_1 + w_2 = 1 \tag{3.47}$$

If no additional information is available, a reasonable decision is probably to choose $w_1 = w_2 = 0.5$ and simply average the two observation time series. A better solution is obtained if the error of each instrument can be estimated. Of course, the error [i.e., the difference between $T_1(t)$ or $T_2(t)$ and $T_t(t)$] is not known at each time $t$; otherwise, the true temperature $T_t(t)$ would be readily obtained. Besides, the statistics of the error can be known, for instance, by measuring the accuracy of each instrument. This can be quantified by the variances $\sigma_1^2$ and $\sigma_2^2$ of the error of $T_1$ and $T_2$, respectively:

$$\begin{aligned} \sigma_1^2 &= E\left[(T_1 - T_t)^2\right] \\ \sigma_2^2 &= E\left[(T_2 - T_t)^2\right] \end{aligned} \tag{3.48}$$

where $E(a)$ represents the expected value of $a$, that is, the value that would be obtained if many observations of $a$ were averaged. The precision of the instrument then can be given by the inverse of the variance of the observational error $1/\sigma_1^2$.

On this basis, we can determine the values of the weights which minimise $\sigma_a^2$, the variance of the error of $T_a(t)$:

$$\sigma_a^2 = E\left[(T_a - T_t)^2\right] = E\left\{\left[w_1(T_1 - T_t) + w_2(T_2 - T_t)\right]\right\}^2 \tag{3.49}$$

where we have applied Eq. (3.47).

The problem can be solved using a classical **least-squares method** which gives

$$\begin{aligned} w_1 &= \frac{1/\sigma_1^2}{1/\sigma_1^2 + 1/\sigma_2^2} \\ w_2 &= \frac{1/\sigma_2^2}{1/\sigma_1^2 + 1/\sigma_2^2} \end{aligned} \tag{3.50}$$

with the variance of the error of $T_a$

$$\frac{1}{\sigma_a^2} = \frac{1}{\sigma_1^2} + \frac{1}{\sigma_2^2} \tag{3.51}$$

Equation (3.50) simply states that the weight of each time series is inversely proportional to the variance of its error (or that it is proportional to the precision of the instrument). In other words, we should give more weight to the most accurate measurement, as expected. Additionally, Eq. (3.51) implies that the error variance of the analysis is always smaller than the error variance of any of the measurements, and thus, that the precision of the analysis is always improved compared to the individual time series. The goal of data assimilation is thus well achieved.

The conclusions derived from this simple case are also valid for more complex situations (see, e.g., Kalnay 2003). In particular, they underline the importance of an accurate representation of the errors on all the sources of information to have an optimal estimate of the state of the system through data assimilation.

To be adapted to more realistic problems, the formalism has to be modified. First, rather than considering equivalent time series coming from various instruments as in Eq. (3.46), it is convenient to separate the sources of information between a model estimate called the 'background state' $x_b(t)$ and the observations $y(t)$:

$$x_a(t) = w_1 x_b(t) + w_2 y(t) \tag{3.52}$$

which can be rewritten using Eq. (3.47) as

$$\begin{aligned} x_a(t) &= (1 - w_2) x_b(t) + w_2 y(t) \\ &= x_b(t) + w_2 \left[ y(t) - x_b(t) \right] \end{aligned} \tag{3.53}$$

Second, in climate science, the interest is generally not in time series at one location but in the time development of spatial fields. All the corresponding variables, at all the locations of interest, can be grouped in a **state vector** $\mathbf{x}(t)$ of dimension $n$ (i.e., a column matrix) representing the state of the system at a time $t$. For instance, consider that a model includes $i$ different **state variables** (e.g., temperature and humidity) on a **grid** characterised by $j$ points in latitude, $k$ points in longitude and $l$ vertical levels. This lead to $n = i \cdot j \cdot k \cdot l$, meaning that the size of $\mathbf{x}_b(t)$ could be very big.

Third, the size of $\mathbf{x}_a$ should be the same as $\mathbf{x}_b$, but observations are not available at all the model grid points. Consequently, $p$, the size of the vector $\mathbf{y}(t)$, including all the time series of observations, is much smaller than $n$. It is thus necessary to introduce the 'forward observational operator' $\mathbf{H}$, also called the 'observation operator'. $\mathbf{H}$ transforms the model results so that they can be compared to the observation $\mathbf{y}(t)$. It could represent a simple selection of the observed variables among all the model variables and an interpolation to the location of the measurements. Besides, the variable observed may have no direct equivalent in the model, and a complex treatment of the model results may be required before any comparison with measurements (see, e.g., Section 5.3.1).

Applying Eq. (3.53) to the estimation of the state vector $\mathbf{x}_a$ then leads to

$$\mathbf{x}_a(t) = \mathbf{x}_b(t) + \mathbf{W}\left\{\mathbf{y}(t) - \mathbf{H}\left[\mathbf{x}_b(t)\right]\right\} \tag{3.54}$$

where $\mathbf{W}$ is a weight matrix of size $n \times p$.

An interpretation of Eq. (3.54) is to consider that the model gives a first esti-mate $\mathbf{x}_b$ of the state of the system which is corrected at the analysis stage thanks to available observations. This is why it is often said that observations constrain or guide the model through data assimilation. Because of the additional term, the analysis $\mathbf{x}_a$ is not necessarily a solution of the model equations. This is fine because the model itself is not assumed to be perfect, but this should be kept in mind when analysing the results of simulations with data assimilation.

The correction applied to the background state $\mathbf{x}_b$ is proportional to $\mathbf{d} = \left\{\mathbf{y}(t) - \mathbf{H}[\mathbf{x}_b(t)]\right\}$, the difference between observations and the background model estimate or, in other words, to the model misfit. $\mathbf{d}$ is called the 'innova-tion' or 'observational increment'. Equation (3.54) also shows that observations of some variables at a few locations potentially influence the analysis $\mathbf{x}_a$ at all the model points and for all the variables. Data assimilation then can be viewed as a sophisticated way to interpolate observations in regions where no data are avail-able or to obtain information consistent with existing data and model physics on variables that cannot be observed directly.

Equation (3.54) is the basis of what is called 'sequential data assimilation', as routinely applied, for instance, in weather forecasts (Figure. 3.26). Starting from an estimate of initial conditions for a particular day, a forecast is performed for the next one (day + 1), providing $\mathbf{x}_b$. When observations $\mathbf{y}$ are available at day + 1, the forecast is corrected using Eq. (3.54) to obtain $\mathbf{x}_a$. This analysis of the state of the system for day + 1 is then used as the initial conditions in the forecast for day + 2 and so on. Note that the interval of one day in this example between two analyses may be much larger than the time step required by the resolution of the model equations (see Section 3.4), and thus, data assimilation does not necessar-ily occur at each model time step.

Different options are available to specify the matrix $\mathbf{W}$ in Eq. (3.54). The sim-plest one is to update during the analysis only the variables which are observed, choosing $\mathbf{W}$ as a diagonal matrix. All the non-zero terms could have different values, but assume for convenience here that they are all identical and equal to $k_n$. If only the surface temperature $T_s$ is observed, Eq. (3.54) then can be reduced to

$$T_{s,a}(t) = T_{s,b}(t) - k_n\left(T_{s,\text{obs}} - T_{s,b}\right) \tag{3.55}$$

with the equation only applied at model grid points for which observations are available. This method is called a 'nudging' or 'relaxation'. For $k_n = 1$, observed values simply replace the corresponding model results. Because of the simplic-ity of the approach, the term $-k_n(T_{s,\text{obs}} - T_{s,b})$ also can be directly included in the prognostic equations of the model [e.g., in Eq. (3.15)] to continuously nudge the solution towards the observations. This is technically more convenient as the

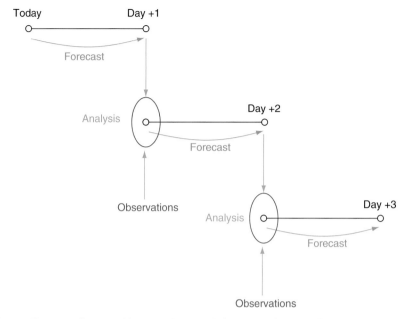

Today                    Day +1

Forecast

Analysis

Day +2

Forecast

Observations

Analysis

Day +3

Forecast

Observations

**Fig. 3.26**    Schematic illustration of sequential data assimilation in which a previous forecast performed with a model is corrected during the analysis step using observations to provide the initial conditions for the next forecast.

correction can be applied at each model time step without the need to have an explicit analysis phase. In this framework, $k_n$ can be related to the inverse of the time scale with which the model solution converges towards observations (see, e.g., Cushman-Roisin and Beckers 2011).

Nudging is far from being an optimal solution. There is no a priori rule to specify of $k_n$. Furthermore, nudging is a 'univariate method', meaning that the observation of one variable is used to directly update only that variable. The other fields are not modified at the analysis step, and it is assumed that the propagation of information from the observed variable through the variables which are not observed will be achieved by the model during the forecast step.

Nevertheless, a direct link between various variables can be reasonably assumed. For instance, a warmer sea surface temperature might be associated with a reduced sea-ice concentration or with more evaporation and thus a higher salinity. Consequently, different techniques have been proposed to determine $\mathbf{W}$, taking advantage of the relationships between the various variables. The formulation of $\mathbf{W}$ also must account for the estimates of the errors on those relationships, on the model results, on the measurements themselves and on the error of representativeness. The general justification is the same as the one leading to Eqs. (3.50) and (3.51). In the so-called optimal interpolation, $\mathbf{W}$ is assumed to be constant through time, whilst in Kalman filtering, $\mathbf{W}$ depends on the state of the system and thus is time dependant. In the latter method, $\mathbf{W}$ is denoted $\mathbf{K}$, the 'Kalman gain', and the relationships between the various variables (more exactly, the error **covariance**) are often determined via an ensemble of model simulations, leading to what is called the 'ensemble Kalman filter'. The mathematical formulation of these techniques will not be developed here. Interested readers are referred to Kalnay (2003) and Cushman-Roisin and Beckers (2011).

The sequential data assimilation described by Figure 3.26 and Eq. (3.54) is just one widely used example among a whole range of data-assimilation methods. Particle filtering (see, e.g., van Leeuwen 2009) is based on a similar approach, but rather than updating the model state with a term proportional to the innovation, the data are used to select, among a large ensemble of simulations, the members that are the closest to observations. The advantage of this approach is that some assumptions about the linearity of the system included in Eq. (3.54) can be relaxed. This comes, however, with the sometimes prohibitive requirement that a very large ensemble is needed to find among its members a sufficiently good analogue to the observed state.

For technical reasons, instead of applying Eq. (3.46), it is sometimes more efficient to directly find the state which minimises the mismatch between the various sources of information. For instance, in the simple case of two different observations of the same temperature studied earlier, a solution equivalent to Eq. (3.50) could be obtain by finding $T_a$ corresponding to the minimum of the cost function $J(T)$ (Kalnay 2003):

$$ J(T) = \frac{1}{2}\left[\frac{(T-T_1)}{\sigma_1^2} + \frac{(T-T_2)}{\sigma_2^2}\right] \tag{3.56} $$

This is referred to as a 'variational approach' or, in some cases, an 'inverse method'. In classical sequential data assimilation, the analysis is performed at some specific times only, when some jumps in the solution occur (which actually can be smoothed using some sophisticated techniques, if needed). An advantage of the variational approach is that it can be applied directly to observations available at various times. In particular, sequential data assimilation only uses past observations to reconstruct the state at a particular time. This is fine for an analysis providing initial conditions for forecasts, but for reanalysis over long periods, estimate of the state of the system can be more easily constrained by past and future (compared to the time studied) observations using the variational approach.

## Review Exercises

1. Parameterisations are introduced in models
   a. because many processes are still not sufficiently well known to include their detailed representation in models.
   b. to have a simpler representation of some processes.
   c. because the resolution of the numerical grid is too coarse to represent explicitly small-scale processes.

2. In energy-balance models (EBMs), the greenhouse effect is taken into account through
   a. a modification of the albedo.
   b. a radiative scheme which takes into account explicitly the absorption of some radiations by greenhouse gases.
   c. a prescribed infrared transmissivity of the atmosphere.

3. Simplifications are included in earth models of intermediate complexity (EMICs) to be able to make simulations over longer periods.
   a. True
   b. False

4. Compared to the GCM fields which are used at their boundaries, regional climate models provided an added value because
   a. they have a better representation of complex topography than GCMs.
   b. they are able to represent some details of the circulation that have a too small spatial scale for GCMs.
   c. they use numerical schemes which are more sophisticated thanks to the higher spatial resolution.

5. Applying Newton's second law, the conservation of mass, the first law of thermodynamics and the equation of state is enough to have a set of equations which can be solved numerically to obtain the development of the atmospheric state.
   a. True
   b. False

6. In sea-ice models, the heat conduction can be considered as a vertical process because
   a. the thickness of the ice is much smaller than its horizontal extent.
   b. floes of many different sizes are present in a model grid box.

7. Among the processes cited below, which ones are not taken into account in simple bucket models but are present in more sophisticated models?
   a. Control of evapotranspiration by the amount of available water in soils
   b. Duration of the transport of water from the grid cell to the ocean by rivers.

8. Is it enough to verify that the numerical scheme used to solve the equations is stable, consistent and coherent?
   a. Yes
   b. No

9. A climate model which has passed all the standard tests successfully can be considered to be validated.
   a. True
   b. False

10. Calibration of model parameters should be avoided because this would mask some model deficiencies.
    a. True
    b. False

11. The main interest of model inter-comparison projects is
    a. to have large ensembles of model results easily available for the whole scientific community.
    b. to have a set of experiments which allows estimating the model uncertainties and the causes of disagreement between different models.
12. Correction of model biases can be applied
    a. to errors on the mean state.
    b. to errors on the mean state and the inter-annual variability.
    c. to errors on the mean and the variability at all frequencies (including long-term trends).
13. Data assimilation can be interpreted as adding a term to the background model state which is proportional to the difference between model results and observations to improve the agreement between the different sources of information. The update of model results is made directly only for the variables which are observed.
    a. True
    b. False

# 4 Response of the Climate System to a Perturbation

## OUTLINE

The major types of perturbations influencing the climate system are briefly reviewed in this chapter. The main focus is on the notions of forcing and feedback which provide a widely used framework to interpret the response of the system to changes in external conditions. The standard physical feedbacks are first presented before describing interactions implying jointly the energy-balance, hydrological and biogeochemical cycles.

## 4.1 Climate Forcing and Climate Response

### 4.1.1 Notion of Radiative Forcing

The climate system is influenced by different types of perturbations: changes in the amount of incoming solar radiation, in the composition of the atmosphere, and so on. To compare the magnitude of these perturbations and to evaluate their effect on the climate, it is convenient to analyse their impact on the radiative balance of the Earth. In this framework, a positive **radiative forcing** corresponds to more energy input to the system (or less output).

The climate changes in response to perturbations ultimately reach a new equilibrium (see Section 4.1.3). To have a clearer view of the dynamics of the system, it is important to separate as objectively as possible the forcing from this response. Consequently, the radiative forcing should be evaluated whilst maintaining the **state variables** of the system at their values before the perturbation is applied. In the case of an instantaneous forcing, this can be interpreted as the immediate changes in radiative budget of the Earth imposed by the perturbation before any adjustment of the climate. For some agents, such as the solar forcing (see Section 4.1.2.4), evaluation of the radiative forcing is relatively direct, as this can be done by measuring the variations in incoming solar radiation at the top of the atmosphere. For other agents, such as greenhouse gases, estimates are less straightforward as they should be based on computation of the impact of the changes in composition on the radiative transfer in the whole atmosphere.

The instantaneous forcing is not necessarily representative of the perturbation of the heat budget of the **troposphere** (which is generally the part of the Earth

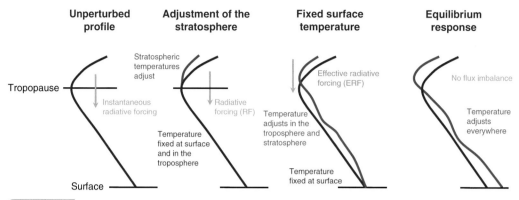

**Fig. 4.1**   Schematic representation of the definition of radiative forcing and effective radiative forcing. (Source: Modified from Hansen et al. 2005.)

that is of interest for climate) on the **time scale** needed for its adjustment. In particular, the stratosphere is nearly in equilibrium with the perturbation after a few months, whilst the surface takes several decades at least to achieve equilibrium (see Section 4.1.5). Consequently, the 'radiative forcing' (RF) is commonly defined as the net change in the Earth's radiative budget at the **tropopause** after the stratospheric temperatures are allowed to reach a new radiative equilibrium and the surface and tropospheric temperatures are fixed at their unperturbed values (Hansen et al. 1985) (Figure 4.1). The forcing at the tropopause thus typically represents the influence of the perturbation on longer time scales than a season to a year.

Many characteristics of the troposphere also respond rapidly to perturbations. This has led to definition of the **'effective radiative forcing'** (ERF), in which the forcing is evaluated whilst maintaining the surface temperature or the temperature of a fraction of the surface, unchanged, with the atmospheric temperature, water vapour and clouds allowed to adjust. Practically, ERF is often estimated in models with only sea surface temperature kept at a constant value as it is not technically possible to fix the ground surface temperature. Rapid land surface temperature changes then are included too in the fast adjustments. Because the atmospheric temperature reaches an equilibrium under the imposed conditions, the ERF is nearly constant over the vertical and should not necessarily be computed at the tropopause (Hansen et al. 1985). It is then often measured at the top of the atmosphere (Myhre et al. 2013).

For most forcing agents, RF and ERF are relatively close to each other. When used in a general way, forcing could refer to both RF and ERF in the following sections and chapters. Nevertheless, ERF generally provides more precise information of the final temperature response (Myhre et al. 2013). Furthermore, ERF is much more adapted for some forcings, such as, for instance, the one related to cloud-aerosol interactions (Section 4.1.2.2). The rapid changes in cloud properties occurring during these interactions cannot be accounted for directly by RF, which excludes by definition any adjustment of the troposphere.

The values generally are given in annual and global means (Figure 4.2). This is instructive when different perturbations are compared, but each of them has a spatial distribution. It is particularly clear for aerosol forcing, for instance (see Section 4.1.2.2), but even the carbon dioxide ($CO_2$) forcing varies spatially because temperature and cloudiness influence the radiative properties of the atmosphere and thus the modifications imposed by a higher concentration of $CO_2$ (see, e.g., Shine and Foster 1999; Feldl and Roe 2013).

## 4.1.2 Major Radiative Forcing Agents

### 4.1.2.1 Greenhouse Gases

The main radiative forcings that affect the Earth's climate can be grouped into different categories. This has classically been done to estimate both the anthropogenic and natural forcings compared to pre-industrial conditions corresponding typically to 1750 (Figure 4.2) (see also Section 5.7). Over the last 250 years, the changes in greenhouse gas concentrations have played a dominant role (note that this also seems to be valid in the more remote past; see Section 5.3). The largest contribution comes from the modification of the atmospheric $CO_2$ concentration, with a radiative forcing of about 1.8 W m$^{-2}$ between 1750 and 2011. Adding to this estimate the forcing resulting from the other well-mixed greenhouse gases, $CH_4$, $N_2O$, and the **halocarbons**, the forcing reaches 2.8 W m$^{-2}$.

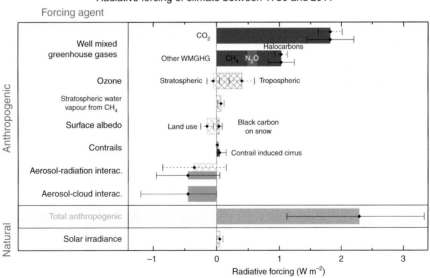

**Fig. 4.2** Radiative forcing (RF, hatched) and effective radiative forcing (ERF, solid) between 1750 and 2011 for individual forcing agents and the total anthropogenic forcing. Uncertainties for RF (dotted lines) and ERF (solid lines) are given as estimated 5–95% confidence range. (Source: Myhre et al. 2013.)

Estimating the radiative forcing $\Delta Q$ associated with the changes in the concentrations of these gases requires a comprehensive radiative transfer model. However, relatively good approximations can be obtained for $CO_2$ from a simple formula (Myhre et al. 1998):

$$\Delta Q = 5.35 \ln\left(\frac{[CO_2]}{[CO_2]_r}\right) \qquad (4.1)$$

where $[CO_2]$ and $[CO_2]_r$ are the $CO_2$ concentrations in parts per million (ppm) for the period being investigated and for a reference period, respectively. The logarithmic dependence in Eq. (4.1) illustrates that a $CO_2$ concentration increase has a larger effect at low $CO_2$ concentrations than at high ones. Consequently, doubling the $CO_2$ concentration from 280 to 560 ppm induces in first approximation the same radiative forcing as an increase from 560 to 1120 ppm, whilst the difference in concentrations is two times larger in the second case. This so-called saturation effect is related to the large absorption of **longwave radiation** by $CO_2$ in specific parts of the electromagnetic spectrum (see Section 2.1.2). When its concentration is high, nearly all the radiation in those frequency bands is already absorbed by the atmosphere, reducing the role of any additional $CO_2$. Nevertheless, a total saturation is not reached on Earth, meaning that an increase in $CO_2$ concentration always induces a radiative forcing. This is due to the multiple absorptions and re-emissions at different levels of the atmosphere and ultimately at the upper atmosphere, where the effect of $CO_2$ on infrared radiation is not saturated, before being radiated to space. Additional processes also contribute to the radiative forcing, for instance, related to the link between the width of the absorption band and the concentration [for more information, see, e.g., Archer (2011) and Pierrehumbert (2010)].

Similar approximations can be made for $CH_4$ and $N_2O$:

$$\Delta Q = 0.036\left(\sqrt{[CH_4]} - \sqrt{[CH_4]_r}\right) \qquad (4.2)$$

$$\Delta Q = 0.12\left(\sqrt{[N_2O]} - \sqrt{[N_2O]_r}\right) \qquad (4.3)$$

where the same notation is used as in Eq. (4.1), the concentrations being in parts per billion (ppb). Some corrections are often added to Eqs. (4.2) and (4.3) which are functions of the concentration of both $CH_4$ and $N_2O$ (Myhre et al. 1998). For halocarbons, a linear expression appears to be valid. When evaluating the radiative forcing since 1750, reference values are classically $CO_2$ 278 ppm, $CH_4$ 722 ppb and $N_2O$ 270 ppb (Myhre et al. 2013).

$CO_2$, $CH_4$, $N_2O$ and halocarbons are long-lived gases that remain in the atmosphere for a decade to many centuries. Their geographical distribution is thus quite homogeneous, and they can be characterised by their large values, although some small differences are observed between the two hemispheres or close to some production regions. Other greenhouse gases such as $O_3$ (**ozone**) have a shorter life. As a consequence, their concentration, and thus the associated radiative forcing, tends to be higher close to areas where they are produced. **Tropospheric** ozone is formed mainly through photochemical reactions driven by the emission of nitrogen oxides, carbon monoxide, methane and some organic compounds. Globally, the impact of the increase in ozone concentration is estimated to induce a radiative forcing of around $0.40$ W m$^{-2}$. However, the forcing is higher close to industrial regions, where the gases leading to ozone production are released.

By contrast, stratospheric ozone has decreased since pre-industrial time, leading to a global average radiative forcing of $-0.05$ W m$^{-2}$. The stratospheric ozone changes are particularly large in polar regions as the reactions responsible for the destruction of ozone in the presence of certain chemical compounds (e.g., the **chlorofluorocarbons**, which are a subset of the halocarbons) are more efficient in the presence of polar stratospheric clouds, which exist at low temperatures. The ozone-depleting halocarbons thus directly induce a positive forcing as a greenhouse gas and indirectly induce a negative forcing via their influence on stratospheric ozone concentration, the overall effect being a positive radiative forcing.

The largest decrease in stratospheric ozone concentration is observed over the high latitudes of the southern hemisphere. There, the famous ozone hole, discovered in the mid-1980s, is a large region of the stratosphere where about half the ozone disappears in spring. Because the Montreal Protocol banned the use of chlorofluorocarbons, the concentration of these gases in the atmosphere is no longer increasing, and the stratosphere is expected to recover in the coming decades (WMO 2014). An additional forcing due to stratospheric processes, estimated at about $0.07$ W m$^{-2}$, is due to the oxidation of methane [Eq. (2.53)] above the tropopause, which leads to an increase in the vapour content of the stratosphere. Finally, the contribution of contrails, that is, of clouds directly induced by human activities, such as water emission or turbulence caused by aircraft, is small at global scale.

### 4.1.2.2 Aerosols

Many atmospheric **aerosols** are natural: they may be generated by evaporation of sea spray, by wind blowing over dusty regions, by forest and grassland fires, by living vegetation (e.g., the production of sulphur aerosols by phytoplankton), and by volcanoes (see Section 4.1.2.4). Human activities also produce aerosols by the burning of fossil fuels or biomass and by the modification of natural surface cover which influences the amount of dust carried by the wind. Among the anthropogenic aerosols, sulphate aerosols and black carbon have received particular attention in climatology. Sulphate is mainly produced by the oxidation of sulphur dioxide ($SO_2$) in the aqueous phase, with fossil fuel burning, in particular, coal burning, as the main anthropogenic source. Black carbon is the result of incomplete combustion during fossil fuel and biomass burning.

Since most aerosols remain in the atmosphere for only a few days, anthropogenic aerosols are concentrated mainly downwind of industrial areas, close to regions where land-use changes have led to dustier surfaces (desertification) and where slash-and-burn agricultural practices are common. As a consequence, maximum

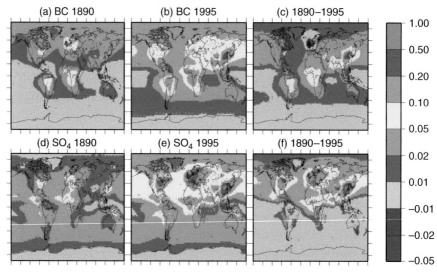

**Fig. 4.3**   Aerosol **optical depths** (i.e., a measure of atmospheric opacity) for black carbon (BC, x10) (a) in 1890 and (b) in 1995. (c) The change between 1890 and 1995. (d–f) The same measures for sulphates. (Source: Koch, D., S. Menon, A. Del Genio, R. Ruedy, I. A. Alienov and G. A. Schmidt (2009). Distinguishing aerosol impacts on climate over the past century. *Journal of Climate* 22, 2659–77. © American Meteorological Society. Used with permission.)

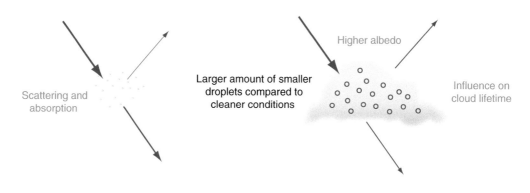

**Fig. 4.4**   Schematic representation of some aerosol-radiation interactions and aerosol-cloud interactions, focussing on the influence on solar radiation. Clouds also affect infrared fluxes (see Section 4.2.3). The rapid adjustments in the atmosphere due to the scattering and absorption of solar radiation, which leads to an additional effective radiative forcing, are not displayed in the figure.

concentrations are found in eastern America, Europe, and eastern Asia, as well as in some regions of tropical Africa and South America (Figure 4.3).

Aerosols directly affect our environment as they are responsible, for example, for health problems and acid rain. They also have multiple effects on the radiative properties of the atmosphere (Figure 4.4), which can be separated into aerosol-radiation and aerosol-cloud interactions (Boucher et al. 2013).

Aerosols absorb and scatter **shortwave** and **longwave** radiation. These aerosol-radiation interactions have been previously referred to as 'direct aerosol effects' (Forster et al. 2007). For sulphate aerosols, the main effect is the net **scattering** of a significant fraction of the incoming solar radiation back to space. This induces a negative radiative forcing, estimated to be around $-0.4$ W m$^{-2}$, on average, across the globe (Boucher et al. 2013). This distribution is highly heterogeneous because of regional variations in the concentration of aerosols (see Figure 4.3). By contrast, the main effect of black carbon is its strong absorption of solar radiation, which contributes to increased local temperatures. The associated positive radiative forcing since 1750 is about $+0.4$ W m$^{-2}$, on average. Taking all the aerosols into account, including organic and nitrate aerosols and mineral dust which each contributes with a forcing close to $-0.1$ W m$^{-2}$, the global average is about $-0.35$ W m$^{-2}$ (Boucher et al. 2013) (see Figure 4.2).

The absorption and scattering of solar radiation by the aerosols modify the air temperature, its humidity, and the vertical stability of the air column, affecting the formation and developments of clouds. This effect, often referred to as the 'semi-direct aerosol effect' (Denman et al. 2007), is relatively rapid and thus is included in the definition of the effective radiative effect due to aerosol-radiation interactions. Consequently, the ERF due to aerosol-radiation interactions is more negative than the RF, with a mean value of about $-0.45$ W m$^{-2}$ (Boucher et al. 2013) (see Figure 4.2). Furthermore, the deposit of black carbon on snow modifies snow's **albedo**, generating an additional small positive forcing ($\sim 0.04$ W m$^{-2}$). The aerosols affect **cloud microphysics**, inducing changes in clouds' radiative properties, their frequency and their lifetimes. These aerosol-cloud interactions cover a wide range of processes. First, aerosols affects the cloud albedo. This has been referred to as the 'first indirect', 'cloud albedo' or 'Twomey effect' (Denman et al. 2007). Many aerosols act as nuclei on which water condenses. A high concentration of aerosols thus leads to clouds that contain many more (and hence smaller) water droplets, with a higher total droplet area, than clouds with the same water content formed in cleaner areas. As such clouds are more highly reflecting (i.e., have a higher albedo), this induces a negative radiative forcing. The size of the droplets also can affect the quantity of water required before precipitation occurs and in this way the cloud lifetime and thus the amount of cloud present. However, the links between aerosol concentration and the development of a cloud are complex and depend on environmental conditions (Boucher et al. 2013). Estimates of the ERF of aerosol-cloud interactions thus are very uncertain, with a mean value of $-0.45$ W m$^{-2}$. As those interactions crucially depend on rapid adjustments in the atmosphere, the radiative forcing is hard to obtain and interpret and thus generally not given (see Figure 4.2).

This leads to a best estimate of the total ERF due to aerosols (excluding the effect of black carbon on the albedo of snow and ice) of $-0.9$ W m$^{-2}$. The likely range, however, with values between $-1.9$ and $-0.1$ (Boucher et al. 2014) is very large. This illustrates that the aerosols represent one of the largest uncertainties in our estimates of past and future changes in radiative forcing. This is true for the twentieth and twenty-first centuries, but aerosols also have played a role, not yet known precisely, in past climate changes. For instance, during the last glacial period (see Section 5.5), the drier conditions led to a higher number of aerosols in the atmosphere, producing a negative radiative forcing, probably larger than 1 W $m^{-2}$, which contributed to amplifying the cold conditions at the time.

### 4.1.2.3 Land-Use and Land-Cover Change

Humans have been modifying their environment for millennia, in particular, through deforestation. Before 1950, this occurred mainly in Europe, North America, India and China, leading to a high fraction of cropland in these areas (Figure 4.5). In the last fifty years, the extension of cropland has been stabilised in many places, some regions even showing an increase in the surface covered by forest. By contrast, deforestation has occurred rapidly over this period in many countries in the tropics.

Deforestation has a direct impact on emissions of $CO_2$ and $CH_4$ (see Section 4.1.2.1), as well as on the production of dust aerosols and aerosols due to biomass burning (see Section 4.1.2.2). Furthermore, the anthropogenic changes in land

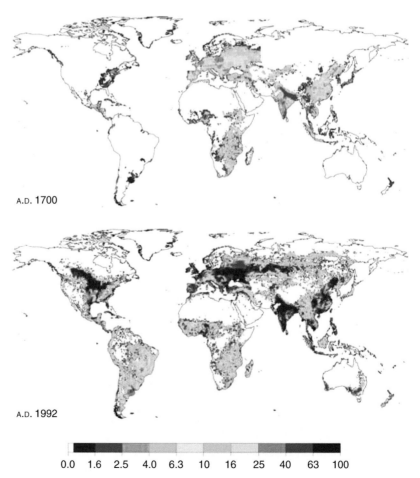

**Fig. 4.5**      The fraction of land occupied by crops in 1700 and 1992. (Source: Pongratz et al. 2008. Reprinted by permission from John Wiley and Sons: *Global Biogeochemical Cycles*; copyright 2008.)

use have altered the characteristics of the Earth's surface, leading to changes in the energy and moisture budgets, in particular, through modification of **evapotranspiration** (see Section 4.2.4). Several of these surface changes cannot be represented adequately by radiative forcing as their impact on the radiative balance of the atmosphere is not direct. However, it is possible to compute a radiative forcing for the modifications of **albedo** associated with land-use changes as this directly affects the radiative balance of the surface (although it is not always easy to separate this forcing from the feedbacks between vegetation and climate; see Section 4.3.2). Forests have a lower albedo than crops or pasture, in particular, when snow is present (see Sections 1.5 and 4.3.2). Consequently, the deforestation since 1750 has induced a radiative forcing which has been estimated to average around $-0.15$ W m$^{-2}$ across the globe (see Figure 4.2). However, the forcing is much higher in regions where deforestation has been the most severe, reaching several watts per square metre in some places (Myhre et al. 2013).

### 4.1.2.4  Solar and Volcanic Forcings

Preceding sections have been devoted mainly to anthropogenic forcings. However, natural forcings such as those associated with explosive volcanoes and changes in **total solar irradiance** (TSI) also affect the Earth's climate. Precise measurements of the TSI have been available from satellites for the last thirty years. They clearly show an eleven-year cycle associated with the well-known periodicity of solar activity. However, the long-term trend since 1980 is very weak (Figure 4.6). Over this period, the amplitude of the changes in TSI has been on the order of 0.1%, corresponding to a mean peak-to-peak difference in TSI of around 1 W m$^{-2}$. Dividing by 4 because of the geometrical properties of the system [see Eq. (2.3)] and multiplying by 0.7 to account for the albedo of the Earth give a value for the radiative forcing between high and low solar activity of about 0.2 W m$^{-2}$. Larger fluctuations in TSI are observed over longer periods, as discussed in Sections 5.4.1 and 5.6.3.

Major volcanic eruptions have a dramatic local impact, causing fatalities and damage to properties, crops, forests, and so on. The ash produced can travel

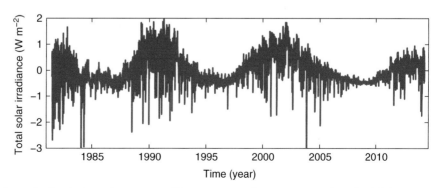

**Fig. 4.6**  **Anomaly** of total solar irradiance estimated from a composite of measurements performed with different satellites. (Source: RMIB TSI composite and updates: Mekaoui and Dewitte 2008.)

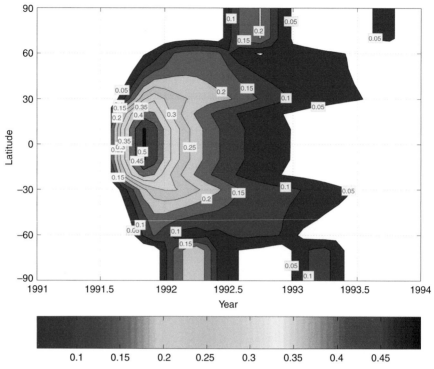

**Fig. 4.7**     Estimate of the volcanic aerosol **optical depth** after the 1991 Pinatubo eruption as a function of latitude and time. (Source: Gao et al. 2008. Reprinted by permission from John Wiley and Sons: *Journal of Geophysical Research*; copyright 2008.)

hundreds of kilometres, altering atmospheric properties for days or weeks and modifying the characteristics of the Earth's surface after its deposition. Explosive volcanic eruptions even can have an influence on a larger spatial scale, affecting the whole of the Earth's climate significantly (e.g., Robock 2000). Indeed, explosive eruptions can transport aerosols (mainly sulphates) directly to the stratosphere, where they remain for a few years and affect nearly all regions (Figure 4.7). As discussed for anthropogenic aerosols (Section 4.1.2.2), the presence of sulphate aerosols in the stratosphere induces both a local warming in the stratosphere (mainly because of the enhanced absorption of solar radiation) and a cooling below (associated with the scattering of some radiation back to space). For major eruptions, the net radiative forcing reaches an average of several watts per square metre across the globe in the year following the eruption and takes a few years to decrease to nearly zero (see Section 5.6.3).

### 4.1.3  Equilibrium Response of the Climate System: A Definition of Feedback

In response to a radiative forcing $\Delta Q$, the variables characterising the state of the climate system will change. This modifies, in turn, the radiative fluxes, inducing a **feedback** which can amplify (positive feedback) or dampen (negative feedback) the direct influence of the perturbation. These modifications involve complex

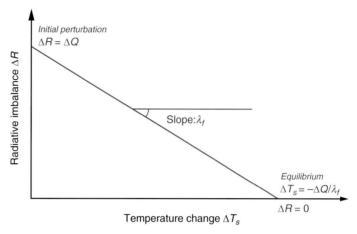

**Fig. 4.8** Schematic representation of Eqs. (4.4) and (4.5) illustrating the role of the climate feedback parameter $\lambda_f$. In particular, $\lambda_f$ must be negative to have a solution at equilibrium corresponding to a positive $\Delta T_s$ for a positive radiative forcing. (Source: Adapted from Gregory et al. 2004.)

mechanisms. However, insights can be gained by assuming that the changes in the radiative fluxes at the **tropopause** or at the top of the atmosphere due to the climate response can be estimated as a function of the changes in global mean surface temperature $\Delta T_s$. If we use $\Delta R$ to denote the imbalance in the radiative budget, we can write (see, e.g., Hansen et al. 1984; Bony et al. 2006)

$$\Delta R = \Delta Q + \lambda_f \Delta T_s \qquad (4.4)$$

where $\lambda_f$ is called the 'climate **feedback** parameter' (expressed in W m$^{-2}$ K$^{-1}$). In Eq. (4.4), downward fluxes are assumed to be positive.

If the perturbation lasts a sufficiently long time, the system eventually will reach a new equilibrium corresponding to $\Delta R = 0$ (Figure 4.8). This can be used to compute the global mean equilibrium temperature change in response to $\Delta Q$ as

$$\Delta T_S = -\frac{1}{\lambda_f} \Delta Q \qquad (4.5)$$

where $-1/\lambda_f$ is called the 'climate sensitivity parameter' (expressed in K m$^{-2}$ W$^{-1}$). It gives the change in the global mean temperature at equilibrium in response to a radiative forcing of 1 W m$^{-2}$. The climate sensitivity parameter must be positive; otherwise, an equilibrium cannot be reached if the climate warms, as expected, in response to a positive radiative forcing. In other words, the climate feedback parameter $\lambda_f$ must be negative. As discussed later, the value of $\lambda_f$ results from the combined effect of positive and negative feedbacks, but the latter ones must dominate in a stable system. The larger the value of $\lambda_f$ in absolute value (i.e., the stronger the negative feedbacks), the more stable is the system, and the smaller is the temperature change for a given radiative forcing $\Delta Q$.

Since the temperature increase $\Delta T_s = -1/\lambda_f$ caused by $\Delta Q = 1$ W m$^{-2}$ is a bit abstract, the '**equilibrium climate sensitivity**' generally is defined as the global mean surface temperature change after the climate system has stabilised in a new equilibrium state in response to a doubling of the $CO_2$ concentration in the atmosphere (unit °C). This corresponds to a radiative forcing $\Delta Q = 3.7$ W m$^{-2}$ (see Section 4.1.2.1).

As equilibrium climate sensitivity links in a simple way the forcing and the response, many studies have been devoted to its estimation. In theory, knowing $\Delta Q$ and $\Delta T_s$ for any past period close to equilibrium can be used following Eq. (4.5). If the system is in a transient state (like the last centuries) and the imbalance $\Delta R$ can be measured, the equilibrium climate sensitivity still can be given by Eq. (4.4) (Andrews et al. 2012). Nevertheless, such estimates are relatively uncertain because the forcing or temperature changes are often not known accurately or because **internal variability** (see Section 5.2) can affect the value obtained if short periods are selected for the availability of good observations. Consequently, it is difficult to have strong constrains on the range of the equilibrium sensitivity. Based on the most recent evidences, its value is likely between 1.5 and 4.5°C, a range which is close to that given by current global climate models of between 2.1 and 4.7°C (Collins et al. 2013).

The concepts of radiative forcing, climate feedback and climate sensitivity are very useful in providing a general overview of the behaviour of the system. However, when using them, we must bear in mind that the framework described earlier is a greatly simplified version of a complex three-dimensional (3-D) system.

Firstly, the global surface temperature is an interesting measure of the general state of the system, and changes in many characteristics of the climate in a response to a perturbation can be more or less strongly related to it (see Chapter 6). Nevertheless, the formalism behind Eq. (4.4) does not provide explicit information on important variables such as the spatial distribution of precipitation or the probability of extreme events such as storms or hurricanes, for instance.

Secondly, Eq. (4.4) assumes that the modifications of radiative flux in response to the forcing can be expressed as a linear function of the surface temperature. Non-linearities, however, are also present. A clear example is the occurrence of rapid events (see Sections 5.5.4 and 6.3.3) where a threshold is crossed because of the perturbation. In such cases, the climate development is mainly due to the internal dynamics of the system and is only weakly related to the magnitude of the forcing. The validity of the assumptions leading to Eq. (4.4) then becomes questionable, but in some other cases, adding non-linear terms in Eq. (4.4) already improves the adequacy of the simple model (Feldl and Roe 2013; Vial et al. 2013).

Thirdly, the magnitude of the climate feedbacks and the equilibrium climate sensitivity depends on the forcing applied, some forcings being more 'efficient' than others for the same global mean radiative forcing. This efficacy is measured as the ratio of the surface temperature change in response to this forcing versus the change from an increase in $CO_2$ concentration with the same mean forcing. It is generally closer to 1 when considering the **effective radiative forcing** than the **radiative forcing** (Hansen et al. 2005) as the later includes rapid adjustments that differ between forcing agents.

Fourthly, the feedbacks depend on the mean state of the climate system. For instance, it is obvious that feedbacks related to the **cryosphere** (see Section 4.2.2) play a larger role in relatively cold periods, where large amounts of ice are present at the surface, than in warmer periods. As a consequence, assuming that the equilibrium climate sensitivity is a constant is an approximation whose validity is generally fine for small perturbations but must be evaluated carefully when dealing with larger changes.

Fifthly, the equilibrium climate sensitivity is a function of the time scale investigated. Focussing on decadal to centennial time scales, most estimates deal only with the response of the atmosphere, land surface, ocean and sea ice. Over a longer term, vegetation, carbon cycle and the ice sheet must be taken into account too (see, e.g., Hansen et al. 2008; Paleosens Project Members 2012). To include all these elements, the notion of '**Earth system sensitivity**' has been introduced in addition to the notion of equilibrium climate sensitivity.

Before closing this section, it is important to also mention that some changes can be classified as either forcing or response depending on the particular focus of the investigator. For instance, in a study of glaciation during the late Eocene, the building of ice sheets can be considered to be a response of the system to forcing, implying powerful **feedbacks** (see Section 5.4). However, if an investigator is mainly interested in atmospheric and oceanic circulation at a particular time during that period, the ice sheets could be treated as boundary conditions and their influence on the Earth's radiative balance (in particular, through their **albedo**) as a radiative forcing. This distinction between forcing and response in some cases, can be even more subtle. It is thus important in climatology, as in many other disciplines, to define precisely what we consider the system we are studying to be and what the boundary conditions and forcings are.

## 4.1.4  Direct Physical Feedbacks

The changes in surface temperature $T_s$ are associated with modifications of many other variables which also affect the global heat budget. If we consider an ensemble of $n$ variables $x_i$ that affect $R$, neglecting the second-order terms, we can express $\lambda_f$ as a function of those variables as

$$\lambda_f = \frac{\partial R}{\partial T_s} = \sum_{i=1,}^{n} \frac{\partial R}{\partial x_i} \frac{\partial x_i}{\partial T_s} \qquad (4.6)$$

Thus $\lambda_f$ can be represented by the sum of the feedback parameters $\lambda_i$ associated with each variable $x_i$. This is again a linear approximation, and the non-linear terms resulting from the interactions between various feedbacks can be included, if needed. The analyses based on Eq. (4.4) focus on the variables which directly affect the radiative balance at the top of the atmosphere via physical processes. For simplicity, we will call these the 'direct physical feedbacks' (see Section 4.2). The standard decomposition of $\lambda_f$ then involves the separation into feedbacks related to the temperature ($\lambda_T$), the water vapour ($\lambda_w$), the clouds ($\lambda_c$) and the surface **albedo** ($\lambda_\alpha$). The temperature feedback is itself further split as $\lambda_T = \lambda_0 + \lambda_L$.

In evaluating $\lambda_0$, we assume that the temperature changes are uniform through-out the troposphere. $\lambda_L$ is called the '**lapse-rate** feedback' (see Section 4.2.1) and is the feedback due to the non-uniformity of temperature changes over the vertical. Thus

$$\lambda_f = \sum_i \lambda_i = \lambda_0 + \lambda_L + \lambda_w + \lambda_c + \lambda_\alpha \qquad (4.7)$$

Although many other feedbacks (in which the dominant processes cannot be simply related to the radiative balance at the top of the atmosphere) also can play an important role, they are excluded from this decomposition. They will be con-sidered separately in Sections 4.2.4 (soil moisture feedbacks), 4.2.5 (advective feedback in the ocean) and 4.3 (biogeochemical feedbacks), whilst some addi-tional feedbacks will be mentioned briefly in Chapters 5 and 6.

The feedback parameter $\lambda_0$ can be evaluated relatively easily because it simply represents the dependence of the outgoing longwave radiation on temperature through the **Stefan-Boltzmann law**. Using the integrated balance at the top of the atmosphere (see Section 2.1)

$$\Delta R = (1 - \alpha)\frac{S_0}{4} - \sigma T_e^4 \qquad (4.8)$$

and assuming the temperature changes to be homogeneous in the troposphere

$$\Delta T_s = \Delta T_e \qquad (4.9)$$

we obtain

$$\lambda_0 = \frac{\partial R}{\partial T_e}\frac{\partial T_e}{\partial T_s} = -4\sigma T_e^3 \qquad (4.10)$$

For a spatially homogeneous effective emission temperature $T_e = 255$ K, this pro-vides a value of $\lambda_0 \sim -3.8$ W m$^{-2}$ K$^{-1}$. Estimates derived from climate models using a vertically homogeneous (but spatially varying) temperature change, taking into account the radiative properties of the present-day atmosphere, give a slightly dif-ferent global mean value of $\lambda_0$ of around $-3.2$ W m$^{-2}$ K$^{-1}$ (see, e.g., Soden and Held 2006). We then can compute the equilibrium temperature change in response to a perturbation if this feedback was the only one active as

$$\Delta T_{s,0} = -\frac{\Delta Q}{\lambda_0} \qquad (4.11)$$

For a radiative forcing due to a doubling of the $CO_2$ concentration in the atmo-sphere corresponding to about 3.7 W m$^{-2}$ [see Eq. (4.1)], Eq. (4.11) leads to an equilibrium climate sensitivity which would be slightly greater than 1°C. This is often referred to as the 'blackbody response' of the system and $\lambda_0$ as the 'Planck feedback'. If now we include all the feedbacks, we can write

$$\Delta T_s = -\frac{\Delta Q}{\sum_i \lambda_i} = -\frac{\Delta Q}{\lambda_0 + \lambda_L + \lambda_w + \lambda_c + \lambda_\alpha} \tag{4.12}$$

The changes associated with the other feedbacks are often compared to the blackbody response, considered as a reference which is modified by the action of the lapse rate, water vapour, cloud and albedo feedbacks. This leads to

$$\begin{aligned}
\Delta T_s &= -\frac{1}{\left(1 + \dfrac{\lambda_L}{\lambda_0} + \dfrac{\lambda_w}{\lambda_0} + \dfrac{\lambda_c}{\lambda_0} + \dfrac{\lambda_\alpha}{\lambda_0}\right)}\left(\frac{\Delta Q}{\lambda_0}\right) \\
&= \frac{1}{\left(1 + \dfrac{\lambda_L}{\lambda_0} + \dfrac{\lambda_w}{\lambda_0} + \dfrac{\lambda_c}{\lambda_0} + \dfrac{\lambda_\alpha}{\lambda_0}\right)}\Delta T_{s,0} \\
&= f_f \Delta T_{s,0}
\end{aligned} \tag{4.13}$$

Here $f_f$ is called the 'climate feedback factor'. If $f_f$ is larger than 1, it means that the equilibrium temperature response of the system is larger than the response of a blackbody. The ratio $g_i = -\lambda_i/\lambda_0$ is classically referred to in the climate literature as the 'feedback gain' for feedback $i$, and $g_T = \sum_{i \neq 0} g_i$, with the sum applied for all the feedback parameters except $\lambda_0$, is the 'total feedback gain' (Hansen et al. 1984; Bony et al. 2006). As $\lambda_0$ is negative, $g_i$ has the same sign as $\lambda_i$. Equation (4.13) shows that if a feedback parameter $\lambda_i$ or gain factor $g_i$ is positive, the corresponding feedback tends to amplify the temperature change (positive feedback) as this is associated with a decrease in the denominator of Eq. (4.13). If $\lambda_i$ is negative, it damps the changes (negative feedback). Note that this nomenclature is reversed in climatology compared to many other fields, where the equivalent of $f_f$ is called a 'gain' and the equivalent of $g_i$ is a 'feedback factor' [for more details, see, e.g., Roe (2009)].

An alternative way to interpret Eq. (4.13) is to consider that the reference blackbody response would lead in the absence of any additional feedback to a response $\Delta T_{S,0}$. However, because of those feedbacks, the surface temperature change modifies the forcing (Figure 4.9). This link between the temperature change and the radiative fluxes at the tropopause is expressed via the feedback parameters [Eqs. (4.4) and (4.7)], in which a positive $\lambda_i$ corresponds to a flux which reinforce the perturbation. We should exclude $\lambda_0$ for the computation of this additional forcing as the blackbody response is already included in the reference. The additional forcing $\Delta Q'$ resulting from the temperature change associated with the blackbody response then can be written as

$$\begin{aligned}
\Delta Q' &= \sum_{i \neq 0} \lambda_i \Delta T_{s,0} = -\sum_{i \neq 0} \lambda_i \left(\frac{\Delta Q}{\lambda_0}\right) = \sum_{i \neq 0} g_i \Delta Q \\
&= g_T \Delta Q
\end{aligned} \tag{4.14}$$

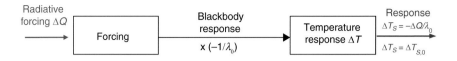

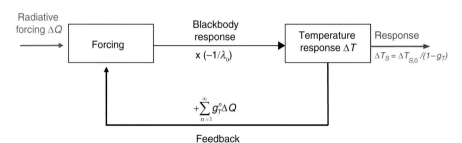

Feedback

**Fig. 4.9** Diagram illustrating (top) the blackbody response in which the forcing is divided by $-\lambda_0$ to obtain the surface temperature change $\Delta T_{s,0} = -\Delta Q/\lambda_0$. In the presence of additional feedbacks (bottom), this surface temperature change leads to an increase in the forcing of $\sum_{n=1}^{\infty} g_T^n \Delta Q$ and finally a warming $\Delta T_s = \Delta T_{s,0}/(1-g_T)$.

After this first step, the total forcing is then $(1 + g_T)\Delta Q$, which induces a total temperature change of $\Delta T_s' = (1 + g_T)\Delta T_{s,0}$. Again, the subsequent warming of $g_T \Delta T_{s,0}$ modifies the forcing by

$$\Delta Q'' = g_T \Delta Q' = g_T^2 \Delta Q \tag{4.15}$$

and so on. The total forcing thus follows a **geometrical series** $(1 + g_T + g_T^2 + \cdots)$ $\Delta Q = \sum_{n=0}^{\alpha} g_T^n \Delta Q$ and then

$$\Delta T_s = -\frac{\sum_{n=0}^{\infty} g_T^n \Delta Q}{\lambda_0} = \sum_{n=0}^{\infty} g_T^n \Delta T_{s,0} = \frac{\Delta T_{s,0}}{1-g_T} \tag{4.16}$$

as given by Eq. (4.13). The series converge only if $g_T < 1$ [which is equivalent to stating that $\lambda_f < 0$; see Eqs. (4.7) and (4.13)], corresponding to situations for which amplification of the initial radiative perturbation by the various feedbacks is not able to destabilise the system (see Figure 4.8).

This reasoning makes the feedback loop clearer (Gregory et al. 2009) and justifies why some authors consider that the blackbody response should not be seen as a feedback but rather as the reference response which is amplified or damped by other processes which could more properly be termed 'feedbacks' (see, e.g., Roe 2009).

### 4.1.5 Transient Response of the Climate System: Ocean Heat Uptake

Because of the thermal inertia of the Earth (see Section 2.1.5), the equilibrium response described in Section 4.1.3 is achieved only when all the components of the system have adjusted to the new forcing. It can take years for the atmosphere and centuries or millennia for the oceans and ice sheets to reach this new equilibrium.

Because nearly all the heat capacity of the system resides in the ocean on multi-annual time scales, the net radiative flux at the top of the atmosphere $\Delta R$ is nearly equal to the net heat flux at the ocean surface. For simplicity, we will assume here that the thermal inertia of the climate system can be represented at the first order by an ocean slab with homogeneous temperature $T_s$ and heat capacity $C_s$. This is an approximation as the heat capacity $C_s$ should vary with time as the perturbation penetrates to deeper oceanic layers (Watterson 2000).

Using the same notation and the same approach as for Eq. (4.4), the energy balance of the system can be written as

$$C_s \frac{d\Delta T_s}{dt} = \Delta Q + \lambda_f \Delta T_s \qquad (4.17)$$

The left-hand side of Eq. (4.17) represents the heat uptake by the ocean, which plays a central role in the transient response of the system to a perturbation (see Sections 5.7, 6.2.3 and 6.3.1).

If we assume that the radiative forcing $\Delta Q$ is equal to 0 for $t < 0$ and is constant for $t \geq 0$, this equation can be easily solved, leading to

$$\Delta T_s = -\frac{\Delta Q}{\lambda_f}\left(1 - e^{-t/\tau}\right) \qquad (4.18)$$

with

$$\tau = -\frac{C_s}{\lambda_f} \qquad (4.19)$$

When $t$ is very large, we obtain, as expected, the equilibrium solution described by Eq. (4.5). $\tau$ represents a time scale, and when $t = \tau$, the temperature change has reached 63% of its equilibrium value. $\tau$ depends on the heat capacity of the system $C_s$ and the strength of the feedbacks. This implies that with larger values of $-1/\lambda_f$ (i.e., a greater **equilibrium climate sensitivity** or a smaller climate feedback parameter in absolute value), the time taken to reach equilibrium will be longer for a given heat capacity. This is an important characteristic of the climate system which also holds when much more sophisticated representations of the climate system than that shown in Eq. (4.17) are used (Hansen et al. 1985).

This behaviour can be clearly illustrated by an example (Figure 4.10). Let us choose for the equilibrium climate sensitivity values of 2 and 4°C (equivalent to values of $\lambda_f$ equal to –1.85 and –0.925 W m$^{-2}$ K$^{-1}$, respectively), a heat capacity corresponding to a depth of 200 m of water distributed evenly over the whole globe ($C_s = 4,180 \times 10^3 \times 200 = 8.36 \times 10^8$ J K$^{-1}$ m$^{-2}$) and a radiative forcing corresponding to a doubling of the $CO_2$ concentration in the atmosphere ($\Delta Q = 3.7$ W m$^{-2}$). As expected, the two different climate sensitivities produce a factor of 2 between the equilibrium responses. However, the two curves are virtually identical during the first fifteen years. It can be easily demonstrated that the

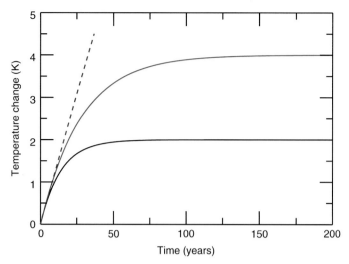

**Fig. 4.10** Temperature changes obtained as a solution of Eq. (4.17) using a forcing $\Delta Q$ of 3.7 W m$^{-2}$, $C_s$ equal to $8.36 \times 10^8$ J K$^{-1}$ m$^{-2}$ and values of $\lambda_f$ of $-1.85$ (black) and $-0.925$ W m$^{-2}$ K$^{-1}$ (red). The dashed blue line indicates the slope at $t = 0$, which has a value of $\Delta Q / C_s$.

slope of the curve at $t = 0$ is independent of the equilibrium climate sensitivity. As a consequence, knowing the temperature changes in the years immediately following application of the perturbation does not necessarily provide clear information on the long-term evolution of the system.

The long adjustment of the climate system to the forcing caused by the oceanic heat uptake has led to the definition of the '**transient climate response**' (TCR). It is measured by the global annual mean temperature change averaged over the years 60 to 80 in an experiment in which the $CO_2$ concentration is increased by 1% per year until year 70 (by which time it is double its initial value). The TCR values derived from models and data range between 1.0 and 2.5°C (Collins et al. 2013); that is, TCR is a bit less than a factor of 2 smaller than the equilibrium climate sensitivity.

An even simpler representation of the oceanic heat uptake is to assume that it is proportional to $\Delta T_s$. The heat balance [equivalent of Eq. (4.17)] is then

$$\kappa_c \Delta T_s = \Delta Q + \lambda_f \Delta T_s \qquad (4.20)$$

where $\kappa_c$ is a constant called the 'ocean heat uptake efficiency' (in W m$^{-2}$ K$^{-1}$). This approximation is, of course, not valid when the system is close to a new equilibrium because ocean heat uptake should tend to 0 in this case for a finite value of $\Delta T_s$. Nevertheless, it seems adequate during periods when the forcing is steadily increasing or decreasing (Gregory and Foster 2008). Equation (4.20) can be interpreted as if a positive radiative perturbation $\Delta Q$ is balanced by extra heat loss to space ($-\lambda_f \Delta T_s$) and to the deeper oceanic levels ($\kappa_c \Delta T_s$). Both of these heat losses are considered proportional to the temperature of the surface layer, which has itself a small heat capacity.

Another advantage of Eq. (4.20) is that the ocean heat uptake has the same mathematical form as the changes in the radiative fluxes at the top of the atmosphere in response to the perturbation. The ocean heat uptake efficiency has the same units as the climate feedback parameter (be careful of the sign as $\kappa_c \Delta T_s$ appears on the other side of the equation from $\lambda_f \Delta T_s$ ). Consequently, the oceanic heat uptake can be interpreted formally in a similar way as the direct physical feedbacks investigated in Section 4.1.4. The magnitude of those effects also can be compared through the values of the climate feedback parameter and the ocean heat uptake efficiency $\kappa_c$ is generally smaller than $\lambda_f$ in absolute value, $\kappa_c$ being, for instance, around $0.6$ W m$^{-2}$ K$^{-1}$ compared to a value of $\lambda_f$ of $-1.4$ W m$^{-2}$ K$^{-1}$ in the models investigated by Gregory and Foster (2008). In this framework, which we recall is not valid for all cases, the ocean heat uptake thus is equivalent to a negative feedback.

Equation (4.20) then gives the temperature $\Delta T_s$ as

$$\Delta T_s = \frac{\Delta Q}{\kappa_c - \lambda_f} = \frac{\Delta Q}{\rho_f} \tag{4.21}$$

where $\rho_f$ is the 'climate resistance' (in W m$^{-2}$ K$^{-1}$) (see Gregory et al. 2009). As $\lambda_f$ must be negative and $\kappa_c$ positive, $\rho_f$ is always positive, as expected. Equation (4.21) is equivalent to Eq. (4.5) for a transient situation. It then can be used to estimate the TCR if $\Delta Q$ corresponds to the forcing due to a $CO_2$ doubling.

## 4.2  Physical Feedbacks

### 4.2.1  Water-Vapour Feedback and Lapse-Rate Feedback

According to the **Clausius-Clapeyron equation**, the **saturation vapour pressure** and the **specific humidity** at saturation are quasi-exponential functions of temperature (see Section 1.2.1). Furthermore, observations and numerical experiments consistently show that the **relative humidity** tends to remain more or less constant in response to climate change. A warming thus produces a significant increase in the amount of water vapour in the atmosphere. As water vapour is an efficient greenhouse gas, this leads to a strong positive feedback (Figure 4.11). The radiative effect is larger in places where specific humidity is relatively low in unperturbed conditions and strongly increases in response to the perturbation, such as in the upper troposphere in the tropics (Soden and Held 2006; Zelinka and Hartmann 2012). A wrong interpretation of the feedback loop presented in Figure 4.11 would be to consider that such a positive feedback ultimately leads to an instability as temperature and humidity changes always reinforce each other. Equation (4.13) shows clearly that it is not the case because of the presence of other processes, in particular, the blackbody response, which stabilises the system as long as $\lambda_f = \sum_i \lambda_i < 0$ [see Eqs. (4.4) and (4.7)].

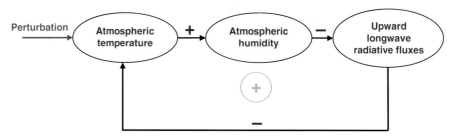

Simplified signal-flow graph illustrating water-vapour feedback. A positive sign on an arrow means that the sign of the change remains the same when moving from the variable at the origin of the arrow (on the left in the top row) to the one pointed to by the arrow (on the right in the top row), whilst a negative sign implies that an increase (decrease) in one variable induces a decrease (increase) in the next one. The positive sign in a circle indicates that the overall feedback is positive.

The feedback parameters can be estimated by different methods. One that is used widely is the 'kernel method' (Soden et al. 2008), which is derived directly from Eq. (4.6):

$$\lambda_i = \frac{\partial R}{\partial \mathbf{x}_i} \frac{\partial \mathbf{x}_i}{\partial T_s} = \mathbf{K}_{x_i} \frac{\partial \mathbf{x}_i}{\partial T_s} \qquad (4.22)$$

where $\mathbf{K}_{x_i}$ is kernel for the variable $\mathbf{x}_i$ (here the water vapour). In Eqs. (4.9) and (4.10), obtaining the climate feedback parameter from Eq. (4.6) was quite straightforward because $\Delta R$ was assumed to depend only of one globally homogeneous variable. More generally, the radiative balance is influenced by the variations in water vapour or of temperature at all the locations, grouped here in a vector $\mathbf{x}_i$. The kernel $\mathbf{K}_{x_i}$ is thus also a vector function of the geographical position, the time of year and the time of day, which relates local changes in the variable of interest to the radiative balance at the top of the atmosphere. It can be obtained by performing a set of numerical experiments in which this variable is perturbed at each location and each model layer and estimating the impact of the change on the radiative flux at the tropopause. To obtain $\lambda_i$, the kernel is multiplied by the changes in the variable simulated in a particular experiment, generally an instantaneous modification of the atmospheric $CO_2$ concentration, as a function of the changes in $T_s$.

The recent estimates provide a value of $\lambda_W$ of around $1.6 \, \mathrm{W \, m^{-2} \, K^{-1}}$ (Soden and Held 2006; Flato et al. 2013; Vial et al. 2013). This means that in the absence of any other feedback, the surface temperature change due to a perturbation would be about twice as large as the blackbody response [see Eq. (4.11)] because of this amplification associated with the water vapour. This makes the water-vapour feedback the largest of all the direct physical feedbacks. It is also one of the best understood.

The vertical variations in the temperature change also have a climatic effect through the **lapse-rate feedback** $\lambda_L$. The models predict enhanced warming in the upper troposphere of tropical regions in response to an increase in the

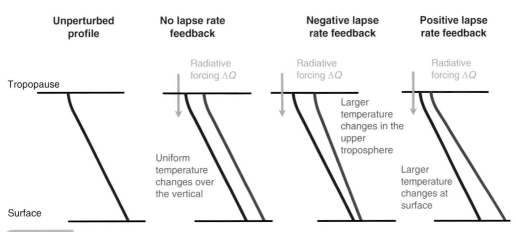

| Unperturbed profile | No lapse rate feedback | Negative lapse rate feedback | Positive lapse rate feedback |

**Fig. 4.12**    Schematic representation of positive and negative lapse-rate feedbacks.

concentration of greenhouse gases. If we assume that the temperature there roughly follows a moist **adiabatic**, this can be directly interpreted as a decrease in the moist lapse rate in response to an increase in temperature (see Section 1.2.1). Because of this change in the lapse rate, temperature at high altitude will be larger than for a homogeneous temperature change over the vertical. Consequently, the emission of outgoing **longwave** radiation by the upper levels, which provides a large contribution to the total emission of the Earth (see Sections 2.1.2 and 2.1.6), will be higher. The system then will lose more energy, so inducing a negative feedback (Figure 4.12). At high latitudes, a higher low-level warming is projected as a response to the positive radiative forcing, in particular, because of the temperature inversion which traps the warming close to the surface (see Section 6.2.4). This then provides a positive feedback (see Figure 4.12). The global mean value of $\lambda_L$ thus depends on the relative magnitude of those two opposite effects. On average, the influence of the tropics dominates, leading to a value of $\lambda_L$ of around $-0.6$ W m$^{-2}$ K$^{-1}$ (Soden and Held 2006; Flato et al. 2013; Vial et al. 2013) in recent models.

    The water-vapour feedback and the lapse-rate feedback are generally negatively correlated between different models (Soden and Held 2006; Ingram 2013). If the temperature increases more in the upper **troposphere**, causing a negative lapse-rate feedback, the warming also will be associated with higher concentrations of water vapour by the Clausius-Clapeyron equation in a region where it has a large radiative impact, leading to an additional positive water-vapour feedback. The exact changes in temperature and humidity at high altitudes in response to a perturbation are not well known. However, as the effects of the two feedbacks discussed in this sub-section tend to compensate each other, the uncertainty in the sum $\lambda_L + \lambda_W$ is smaller than in the two feedbacks individually. Furthermore, because of this partial cancellation between the two feedbacks, it has been proposed to decompose the total feedback in a different way using, for instance, the relative humidity rather than the specific humidity in Eqs. (4.6) and (4.7) (Held and Shell 2012; Ingram 2013).

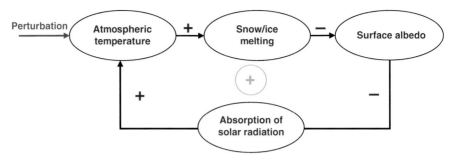

**Fig. 4.13** Flow graph illustrating the snow-and-ice–albedo feedback. A positive (negative) sign on an arrow means that an increase in one variable produces an increase (decrease) in the one pointed to by the arrow. The positive sign inside a circle indicates that the overall feedback is positive.

## 4.2.2 Cryospheric Feedbacks

The most important feedback associated with the **cryosphere** is that due to the high **albedo** of snow and ice (see Table 1.3). If the temperature increases in response to a perturbation, snow and ice will tend to melt, leading to a reduction in the surface albedo (Figure 4.13). As a consequence, the fraction of incoming solar radiation absorbed by the Earth will increase, leading to greater warming. The quantification of this feedback, which is referred to as the 'snow-and-ice–albedo feedback' or the 'temperature-albedo feedback' depends on the exact change in the surface albedo in response to the temperature change. The albedo can be influenced by changes in the surface covered by ice or snow (and thus also by **leads**) or by modifications of the snow and ice surface properties caused by surface melting (in particular, the formation of **melt ponds** at the sea-ice surface).

The snow-and-ice–albedo feedback has a significant impact on a global scale, with an estimated value based on recent model simulations of $\lambda_\alpha = 0.3 \text{ W m}^{-2} \text{ K}^{-1}$ (Flato et al. 2013; Vial et al. 2013). However, its influence is larger at high latitudes, where most of the cryospheric changes are observed. For snow over land, the impact is particularly strong in spring, when warming can produce fast disappearance of the snow cover, leading to large albedo changes at a time when the incoming solar radiations are intense.

The snow-and-ice–albedo feedback is also important in producing the greater surface temperature changes at high latitudes than at middle and lower latitudes in response to a radiative perturbation. This polar amplification of climate change is a robust characteristic of climate model simulations. It is also strongly influenced by changes in poleward energy transport and by water-vapour, lapse-rate and cloud feedbacks at high latitudes (see Section 6.2.4).

Another important cryospheric feedback is related to the thermal insulation effect of sea ice (see also Section 3.3.3). If the temperature increases in polar regions, sea-ice thickness will be reduced. As a consequence, in winter, the conductive heat flux from the relatively warm ocean to the cold atmosphere via sea ice will increase. This will lead to more oceanic heat losses and thus more sea-ice production which will moderate the initial thinning of the ice. This 'negative-growth-thickness feedback' (Bitz and Roe 2004) is more efficient for thin ice than for thick ice. Indeed, for an ice thickness of 3 m or more, the conduction heat flux is very low, and a change in thickness has a weak impact on this flux. Besides, for thin ice, the dependence of conduction and thus of the growth rate on thickness is higher. This partly explains why sea-ice mass loss in response to a perturbation is larger for thick ice than for thin ice.

On longer time scales, the formation of ice sheets is a formidable amplifier of climate changes and plays a large role in glacial-interglacial cycles (see Section 5.5.2). If the snow does not melt completely in summer, it accumulates from year to year, eventually forming large masses of ice (as currently observed in Greenland and Antarctica; see Section 1.4). When such an ice sheet is formed, it induces an increase in the planetary albedo, as discussed earlier in the framework of the snow-and-ice–albedo feedback. Furthermore, because of the high elevation of the ice sheet, the surface is cold and not prone to melting, further stabilising the ice sheet. Both these effects tend to maintain the cold conditions once they have been initiated by the forcing.

### 4.2.3 Cloud Feedbacks

Clouds affect the Earth's radiation budget in a variety of ways (Figure 4.14). On the one hand, they trap a significant fraction of the **longwave** flux coming from the surface (and re-emit part of it downward). The flux emitted at cloud tops towards space is lower than the surface flux because the temperature at the corresponding altitude is lower. Consequently, the presence of clouds tends to reduce the longwave emission from the Earth. This explains why cloudy nights are generally warmer than clear nights as clouds reduce the radiative heat losses of the surface. However, clouds reflect a significant amount of the incoming solar radiation, resulting in a net decrease in the amount of solar radiation absorbed by the Earth (see Section 2.1.6). This explains why cloudy days are colder than sunny days.

These two effects are often referred to as 'longwave' and '**shortwave** cloud radiative forcing' or the 'cloud radiative effect' (CRE). The shortwave CRE appears to be dominant, on average, in present-day conditions with a value of about $-50 \text{ Wm}^{-2}$ compared to the longwave CRE of about $30 \text{ W m}^{-2}$. The clouds thus induce a reduction in the net downward radiation flux at the top of the atmosphere which is estimated to be around $20 \text{ W m}^{-2}$ (Kiehl and Trenberth 1997; Boucher et al. 2013).

However, the CRE varies strongly with location and season. It is also very different for the various cloud types (see Figure 4.14). For instance, low-level clouds tend to be relatively warm and thus have a small impact on the upward longwave flux, whilst they generally have a large albedo. They are thus associated with a reduction in the net radiative flux at the top of the atmosphere. By contrast, most high-level clouds are associated with a higher net radiative flux because they are cold and have a lower albedo.

As a consequence, an analysis of cloud feedback must take into account, for the different regions of the globe, the changes in the amount of different cloud types, cloud temperatures (in particular, related to cloud height as it influences the

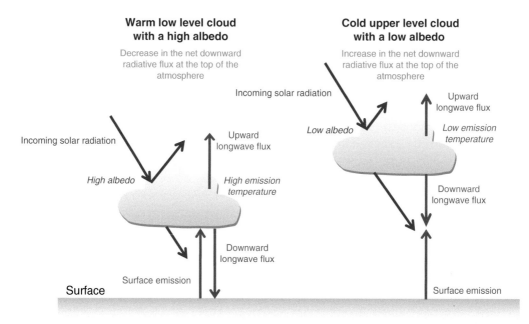

**Warm low level cloud with a high albedo**

Decrease in the net downward radiative flux at the top of the atmosphere

Incoming solar radiation

Incoming solar radiation

High albedo

Upward longwave flux

High emission temperature

Downward longwave flux

Surface emission

**Surface**

**Cold upper level cloud with a low albedo**

Increase in the net downward radiative flux at the top of the atmosphere

Low albedo

Upward longwave flux

Low emission temperature

Downward longwave flux

Surface emission

**Fig. 4.14**     Schematic illustration of the influence of the different types of clouds on the Earth's heat budget. Shortwave radiative fluxes are shown in blue and longwave radiative fluxes in red.

temperature and thus the emission of radiation at their tops) and cloud radiative properties (**optical depth**) as a response to a perturbation. All these changes can be caused by direct thermodynamic effects, by the large-scale dynamics which influence changes in temperature and humidity and by small-scale processes occurring in the clouds themselves (**cloud microphysics**).

Despite many uncertainties, a positive cloud feedback in the longwave emission at low and middle latitudes is a robust feature in current climate models. The tops of tropical high clouds rise in response to the temperature increase, the temperature of the clouds thus decreases and the heat loss due to emission is reduced compared to a situation where the altitude of the cloud tops would have remained the same (cloud-top feedback) (Zelinka and Hartmann 2010, 2012). On the shortwave emissions, the magnitude of the feedbacks vary more between models but is generally negative at high latitudes (Boucher et al. 2013).

Cloud feedback thus is still one of the less well understood feedbacks, and this uncertainty is responsible for a significant fraction the spread of **equilibrium climate sensitivity** in the present generation of climate models. The mean value provided by these models for the feedback parameter is $\lambda_C = 0.3$ W m$^{-2}$ K$^{-1}$ (Flato et al. 2013; Vial et al. 2013), but it goes from less than 0 (i.e., a negative feedback) to more than 1 W m$^{-2}$ K$^{-1}$ (90% uncertainty range of ±0.7 W m$^{-2}$ K$^{-1}$, i.e., more than twice the range for water-vapour feedback).

The preceding analyses have focussed on the global mean value of the climate feedback parameters. Nevertheless, the feedbacks modify the heat radiative balance at the top of the atmosphere locally [in the same way as expressed by Eq. (4.4)]. Any local imbalance in the radiative fluxes must be compensated for by a heat transport at equilibrium (see Section 2.1.5). Consequently, the spatial variations of the direct physical feedback are responsible for changes in this transport. In particular, the strong water-vapour and cloud feedbacks at low latitudes are associated with an increase in the radiative fluxes at the top of the atmosphere in those regions which must be compensated for by an export to higher latitudes (Zelinka and Hartmann 2012; Feldl and Roe 2013).

### 4.2.4　Soil-Moisture Feedbacks

The feedbacks that involve variations in soil moisture are in general impossible to describe via their impact on the global heat balance of the Earth because they deal mainly with exchanges between the surface and the atmosphere. Thus they are not included in the direct physical feedback introduced in the framework of Eq. (4.7). Nevertheless, they play an important role in the surface heat balance (see, e.g., Seneviratne et al. 2010).

When the temperature rises, if water is available, **evapotranspiration** increases as a consequence of the **Clausius-Clapeyron equation**. This cools the surface, providing a negative feedback. However, evapotranspiration induces a drying of the soils. The remaining water may be located at greater depths or may be more tightly bound to soil particles (see Section 2.2.2) and thus less easily extracted by plants. As a consequence, evapotranspiration decreases, finally leading to an increase in temperature and thus to a positive feedback (Figure 4.15). Such a positive feedback is possible only in regions where evapotranspiration is controlled by the soil moisture content (see Section 2.2.2). It does not operate if the soil is too dry and evapotranspiration is very small or if the soil is so wet that sufficient water is available to plant roots whatever the conditions.

The soil moisture content also can affect precipitation, potentially leading to other feedbacks. An increase in soil moisture favours a higher evapotranspiration rate and then a higher atmospheric humidity that may induce condensation and more precipitation. These higher precipitation rates replenish the soil reservoir, closing the loop to generate a positive feedback, but we, of course, also have to take into account that more evapotranspiration implies a direct decrease in soil moisture content, as discussed earlier. Furthermore, the contribution of

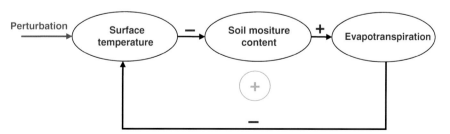

**Fig. 4.15**　Flow graph illustrating the soil-moisture–temperature feedback. A positive (negative) sign on an arrow means that an increase in one variable produces an increase (decrease) in the one pointed to by the arrow. The positive sign inside a circle indicates that the overall feedback is positive.

evapotranspiration to local precipitation depends on many factors. The moisture transferred from the soil to the atmosphere can precipitate in distant locations because of atmospheric transport. The impact of surface conditions on the stability of the **atmospheric boundary layer** often plays a more important role in precipitation than the magnitude of the local moisture transfer itself. For instance, dry soils may be associated with a large surface warming which destabilises the air column above, inducing **convection**, cloud formation (see Section 1.2.1) and then precipitation. The efficiency of this local water recycling between the atmosphere and the soil therefore must be evaluated case by case.

## 4.2.5  Advective Feedback in the Ocean

The **thermohaline circulation** in the North Atlantic is characterised by a northward surface transport of relatively warm and salty waters to high latitudes, where they cool, sink because of their high density and are exported southward at depth (see Section 1.3.2). This **meridional overturning circulation** thus contributes to maintaining the high salinity and high density at high latitudes which drive the flow. This leads to a positive feedback, usually referred to as the 'ocean advective feedback' or 'salinity advection feedback'. If the circulation strength decreases because of a perturbation, the salinity transport towards high latitudes is reduced. This induces a decrease in the density there and thus in the intensity of the flow which will reinforce the initial perturbation.

As initially proposed by Stommel (1961), this oceanic advective feedback can be illustrated by a simple model consisting of two well-mixed boxes of equal volume. Here we will follow the formalism described in Marotzke (2000). The first box represents the high latitudes. It has a temperature $T_1$ and a salinity $S_1$. The second box corresponds to the lower latitudes. It has a salinity $S_2$ and a temperature $T_2$ (Figure 4.16).

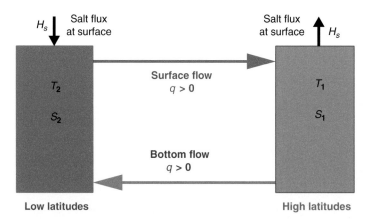

**Fig. 4.16**   Stommel conceptual model of the meridional overturning circulation. (Source: Redrawn from Marotzke 2000.)

It is assumed that the flow between the two boxes is proportional to the density difference between them; that is,

$$q = k_q \left( \rho_1 - \rho_2 \right) \tag{4.23}$$

where $k_q$ is a hydraulic constant. The density $\rho$ is approximated by a linear function of the salinity $S$ and temperature $T$:

$$\rho = \rho_0 - \alpha_T \left( T - T_0 \right) + \beta_S \left( S - S_0 \right) \tag{4.24}$$

where the $\rho_0$, $T_0$ and $S_0$ are the density, temperature and salinity of a reference state, respectively, and $\alpha_T$ and $\beta_S$ are the thermal expansion and haline contraction coefficients, respectively (both positive). The flow then can be written as

$$\begin{aligned} q &= k_q \left[ \alpha_T \left( T_2 - T_1 \right) - \beta_S \left( S_2 - S_1 \right) \right] \\ &= k \left( \alpha_T \Delta T - \beta_S \Delta S \right) \end{aligned} \tag{4.25}$$

If $q > 0$, the flow is from low to high latitudes at the surface and from high to low latitudes at depth, as presently observed in the North Atlantic (see Section 1.3.2). This corresponds to a flow driven by the temperature difference as it implies that $\alpha_T \Delta T > \beta_S \Delta S$.

The temperatures $T_1$ and $T_2$ are assumed to be fixed and controlled by the atmosphere with, as expected from the definition of the problem, $T_2 > T_1$ (i.e., $\Delta T > 0$). Additionally, a freshwater flux is imposed at the surface of each box. The input is positive at high latitudes and negative at low latitudes because of the higher evaporation there (see Section 2.2.3). Since the **state variable** of the model is salinity, the freshwater flux is expressed as an equivalent salinity flux $H_S$ which mimics the effect of dilution on the salinity. In this framework, a positive freshwater flux corresponds to a negative salt flux. The second term in the salt balance, representing transport, is simply given by $q$ times the difference in salinity between the boxes. The equations for $S_1$ and $S_2$ are then

$$\frac{dS_1}{dt} = -H_S + |q| \left( S_2 - S_1 \right) \tag{4.26}$$

and

$$\frac{dS_2}{dt} = +H_S - |q| \left( S_2 - S_1 \right) \tag{4.27}$$

The absolute value of $q$ is used in Eqs. (4.26) and (4.27) because the salt balance of the boxes is the same whatever the flow is northward at the top and southward at the bottom or vice versa.

The steady-state solution can be easily obtained by combining Eq. (4.25) and (4.26):

$$0 = -H_S + \left| k_q \left( \alpha_T \Delta T - \beta_S \Delta S \right) \right| \left( \Delta S \right) \tag{4.28}$$

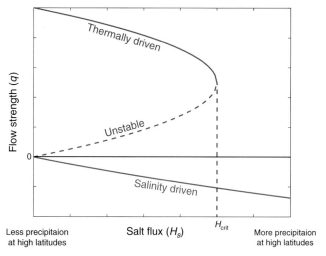

The flow strength $q$ in the Stommel model as a function of salt flux. High values of $H_s$ correspond to a larger freshwater flux at high latitudes. The thermally driven flow is in blue ($q > 0$), whilst the flow driven by the salinity ($q < 0$) is in red. The unstable solution is dashed.

This leads to a quadratic equation whose solutions for $q > 0$ are

$$\Delta S = \frac{\alpha_T \Delta T}{\beta_S} \left[ \frac{1}{2} \pm \sqrt{\frac{1}{4} - \frac{\beta_S H_S}{k_q (\alpha_T \Delta T)^2}} \right], \qquad q > 0 \qquad (4.29)$$

It can be shown that the positive root leads to an unstable solution because any modification of the freshwater flux, for instance, would lead to an anomalous flow that would reinforce the initial perturbation until the system reaches another equilibrium state.

Figure 4.17 clearly shows two consequences of the advective feedback. Firstly, the decrease in the flow function of the freshwater flux in this regime, driven by the temperature difference, is faster than in the absence of feedback. If the flow itself had no effect on the salinity and thus on the density, this decrease would simply follow Eq. (4.25) and would be linear. Secondly, a positive flow cannot be sustained if $4\beta_S H_S > k_q (\alpha_T \Delta T)^2$ [which would induce a negative square root in Eq. (4.29)]. For the value of $H_s$ corresponding to this threshold ($H_{crit}$), the flow still has a strength equal to half the maximum value obtained for $H_S = 0$. Consequently, a small increase in $H_S$ above $H_{crit}$ will lead to a transition from a significantly positive flow to a very different state (this is a bifurcation point). The new state is characterised by a negative value for $q$ because Eq. (4.28) admits an additional solution for $q < 0$:

$$\Delta S = \frac{\alpha_T \Delta T}{\beta_S} \left[ \frac{1}{2} + \sqrt{\frac{1}{4} + \frac{\beta_S H_S}{k_q (\alpha_T \Delta T)^2}} \right], \qquad q < 0 \qquad (4.30)$$

The second root has been discarded because it leads to a value of $q > 0$ which contradicts the hypothesis used to solve the equation. The solution with $q > 0$ corresponds to conditions where the density is higher at high latitudes because of its lower temperature. By contrast, when $q < 0$, the density difference is dominated by the salinity. Dense salty water from the low-latitude box is transported at depth northward, whilst a surface transport from high to low latitudes occurs at surface. Since the solution of Eq. (4.30) is valid for any positive $H_S$, two equilibrium solutions for the flow coexist for values of $H_S$ between 0 and $H_{crit}$, and a perturbation can induce a transition from one to the other (see Figure 4.17).

The Stommel model illustrates in a simple and straightforward way the role of advective feedbacks in the ocean and the possibility of multiple equilibria in the ocean circulation. More complex representations would require taking into account the exchanges between the hemispheres, a key characteristic of the meridional overturning circulation (see, e.g., Rahmstorf 1996). The Stommel model also neglects the contribution of the wind-driven circulation in heat and freshwater transport in the ocean. Furthermore, the hypothesis that the temperatures of the boxes are independent of the flow masks a negative feedback: a weaker positive flow reduces the heat transport from low to high latitudes, decreasing the temperature and the density at high latitudes and thus partly compensating for the positive salt advection feedback.

Nevertheless, many climate models agree surprisingly well with some of the characteristics of the Stommel model. The circulation driven by temperature differences ($q > 0$) displays a non-linear decrease in flow as a function of the freshwater input, and a threshold is present above which no flow is possible. Besides, the second solution of the Stommel model, characterised by a circulation dominated by the salinity differences ($q < 0$) and a poleward flow at depth, is not reproduced by climate models in conditions which are close to the present-day situation. The second equilibrium in those models, when present, is a state without a deep overturning circulation (referred to as a 'shutdown' or 'collapse' of the thermohaline circulation). This has been illustrated in several Earth models of intermediate complexity (EMICs; see Section 3.2.2) driven by slowly varying freshwater perturbations in the North Atlantic (Figure 4.18). They all display a similar behaviour, but the freshwater flux required to induce a shutdown of the meridional overturning circulation differs between models. Furthermore, the present-day climate is reproduced in some models in the regime for which two equilibrium states are stable (characterised by strong and absent formation of North Atlantic deep water, respectively), whilst in other models only the active state is stable. These conclusions have been confirmed by experiments performed with some general circulation models (GCMs) using the same experimental design, but such experiments are very expensive with this type of model.

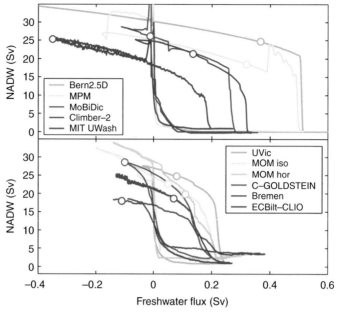

**Fig. 4.18** Response of various EMICs to slowly varying freshwater perturbations. The y-axis represents the maximum of the meridional overturning circulation in the North Atlantic, a measure of the formation of North Atlantic deep water (NADW) in the models. All the models have been aligned so that the rapid transition from a shutdown circulation to an active overturning corresponds to 0 on the x-axis. Circles indicate the present-day climate state of each model. Both axes are expressed in sverdrups (1 Sv $= 10^6$ m$^3$ s$^{-1}$). The bottom panel shows coupled models with 3-D global ocean models; the top panel shows coupled models with simplified ocean models. (Source: Rahmsorf et al. 2005. Reprinted by permission from John Wiley and Sons: *Geophysical Research Letters*, copyright 2005.)

## 4.3    Geochemical, Biogeochemical and Biogeophysical Feedbacks

In addition to the physical feedbacks discussed in Section 4.2, many feedbacks are related to the chemical and biological processes occurring within the climate system, implying, for instance, changes in carbon uptake by the ocean, vegetation on land, aerosols and dust (see, e.g., Arneth et al. 2010; Runyan et al. 2012). The goal of this section is not to be exhaustive but to illustrate some of the dominant interactions.

### 4.3.1  Concentration-Carbon and Climate-Carbon Feedbacks

#### 4.3.1.1  Concentration-Carbon Feedbacks

Any change in the atmospheric $CO_2$ concentration, for instance, in response to a perturbation due to anthropogenic activities, modifies the atmosphere-ocean and

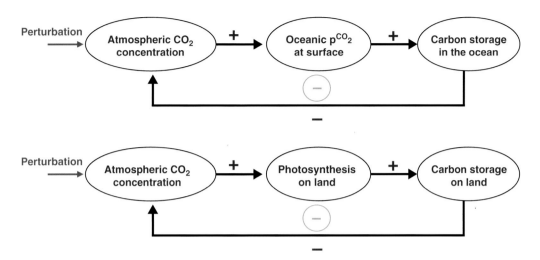

Fig. 4.19 Flow graphs illustrating some important concentration-carbon feedbacks for recent and future changes. A positive (negative) sign on an arrow means that an increase in one variable produces an increase (decrease) in the one pointed to by the arrow. The sign inside a circle indicates the overall sign of the feedback.

atmosphere-land $CO_2$ fluxes. This leads to the so-called concentration-carbon feedback (Figure 4.19).

Firstly, a fraction of the carbon released to the atmosphere is transferred to the ocean because of the higher atmospheric partial pressure of $CO_2$ [see Eq. (2.41)], contributing to an increase in the carbon content of the ocean and a decrease in the carbon content of the atmosphere. In the ocean, the $CO_2$ reacts with water to form $H_2CO_3$ and with carbonate ions ($CO_3^{2-}$) to form bicarbonate ions ($HCO_3^-$), the dominant form of inorganic carbon in the ocean (see Section 2.3.2):

$$H_2CO_3 + CO_3^{2-} \rightleftharpoons 2HCO_3^- \qquad (4.31)$$

This implies that any $CO_2$ uptake by the ocean tends to decrease the availability of carbonates ions and thus reduces the efficiency of reaction (4.31) to form bicar-bonates from $CO_2$. A larger fraction of the dissolved inorganic carbon (DIC) then remains as $H_2CO_3$, increasing the partial pressure of $CO_2$ in the ocean. As a con-sequence, the fraction of $CO_2$ released to the atmosphere and stored in the ocean decreases when total emissions increase. This sensitivity of the oceanic $p^{CO_2}$ to changes in the DIC is often measured by a coefficient referred to as the 'buffer factor' or the 'Revelle factor' (see, e.g., Sarmiento and Gruber 2006). However, we must insist that this process just reduces the rate of transfer from the atmo-sphere to the ocean. The transfer remains positive as long as a new equilibrium is not achieved. The feedback thus is always negative, but its magnitude changes with time.

Over land, the increase in $CO_2$ concentration in the atmosphere generally implies more assimilation and sequestration of carbon by the terrestrial biosphere via photosynthesis [see Eq. (2.49)]. This $CO_2$ **fertilisation** effect is partly due to a direct effect of the atmospheric $CO_2$ concentration on the absorption and use of carbon by plants. It is also due to a better regulation of the plant-atmosphere gas exchanges through **stomata**. With high levels of $CO_2$, smaller gas exchanges are required for the same $CO_2$ uptake, implying less transpiration. This increase in the efficiency of plants to use water then induces a higher production in regions of water stress. In addition to the impact on the carbon cycle, the reduced exchanges modify the latent heat losses of the surface and have a potential impact on the atmospheric humidity and cloud cover. However, many factors limit plant growth, including the availability of nutrients. This reduces the efficiency of the $CO_2$ fertilisation. Consequently, its long-term and large-scale influence has not yet been assessed precisely.

### 4.3.1.2 Climate-Carbon Feedbacks

The biogeochemical effects mentioned in Section 4.3.1.1 occur even in the absence of any climate change induced by modifications in the composition of the atmosphere. In addition, the temperature increase due to higher atmospheric $CO_2$ concentrations reduces the oceanic solubility of $CO_2$ (see Section 2.3.2.1). This is one example of a positive climate-carbon feedback (Figure 4.20). Moreover, increased stratification and a slower oceanic circulation, which in many cases are associated with the warming (see, e.g., Section 6.2.6), reduce the exchanges between the surface layers rich in carbon and the deeper layers. The deep waters tend to have a lower carbon content because they have not yet been in contact with an atmosphere presenting a higher $CO_2$ concentration in response to the imposed perturbation. Consequently, a slower renewal of surface waters induces higher levels of DIC at the surface and thus reduces the oceanic uptake of carbon, providing another positive climate-carbon feedback. Changes in marine ecosystems also may lead to some feedback loops, but they are currently not well understood. Present-day models suggest that their role is weak, but modelling of the marine **biosphere** is still relatively simple in Earth system models (ESMs), and more accurate estimates of those effects are required (see, e.g., Cabré et al. 2015).

Temperature and precipitation changes also affect the carbon cycle on land. Warming tends to accelerate decomposition in soils, which release $CO_2$ to the atmosphere. The primary production is enhanced by warming in cold areas and by an increase in precipitation in dry areas. By contrast, in warm, dry areas where water availability is a limiting factor, a decrease in precipitation produces a reduction in productivity and thus in the uptake of $CO_2$ by plants. Moreover, climate changes influence the distribution of **biomes** (see, e.g., Section 4.3.2) as well as the frequency and extent of wildfires which emit substantial quantities of $CO_2$ (see Section 2.3.3). This illustrates that both positive and negative carbon-climate feedbacks are expected over land in different regions.

### 4.3.1.3 Concentration-Carbon Feedback and the Climate-Carbon Feedback Parameters

Concentration-carbon and climate-carbon feedbacks can be formalised on the basis of the carbon mass balance of the atmosphere-ocean-land system (see, e.g., Gregory et al. 2009). Consider that this system is perturbed by the release to the

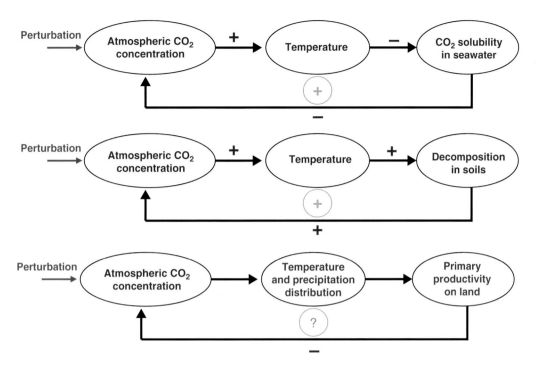

**Fig. 4.20**    Flow graphs illustrating some important climate-carbon feedbacks for recent and future changes. A positive (negative) sign on an arrow means that an increase in one variable produces an increase (decrease) in the one pointed to by the arrow. The sign inside a circle indicates the overall sign of the feedback. The sign of the feedback related to primary productivity is unknown because different signs act in different regions.

atmosphere of an amount of carbon $C_E$. A fraction of this carbon will stay in the atmosphere, changing its content by $\Delta C$, whilst the remaining carbon will be transferred to the ocean and land, inducing a change in their storage given by $\Delta C_O$ and $\Delta C_L$, respectively. This leads to

$$C_E = \Delta C + \Delta C_O + \Delta C_L \qquad (4.32)$$

A standard approximation (see, e.g., Friedlingstein et al. 2006; Gregory et al. 2009) is to assume that the changes in ocean and land storage can be represented as a linear function of the change in atmospheric carbon content $\Delta C$ and of the global mean surface air temperature change $\Delta T_s$; that is,

$$\Delta C_O = \beta_O \Delta C + \gamma_O \Delta T_s$$
$$\Delta C_L = \beta_L \Delta C + \gamma_L \Delta T_s \qquad (4.33)$$

Combining Eqs. (4.32) and (4.33) leads to

$$C_E = \Delta C + \beta_O \Delta C + \gamma_O \Delta T_s + \beta_L \Delta C + \gamma_L \Delta T_s \qquad (4.34)$$

and

$$C_E = \Delta C + \beta_c \Delta C + \gamma_c \Delta T_s \qquad (4.35)$$

where $\beta_c = \beta_O + \beta_L$ and $\gamma_c = \gamma_O + \gamma_L$, referred to here as the 'concentration-carbon feedback parameter' and the 'climate-carbon feedback parameter', respectively. This development has the advantage of representing the concentration-carbon feedback [the second term on the right-hand side of Eq. (4.35)] and the climate-carbon feedback [the third term on the right-hand side of Eq. (4.35)] as a function of two central variables in the system (i.e., the carbon content of the atmosphere and the surface temperature), assuming that the impact of other elements such as precipitation and oceanic circulation changes is directly related to those two variables. For simplicity, the developments are also based on the actual carbon storage in each reservoir, not on the history of the system. As a consequence, $\beta_c$ and $\gamma_c$ depend on the conditions analysed as well as on the type and magnitude of the applied perturbation. They are very useful to analyse a particular situation or to compare the results of different models, but the values estimated in some conditions could not be applied to another case without caution.

Estimates of $\beta_c$ and $\gamma_c$ can be derived from observations by comparing changes in temperature and atmospheric $CO_2$ concentration over specific periods (see, e.g., Frank et al. 2010). It is also possible to obtain the value of $\beta_c$ and $\gamma_c$ from models. To do so, different types of simulations are needed. Firstly, an experiment is performed using a climate model with an interactive carbon cycle including the response to all the feedbacks. In a second experiment, referred to as 'radiatively coupled', changes in atmospheric $CO_2$ concentration affect climate and thus the carbon cycle through climate-carbon feedbacks. The atmospheric concentration is kept fixed for the carbon-cycle model, and its variations thus have no direct influence on the exchange with the land and ocean. Thirdly, in the biogeochemically coupled experiment, the modifications of the atmospheric $CO_2$ concentration have no effect on the radiative forcing, but the carbon-cycle model computes the consequence of its changes on oceanic and land storage. The concentration-carbon feedback plays a dominant role in this biogeochemically coupled experiment, but the climate still responds because a change in atmospheric $CO_2$ concentration modifies, for instance, the latent heat fluxes owing to its impact on gas exchanges through **stomata** (see earlier). Furthermore, the change in $CO_2$ concentration can induce variations in photosynthetic activity and thus in the **leaf-area index** and finally the surface **albedo** (see Section 1.5).

By comparing those three experiments, values of $\beta_c$ and $\gamma_c$ can be derived. They agree well with the ones obtained from observation, but the uncertainty is large on both types of estimates. Nevertheless, it appears robust for the recent past and the twenty-first century that $\beta_c$ is positive, meaning that ocean and land are storing a fraction of the carbon released to the atmosphere, leading to a negative feedback. By contrast, $\gamma_c$ is negative. This implies that the climate change overall reduces the ability of both land and ocean to store carbon. Furthermore, the contribution of the concentration-carbon feedback generally is much larger than that of the climate-carbon feedback [by about a factor of 4 in the experiments analysed by Arora et al. (2013)].

Because of the different units, $\gamma_c$ cannot be compared directly to $\beta_c$ or $\lambda_f$ (the climate feedback parameter; see Section 4.1.3). As a consequence, Gregory et al.

(2009) proposed formally unifying the different approaches by estimating $T_s$ in Eq. (4.35) as a function of $C$. This leads to an expression for $C_E$ that is function of $C$ only and thus parameters which are all dimensionless. This can be done by assuming that $T_s$ is related to the radiative forcing [Eqs. (4.5) and (4.21)]. The radiative forcing is itself a function of the atmospheric $CO_2$ concentration and thus of the atmospheric carbon content [Eq. (4.1)]. This reasoning then can be used to derive the equivalent of Eq. (4.35) in terms of forcing and temperature changes in response to modifications in the carbon cycle. Readers interested in the mathematical details of this should see the paper of Gregory et al. (2009). Thanks to this direct comparison, it is possible to show that carbon-cycle feedbacks have the same magnitude as the direct physical feedbacks discussed in Section 4.2 and that their uncertainty is similar (see also Section 6.2.9).

### 4.3.1.4  Feedbacks Involving Permafrost and Methane

The **permafrost** is the framework of an additional climate-carbon feedback because a surface warming can induce (partial) permafrost thawing. This allows the decomposition by bacteria of some of the organic carbon stored at depth (see Section 2.3.3) and the release to the atmosphere mainly of $CO_2$ but also $CH_4$. Since both methane and $CO_2$ are greenhouse gases, this release contributes to additional warming (Figure 4.21). Because the sub-surface melting relies on the diffusion of heat into the soil, the process is slow (see Section 2.1.5). This is one of the reasons why it is not classically included in the computation of $\gamma_c$, but it may be a significant source of carbon to the atmosphere on a centennial time scale. Furthermore, the decomposition of organic matter by bacteria may provide an additional local source of heat potentially contributing to additional permafrost thawing. In a similar way, methane **hydrates** in oceanic sediments may be destabilised by a warming (see, e.g., O'Connor et al. 2010) because their presence requires low temperature (and high pressure). The subsequent emission of $CH_4$ would, however, be even slower than for permafrost because the warming at the ocean bottom is smaller and delayed compared to the one on the land surface (see, e.g., Sections 6.2.4 and

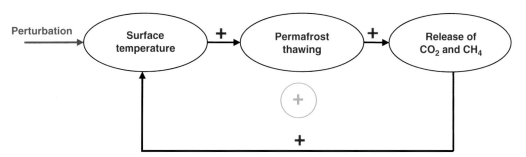

**Fig. 4.21**   Flow graphs illustrating the feedback between temperature change and the release of greenhouse gases as a result of permafrost thawing. A positive (negative) sign on an arrow means that an increase in one variable produces an increase (decrease) in the one pointed to by the arrow. The sign inside a circle indicates the overall sign of the feedback.

6.2.6). Furthermore, a significant part of the methane released below the ocean floor would be oxidised before reaching the atmosphere (see Section 2.3.5).

These processes associated with the permafrost and methane hydrates are not the only ones leading to feedbacks involving the methane cycle. The production of methane in wetlands is also affected by changes in climate and atmospheric $CO_2$ concentration (Ringeval et al. 2011). Firstly, the $CO_2$ fertilisation effect influences the production of organic matter and thus the amount of substrate available for methane production via its decomposition. A $CO_2$ concentration increase thus could lead to more methane release to the atmosphere and therefore additional warming. Furthermore, climate change modifies the distribution of wetlands. However, because it may induce drying in some areas and wetter conditions in other areas, the net effect on methane production is often difficult to estimate.

### 4.3.2 Interactions between Climate and the Terrestrial Biosphere

The terrestrial **biosphere** plays an important role in the global carbon cycle (see Sections 2.3.3 and 4.3.1). Changes in the geographical distribution of the different **biomes**, induced by climate modifications, can modify carbon storage on land and so have an impact on the atmospheric concentration of $CO_2$. In addition to these **biogeochemical feedbacks**, changes in vegetation have a clear impact on the physical characteristics of the surface, in particular, the surface roughness, the albedo and the water exchanges between the ground and atmosphere (see Section 1.5).

Feedbacks involving physical variables influenced by the terrestrial biosphere are referred to as '**biogeophysical feedbacks**'. One important biogeophysical feedback is the **tundra-taiga** feedback which can be observed at high latitudes (Figure 4.22). The **albedo** of a snow-covered forest is much lower than that of snow over grass (Figure 4.23; see also Table 1.3). As a consequence, if, because of warming, trees start to grow in the tundra (transforming the region in a taiga), the surface albedo will tend to decrease, in particular, in spring (see Section 4.2.2). This will lead more warming and thus a positive feedback.

In the tropics, forests are characterised by strong evapotranspiration because of their high **leaf-area index**, which increases the surface of contact with the atmosphere; their roughness, which may enhance the turbulence and exchanges with the air; and their deep root system, which allows the tree to efficiently extract water from the soil. This high evaporation cools the surface and may contribute to higher precipitation rates. Deforestation in the tropics thus can induce warmer and drier conditions because the decrease in evaporation cooling over balances the albedo changes which are dominant at high latitudes. These modifications of local climate then may prevent the regeneration of a forest after a perturbation, leading to an additional positive feedback (see, e.g., Runyan et al. 2012). The processes involved are complex, however, and may depend on the scale of the perturbation because, as a result of the surface warming, the air may rise over deforested areas, leading to saturation at higher levels and eventually additional precipitation (see Section 4.2.4).

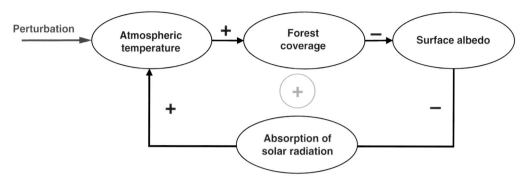

**Fig. 4.22**    Flow graph illustrating the positive tundra-taiga feedback. A positive (negative) sign on an arrow means that an increase in one variable induces an increase (decrease) in the one pointed to by the arrow. The positive sign inside a circle indicates that the overall feedback is positive.

**Fig. 4.23**    The difference in albedo between snow-covered forest and grass. The forest appears darker because it absorbs more of the incoming solar radiation. (Courtesy of Ali Gillet. Reproduced with permission.)

In addition to their direct influence on local climate, vegetation changes also may affect the temperature gradient between the different regions and the large-scale circulation. This will be discussed in the framework of the Sahara greening during the Holocene in Section 5.6.2.

### 4.3.3 Calcium Carbonate Compensation

Nearly all the feedbacks investigated up to this point potentially play a role on decadal to centennial time scales and thus in recent climate changes as well as in those expected over the next decades (see Section 6.2). In addition, other feedbacks are active on longer **time scales**.

As discussed in Section 2.3.4., the burial of $CaCO_3$ in sediments is ultimately compensated for by the input from rivers. Because **weathering** and sedimentation rates appear relatively independent, there is no a priori reason why these two processes should be in perfect balance at any particular time. However, any imbalance between them can lead to large variations in the stock of calcium carbonate (and thus in the alkalinity) [see Eq. (2.48)] in the ocean on millennial to multi-millennial time scales. This would imply significant changes in oceanic $p^{CO_2}$

(see Section 2.3.2) and atmospheric $CO_2$ concentration. Such large shifts have not been observed, at least over the last tens of millions of years, because of a stabilising feedback between the oceanic carbon cycle and the underlying sediment referred to as 'calcium carbonate compensation' (see, e.g., Sarmiento and Gruber 2006).

To understand this feedback, it is necessary to analyse the mechanisms controlling the dissolution of $CaCO_3$. Firstly, let's define the **solubility** $K^{CaCO_3}$ [similar to the solubility of $CO_2$ in Eq. (2.46)] from the equilibrium relationship for the dissolution of $CaCO_3$ [Eq. (2.50)]; that is,

$$K^{CaCO_3} = \left[CO_3^{2-}\right]_{sat} \left[Ca^{2+}\right]_{sat} \tag{4.36}$$

where $\left[CO_3^{2-}\right]_{sat}$ and $\left[Ca^{2+}\right]_{sat}$ are the concentrations when the equilibrium between $CaCO_3$ and the dissolved ions is achieved, that is, at saturation. If at one location in the ocean the product $\left[CO_3^{2-}\right]\left[Ca^{2+}\right]$ is higher than $K^{CaCO_3}$, the water is said to be super-saturated with respect to $CaCO_3$. If the product is smaller than $K^{CaCO_3}$, the water is under-saturated. Since the variations in $Ca^{2+}$ concentration are much smaller than the variations in $CO_3^{2-}$ concentration, saturation is mainly influenced by $\left[CO_3^{2-}\right]$.

Observations indicate that the $CO_3^{2-}$ concentration decreases with depth. The downward transport of inorganic carbon by the soft-tissue pump and the carbonate pump might suggest the opposite. However, recall that the alkalinity is mainly influenced by the concentration of bicarbonate and carbonate ions [see the discussion of Eq. (2.48)]. Alkalinity then can be approximated by

$$\text{Alkalinity} \simeq \left[HCO_3^-\right] + 2\left[CO_3^{2-}\right] \tag{4.37}$$

If we also neglect the influence of carbonic acid on DIC, we can write

$$\text{DIC} \simeq \left[HCO_3^-\right] + \left[CO_3^{2-}\right] \tag{4.38}$$

which leads to

$$\left[CO_3^{2-}\right] \simeq \text{alkalinity} - \text{DIC} \tag{4.39}$$

The dissolution of calcium carbonate releases $CO_3^{2-}$ directly. This is consistent with Eq. (4.39) because the dissolution of 1 mol $CaCO_3$ increases alkalinity by 2 and DIC by 1. By contrast, the remineralisation of organic matter mainly affects the DIC, producing a decrease in $\left[CO_3^{2-}\right]$ according to Eq. (4.39). In current oceanic conditions, the influence of the soft-tissue pump appears to dominate, leading to the observed decrease in $CO_3^{2-}$ at depth. As the solubility of $CaCO_3$ increases in the deep ocean, mainly because of its pressure dependence, the upper ocean tends to be super-saturated, whilst the deep ocean is under-saturated. The depth at which those two regions are separated is called the 'saturation horizon'.

Some of the $CaCO_3$ which leaves the surface layer is dissolved in the ocean water column, but a significant part is transferred to sediment. There, a fraction of the $CaCO_3$ is dissolved and re-injected into the ocean, the rest being

(a) Current state

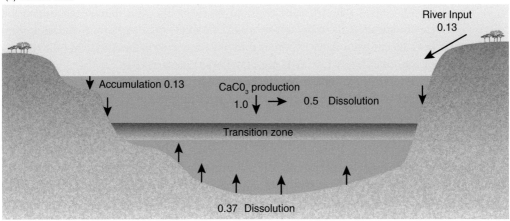

(b) New steady state after a pertrubation

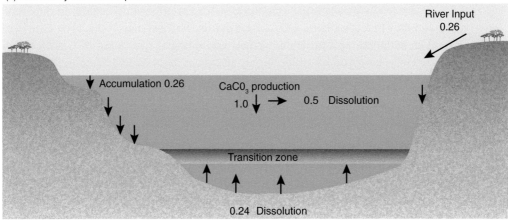

**Fig. 4.24**   (a) The current $CaCO_3$ budget in PgC yr$^{-1}$. 0.13 PgC yr$^{-1}$ comes from the rivers. About 1.0 PgC yr$^{-1}$ of $CaCO_3$ is produced in the ocean, of which 0.5 PgC yr$^{-1}$ is dissolved in the water column and 0.5 PgC yr$^{-1}$ is transferred to sediment. Of the $CaCO_3$ transferred to sediment, 0.37 PgC yr$^{-1}$ is dissolved and goes back to the deep ocean, and 0.13 PgC yr$^{-1}$ accumulates in the sediment, so balancing the input from rivers. (b) If the river input doubles, $\left[CO_3^{2-}\right]$ concentration increases, surface layers are more supersaturated and the saturation horizon deepens. Less $CaCO_3$ dissolves, more accumulates in sediment and a new balance is reached. (Source: Based on Sarmiento and Gruber 2006.)

buried in the sediment on a long time scale. In order to describe these processes, the 'lysocline' can be defined as the depth up to which very little dissolution of $CaCO_3$ occurs in sediments, whilst below the 'calcite compensation
depth' (CCD), nearly all the calcite is lost from the sediment by dissolution.
The region between the lysocline and the CCD is called the 'transition zone'
(Figure 4.24).

   The position of the transition zone depends on several factors, in particular,
the presence of organic material in the sediment. It is significantly influenced

by the saturation of the waters above the sediment: if they are under-saturated, dissolution in the sediment tends to be relatively high, but if the waters are super-saturated (as in the upper ocean), dissolution is very low. If the saturation horizon changes, the position of the transition zone is modified, and the regions of the ocean where $CaCO_3$ is preserved in sediment change.

These shifts in the saturation horizon are responsible for the stabilisation of ocean alkalinity. Imagine, for instance, that the input of $CaCO_3$ from rivers doubles because of more intense weathering on continents (see Figure 4.24). The alkalinity of the ocean and the $\left[CO_3^{2-}\right]$ increase [Eq. (4.39)]. As a consequence, the saturation horizon goes down. The fraction of the ocean floor which is in contact with under-saturated waters will decrease. This would lead to a higher accumulation of $CaCO_3$ in sediments and thus to a larger net flux of $CaCO_3$ from the ocean to the sediment. This deepening in the saturation horizon will continue until a new balance is achieved between the increased input of $CaCO_3$ from rivers and the greater accumulation in sediment.

This feedback limits the amplitude of the variations in alkalinity in the ocean and thus in atmospheric $CO_2$ concentration. For instance, it has been estimated that, in the example presented earlier, a doubling of river input of $CaCO_3$ would lead to a change on the order of only 30 ppm in the atmospheric concentration of $CO_2$ (see Sarmiento and Gruber 2006).

### 4.3.4　Interaction among Plate Tectonics, Climate and the Carbon Cycle

The chemical **weathering** of rocks [see, e.g., Eq. (2.52)] is influenced by a large number of processes. In particular, high temperatures and the availability of water at the surface tend to induce higher weathering rates, at least for some rocks such as the basalts. Since the weathering consumes atmospheric $CO_2$ (see Section 2.3.4), this potentially can lead to a negative feedback which can regulate the long-term variations in atmospheric $CO_2$ concentration (Figure 4.25). For example, more intense tectonic activity might cause the uplift of a large amount of calcium-silicate rocks to the surface, leading to an increase in the weathering rate. The atmospheric $CO_2$ concentration then would decrease, producing a general cooling of the climate system. This would lead to lower evaporation, less precipitation and lower water availability. These climate changes then would reduce the weathering rates, so moderating the initial perturbation (Walker et al. 1981). This is often referred as the 'weathering-$CO_2$ thermostat' (see, e.g., Archer 2010).

In a second hypothesis, we can postulate that the rate of weathering is mainly influenced by mechanical erosion, which increases the exposition of fresh, new rocks to the atmosphere, and these then can be altered more efficiently. If the temperature decreases because of a higher weathering rate, glaciers and ice sheets can cover a larger surface. Because they are very active erosion agents, this would increase the mechanical weathering and so the chemical weathering, amplifying the initial perturbation (Figure 4.25).

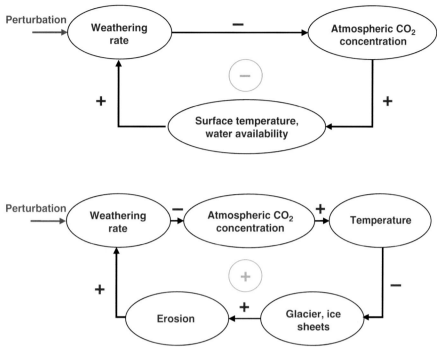

**Fig. 4.25** Flow graph illustrating two mechanisms affecting weathering, $CO_2$ concentration and climate change. A positive (negative) sign on an arrow means that an increase in one variable induces an increase (decrease) in the one pointed to by the arrow. The positive sign in a circle indicates that the overall feedback is positive, a negative sign the opposite. Note that the perturbation also can occur at other locations in the cycle.

These two feedback mechanisms likely have played a role in the variations in $CO_2$ concentration on time scales of hundreds of thousands to millions of years. Their relative contribution is still uncertain, but the stability of the climate since formation of the Earth (see Section 5.4) strongly advocates for a dominant role of the negative feedback relating temperature and weathering rates.

## 4.4 Summary of the Most Important Feedbacks

Sections 4.2 and 4.3 include descriptions of some of the most important feedbacks acting on different time scales and involving all the components of the climate system. They are summarised in Table 4.1. Many others feedbacks are operating, potentially having a large impact locally or for specific variables and periods, but we cannot list all of them here. Determining the ones which are dominant and how they interplay in the problem investigated is a key issue in climate science, as illustrated in the next two chapters for past and future conditions.

| Table 4.1 | List of the feedbacks discussed in Sections 4.2 and 4.3 | | |
|---|---|---|---|
| Feedback | Sign | Time scale | Section where discussed |
| Longwave (blackbody) | – | Hours-days | 4.1.4 |
| Ocean heat uptake | – | Days-century | 4.1.5 |
| Water vapour | + | Days-month | 4.2.1 |
| Lapse rate | – | Hours-days | 4.2.1 |
| Snow-and-ice albedo | + | Days-year | 4.2.2 |
| Sea-ice growth-thickness | – | Months-year | 4.2.2 |
| Ice-sheet elevation | + | Centuries–thousand years | 4.2.2 |
| Cloud tops (longwave, tropics) | + | Days-month | 4.2.3 |
| Soil moisture–temperature | + | Weeks-year | 4.2.4 |
| Soil moisture–precipitation | ? | Weeks-year | 4.2.4 |
| Ocean advection | + | Years-century | 4.2.5 |
| $CO_2$ fertilisation | – | Years-decade | 4.3.1.1 |
| Concentration-carbon (ocean) | – | Years–thousand years | 4.3.1.1 |
| Ocean $CO_2$ solubility–temperature | + | Years–thousand years | 4.3.1.2 |
| Decomposition soils–temperature | + | Years-decade | 4.3.1.2 |
| Fire | + | Years-decade | 4.3.1.2 |
| Methane in permafrost-temperature | + | Decades–thousand years | 4.3.1.4 |
| Tundra-taïga | + | Decades–thousand years | 4.3.2 |
| Carbonate compensation | – | Thousand years | 4.3.3 |
| Weathering $CO_2$ thermostat | – | Million years | 4.3.4 |

*Note:* The sign represents the global effect of the feedback and may result from regional positive and negative contributions. The time scale is only indicative because some feedbacks act on different time scales directly or because of interactions with other feedbacks. Including the blackbody response and thermal inertia in the list or not depends on the interpretation used and is only valid for steadily increasing or decreasing forcing for the latter (see Sections 4.1.4 and 4.1.5).

# Review Exercises

1. When evaluating the radiative forcing (RF) as the net change in the radiative fluxes at the tropopause, the stratospheric temperature is allowed to adjust to the forcing. The difference with the effective radiative forcing (ERF) is that

    a. when evaluating the ERF, the temperature is maintained everywhere at its value before the perturbation is applied (i.e., no stratospheric adjustment).

    b. when evaluating the ERF, the temperature is allowed to adjust in both the stratosphere and the troposphere, but the surface temperature is kept fixed.

    c. when evaluating the ERF, the temperature is allowed to adjust everywhere (stratosphere, troposphere and surface).

2. The radiative forcing due to a change in atmospheric $CO_2$ concentration is

    a. a linear function of the ratio between the $CO_2$ concentration and a reference value.

    b. a logarithmic function of the ratio between the $CO_2$ concentration and a reference value.

3. The ozone hole over Antarctica leads to a large radiative forcing at global scale.

    a. Yes

    b. No

4. Aerosols affect the Earth energy balance in different ways. Among the following terms, select the ones which induce a positive radiative forcing or positive effective radiative forcing.

    a. The scattering of the incoming solar radiation by sulphate aerosols.

    b. The absorption of solar radiation by black carbon.

    c. The deposit of black carbon on snow.

    d. Aerosols acting as condensation nuclei.

5. At equilibrium, what is the change in the net radiative flux at the tropopause in response to a positive (downward) radiative forcing?

    a. The net radiative flux is positive.

    b. The net radiative flux is equal to zero.

    c. The net radiative flux is negative.

6. The equilibrium climate sensitivity is usually defined as the surface temperature change

    a. at equilibrium when a radiative forcing of 1 W m$^{-2}$ is applied at the tropopause.

    b. at equilibrium in response to a doubling of the $CO_2$ concentration in the atmosphere.

7. The blackbody response of the Earth induces a negative feedback. If it were the only feedback active, the equilibrium climate sensitivity would be equal to about

    a. 1°C.

    b. 2.5°C.

    c. 4°C.

8. The time it takes to reach 65% of the equilibrium response to a perturbation is mainly influenced by

    a. the equilibrium climate sensitivity.

    b. the thermal inertia of the system.

    c. the equilibrium climate sensitivity and the thermal inertia of the system.

9. Water-vapour feedback is a very powerful positive feedback. If, in addition to the blackbody response, it were the only feedback operating, the equilibrium climate sensitivity would be about
   a. 1°C.
   b. 2.5°C.
   c. 4°C.

10. If the lapse rate increases as a response to a positive radiative forcing (i.e., if the temperature response is larger at the surface than in the upper troposphere), this induces
    a. a positive feedback.
    b. no feedback.
    c. a negative feedback.

11. The snow-and-ice–albedo feedback is related to a decrease in the surface albedo when snow and ice melt in response to a warming.
    a. True
    b. False

12. Low-level clouds with a high albedo induce
    a. a positive cloud radiative effect.
    b. nearly no cloud radiative effect.
    c. a negative cloud radiative effect.

13. An increase in temperature reduces the soil moisture content as a result of a higher evapotranspiration rate. This induces
    a. a further warming of the surface because of the lower heat capacity of dry soils.
    b. a further warming because of the decrease in evapotranspiration and thus in latent heat fluxes in drier soils.

14. In the case of an initial reduction in the intensity of the thermohaline circulation in the North Atlantic, a positive feedback is due to
    a. a cooling of the surface as the heat transport by the thermohaline circulation is reduced.
    b. a freshening of the surface as the salt transport by the thermohaline circulation is reduced.

15. Concentration-carbon feedbacks are negative because
    a. a higher atmospheric $CO_2$ concentration induces higher $CO_2$ fluxes towards the ocean.
    b. a higher atmospheric $CO_2$ concentration is associated with a fertilisation effect of the land biosphere.

16. Climate changes caused by the injection of $CO_2$ in the atmosphere have an impact on biogeochemical cycles, inducing climate-carbon feedbacks. Although both positive and negative feedbacks occur, the net effect is positive. In the following list, indicate the positive feedbacks.
    a. A warming affects the oceanic solubility of $CO_2$.
    b. A warming is generally associated with an increased oceanic stratification and a slower oceanic circulation.
    c. A warming influences the decomposition in soils.
    d. Climate changes influence primary production.

17. The tundra-taiga feedback is positive because a warming favours the growth of trees
    a. which have a lower albedo than grass.
    b. which experience less evapotranspiration than grass.
18. A larger influx of calcium carbonate by rivers must be compensated for in the ocean in order to reach a new balance and to avoid unrealistic variations in the atmospheric $CO_2$ concentration. This is achieved by
    a. a stronger carbonate pump which transfers more calcium carbonate to the deep ocean and sediments.
    b. a larger biological production which through the soft-tissue pump induces a larger transport of DIC and thus of carbonate ions to the deep ocean.
    c. a deepening of the saturation horizon which modifies the dissolution of calcium carbonate in sediments.
19. The increase in the weathering rate when temperature and precipitation increase leads to
    a. a positive feedback.
    b. a negative feedback.

# Brief History of Climate: Causes and Mechanisms

## OUTLINE

This chapter analyses climate variations on time scales for one year to billion years, investigating the roles of both external forcing and the internal dynamics. Rather than an exhaustive description of the climate over the whole Earth's history, some key periods and mechanisms are selected. The goal is to illustrate the processes studied in the preceding chapters and to investigate how their interplay drove past climate changes.

## 5.1 Introduction

### 5.1.1 Forced and Internal Variability

We are all familiar with the large day-to-day variations in the weather in the absence of any major change in the energy coming from the Sun or any other external **forcing**. Similar fluctuations are observed at all the **time scales** and are a key element of climate dynamics. Hence, it is common to distinguish the forced variability in response to a forcing (see Chapter 4) from the unforced or **internal variability** which has its origin in the interactions between various elements of the system.

For forced variability, a cause of the observed changes can be found outside the climate system. This is not the case for internal variability. We can identify a chain of events with, for instance, a change in the wind stress which influences tropical temperatures and contributes to the development of El Niño conditions. However, we then should determine the cause of the modification in wind stress, with the full sequence rapidly blurred in the complexity of the system. We thus have no access to an *ultimate cause* of an event but only at best to *proximate causes* immediately contributing to the event.

The development of internal variability is sensitive to small perturbations. Consider, for instance, two situations which are very similar, with just a small difference in one variable which may appear originally negligible. This small difference still slightly perturbs the evolution of the system and, because of various mechanisms and amplifications, leads to states that are clearly different after some time in the two situations. This is often referred to the 'chaotic nature' of the climate (Lorenz 1963), implying that the uncertainty present at one moment, in particular, because of imprecision of observations, grows stronger with time.

This has several fundamental implications. Firstly, it limits considerably the quality of the predictions of the state of the system at a specific date in the future. If a general circulation model (GSM), for instance, is initialised from observations, it will remain close to the observed state only for a short period because it will start to diverge from that state quickly owing to this error growth (see also Sections 3.6.1 and 6.2.2). The statistics of the model and observations can remain similar because they are largely determined by the boundary conditions, but the timing of the events will not be the same. For instance, the model can simulate the spatial and temporal characteristics of the North Atlantic oscillation (NAO) but cannot predict the occurrence of its positive phase during a particular winter several years in advance. Secondly, two GCM simulations initialised from slightly different conditions will display different realisations of the internal variability of the system and thus different time developments (see Figure 5.38). Thirdly, an exact correspondence between climate and observations for the timing of occurrence of unforced individual events is not possible in the long term, except if the model is constrained to follow the observations, as in **reanalysis** (see Section 3.6.2). Compared to GCMs, the situation is different in simpler models, such as energy-balance models (EBMs) and some Earth models of intermediate complexity (EMICs), which generally display no significant internal variability at inter-annual time scales because of the simplification imposed in their representation of the atmospheric dynamics. Consequently, the growth rate of a perturbation is much smaller at inter-annual time scales in those models, but they can include some modes of variability, for instance, associated with ocean circulation or ice-sheet dynamics, leading to a chaotic nature which manifests over longer periods.

The magnitude of internal variability is a strong function of the spatial and temporal scale investigated (see, e.g., Hawkins and Sutton 2012; Jones et al. 2013; Knutson et al. 2013). Because observations always contain both forced and internal variability, it is not straightforward to estimate the contribution of each. This will be discussed in more detail in the various sections of this chapter. Nevertheless, some general characteristics can be investigated here by analysing the control experiments performed with climate models in which the forcing is maintained constant in conditions corresponding to pre-industrial times. In these simulations, all the variability is due to internal dynamics by experimental design (Figure 5.1). The amplitude of inter-annual variations generally is larger over land than over oceans and in the higher latitudes than in the tropics, although peaks are found in equatorial regions such as in the East Pacific. The fluctuations are also much larger at smaller scale than at larger scale. For instance, the standard deviation of the annual mean global surface temperature is between 0.06 and 0.15°C in most of the control runs carried out using the Fifth Coupled Model Inter-Comparison Project (CMIP5) models, with a mean slightly over 0.1°C compared to local values well above 1°C in many regions (see also Figure 5.42).

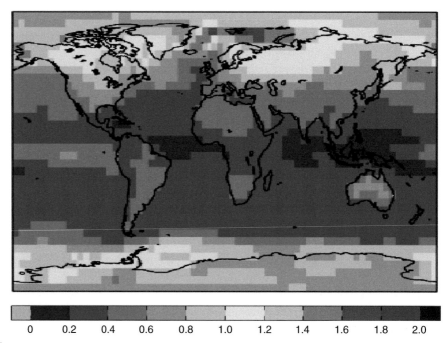

0     0.2     0.4     0.6     0.8     1.0     1.2     1.4     1.6     1.8     2.0

**Fig. 5.1**  Median of the standard deviation of the annual mean surface air temperature (°C) from control simulations performed in the framework of CMIP5. (Source: Courtesy of E. Hawkins; updated from Hawkins and Sutton 2012. Reproduced with permission.)

## 5.1.2  Time Scales of Climate Variations

Since the beginning of Earth's history, climate has varied on all **time scales**. Over millions of years, it has swung between very warm conditions, with annual mean temperatures above 10°C in polar regions, and glacial climates, in which ice sheets covered most of the middle-latitude continents. It has even been postulated that in some past cold periods, the whole surface of the Earth was covered by ice (this is the 'snowball Earth hypothesis'; see Section 5.4.1). At the other end of the spectrum, lower-amplitude fluctuations are observed at inter-annual and decadal time scales, no year being exactly the same as a previous year.

The forcings in part set up the time scale of these variations (Figure 5.2). Because of its own stellar evolution, the energy emitted by the Sun has increased by roughly 30% over the 4.5 billion years of Earth's history. Variations in the **total solar irradiance** (TSI) on shorter time scales have a smaller amplitude, although this amplitude is not precisely known (see Section 5.6.3). Low-frequency changes in the characteristics of the Earth's orbit (see Section 5.5.1) modify the amount of solar energy received in a particular season on every point on the Earth's surface, with the most important fluctuations located in the 10- to 100-ka (1 ka = 1000 years) range. Individual volcanic eruptions produce a general cooling over the years following the eruption (see Section 5.6.3). Furthermore, volcanic activity can be responsible for a low-frequency forcing if large eruptions are grouped in a particular decade or century. On longer time

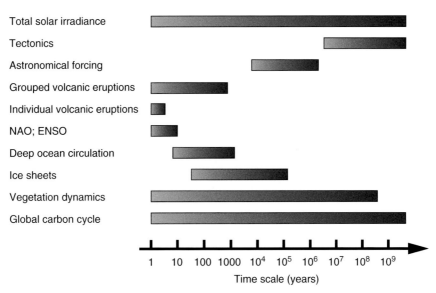

Total solar irradiance
Tectonics
Astronomical forcing
Grouped volcanic eruptions
Individual volcanic eruptions
NAO; ENSO
Deep ocean circulation
Ice sheets
Vegetation dynamics
Global carbon cycle

1    10   100  1000  $10^4$   $10^5$   $10^6$   $10^7$   $10^8$   $10^9$
Time scale (years)

**Fig. 5.2**  Schematic representation of the dominant time scales of selected external forcings and processes related to internal dynamics which affect climate. For each, an indicative range has been plotted here. However, because of mutual interactions, they can exhibit variability on nearly all time scales. For instance, orbital forcing can influence the distribution of temperature and precipitation at the Earth's surface and then induce variations in the oceanic circulation and the El Niño–Southern Oscillation (ENSO) on multi-millennial time scales.

scales, increased volcanic activity related to plate tectonics can perturb the climate over thousands to millions of years.

Internal variability also has some preferred time scales, with, for instance, one El Niño–Southern Oscillation (ENSO) being around three to five years. More generally, internal dynamics shape to a large extent the variability of the Earth's climate. Consequently, the response of the system can involve complex mechanisms which lead to large differences between the characteristics of a forcing and those of the climate changes induced by that forcing. In particular, because of the large inertia of the ocean and the ice sheets, the dominant effect of a perturbation can be related to the integration of the forcing over long time scales, whilst higher-frequency changes are damped. If a forcing excites one mode of internal variability of the system at a particular frequency, leading to a kind of resonance, the magnitude of the response at that frequency will be large, even though the forcing is not particularly intense at that frequency. Small changes in the forcing also can lead to large variations in the climate system if a threshold is crossed and, as a result, the system evolves (nearly) spontaneously from one state to another, possibly quite different, one. Such a transition, involving the deep oceanic circulation, has been proposed to explain some of the abrupt climate changes recorded in Greenland ice cores during the last glacial period (see Section 5.5.4).

The brief overview presented in this section has pointed out a few of the processes which have to be combined to explain past climate changes. Subsequent sections will illustrate some of the most important concepts, starting with the variability due to internal dynamics only. For this, we will focus on inter-annual to

decadal variations because they are the ones for which we have the most information. We will then review past climate changes, starting with very long time scales
and finishing with the last century. The goal is not to extensively describe the climate of all the past periods but rather to focus on some key elements to show how
the mechanisms described in preceding chapters have operated and how models
are able to reproduce them.

## 5.2  Internal Climate Variability

The internal variability displays structures that can be identified by different
methods, including analyses of the correlations between variations observed at
distant locations, referred to as '**teleconnections**'. These structures are described
in terms of modes of variability which are characterised by a spatial pattern and
a climate index. The definition of the index is often given by the **anomalies** of a
specific variable averaged over an area characterised by a large variability or the
difference of anomalies between two regions that are strongly anti-correlated.
These critical regions are often called the 'centres of action of the mode'. The
spatial pattern then can be determined by the correlation between the index of
the mode and the field of temperature or sea-level pressure (SLP). Important
examples are the ENSO, the NAO, the Southern Annular Mode (SAM), the
Atlantic Multi-Decadal Oscillation (AMO) and the Pacific Decadal Oscillation
(PDO). They are discussed in this section, but many others are analysed in the
scientific literature (see, e.g., Hartmann et al. 2013). Note here that the term
'oscillation' is potentially misleading because many modes do not have a dominant period.

These modes allow representation of some of the key characteristics of the
state of the system in a simplified way using a small number of variables. A positive value of ENSO, for instance, is associated with well-known temperature or
precipitation changes. Although very useful, the variations in the index of modes
should be interpreted with caution, however, because any real situation is influenced by many factors and is not necessarily well depicted by those standard features. Two years with a positive index of a particular mode thus could display
slight or more substantial differences. Furthermore, the spatio-temporal characteristics of the modes of variability may vary with time, making the analysis of
any change potentially complex.

### 5.2.1  El Niño–Southern Oscillation

In equatorial regions, the **trade winds** induce a **zonal** transport from the East to
the West Pacific which contributes to maintain the warm sea surface temperatures
(SSTs) and a relatively deep **thermocline** there. The thermocline is shallower in
the East Pacific, and the cold sub-surface waters thus are located closer to the
surface. A central process in the dynamics of these regions is the divergence of

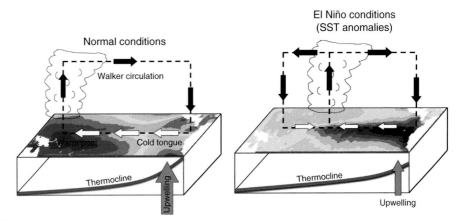

**Fig. 5.3** The Walker circulation in the atmosphere and the position of the thermocline in the ocean under normal conditions and in El Niño years. For normal conditions, red represents warm SST, whilst blue represents cold waters. For El Niño conditions, anomalies of surface temperature are displayed, so red corresponds to higher than normal temperatures and blue colder temperatures. (Source: Christensen et al. 2013.)

upper ocean waters near the equator as a result of the trade winds. The resulting equatorial **upwelling** brings the colder waters located below the thermocline to the surface in the East Pacific and is, to a large extent, responsible for the cold temperatures in this region relative to the western tropical Pacific warm pool area (see Section 1.3 and Figure 5.3).

With SSTs exceeding 28°C over the warm West Pacific, deep atmospheric **convection** and **ascendant** air motion are observed, whereas **subsidence** occurs over the colder East Pacific. This circulation loop is closed by an eastward movement in the upper troposphere and westward atmospheric flow in the lower layers (Figure 5.3). The resulting zonal overturning circulation, referred to as the '**Walker circulation**', thus reinforces the surface easterlies (see Section 1.2).

This situation is associated with a positive **feedback** called the 'Bjerknes feedback'. Imagine an initial eastern equatorial SST perturbation that decreases the long-term mean sea-level pressure **gradient** between the East and West Pacific. This, in turn, weakens the easterly equatorial winds and the Walker circulation. If the situation lasts for a sufficiently long time, the warmer upper ocean western Pacific waters slosh over to the eastern Pacific within two months. The cold-water layer in the eastern Pacific is pushed down, less cold waters get to the surface and the eastern equatorial Pacific warms up. Locally, temperature anomalies can reach up to 5°C. Because of this decrease in the difference in SST between the East and West Pacific, the intensity of the Walker circulation is reduced, leading to an additional reduction in the easterly winds (Figure 5.4).

The warming close to the coast of Peru associated with the reduction of the winds was originally referred to as 'El Niño', but the term is now used to describe all the changes that occur when temperature is anomalously high in the East Pacific Ocean (see Figure 5.3). The opposite condition, corresponding to colder

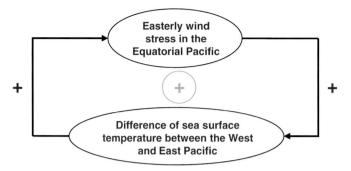

**Fig. 5.4** Flow graph illustrating the Bjerknes feedback. A positive sign on an arrow shows that an increase in one variable produces an increase in the other. The positive sign inside a circle indicates that the overall feedback is positive.

temperatures in the East Pacific, an increased tilt in the ocean thermocline, more upwelling in the East and warm surface waters restricted to the western tropical Pacific is called, by analogy, 'La Niña'.

In summary, through the Bjerknes feedback, the weaker easterly surface wind stress during El Niño conditions reduces the zonal SST gradient, which, in turns, amplifies the initial perturbation. These conditions cannot be maintained for several years, however. In fact, the weaker Walker circulation during El Niño also discharges the heat from the equatorial Pacific. This slow ocean dynamical effect leads to an overall shoaling of the thermocline and the onset of upwelling again in the Eastern Tropical Pacific, thus terminating the El Niño event (see, e.g., Jin 1997; Neelin 2011).

Different indices have been proposed to describe this variability in the tropical Pacific (Figure 5.5). On the atmospheric side, the seesaw in SLP between the East and West Pacific which drives the surface easterlies (Figure 5.6) is measured by the Southern Oscillation index (SOI)

$$\text{SOI} = 10\frac{\text{SLP}'_{\text{Tahiti}} - \text{SLP}'_{\text{Darwin}}}{\sigma_{\Delta\text{SLP}}} \tag{5.1}$$

where $\text{SLP}'_{\text{Tahiti}}$ and $\text{SLP}'_{\text{Darwin}}$ are SLP **anomalies** at Tahiti and Darwin (Australia), respectively, and $\sigma_{\Delta\text{SLP}}$ is the standard deviation of the difference between these two SLP anomalies. The oceanic variations are described by the temperature anomalies averaged over boxes referred to as 'Niño 1' through 'Niño 4', the number increasing westward (see Figure 5.6 for the location of the box Niño 3.4). The strong interactions between oceanic and atmospheric changes and the strong correlation between the indices have led to the choice of ENSO to refer to the whole process.

In addition to their effect on the wind stress, positive SST anomalies related to the positive phase of ENSO induce an eastward migration of the atmospheric **convection** in the Pacific which follows the location of the high SSTs (see Figure 5.3). This causes higher precipitation in the central Pacific and

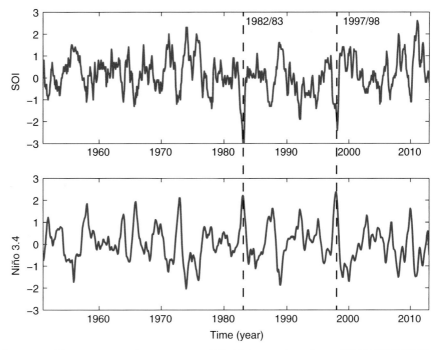

**Fig. 5.5**    Time series of the temperature in the eastern equatorial Pacific (averaged over the area 5°N, 5°S, 170°W, 120°W, the so-called Niño 3.4 index) and the SOI. A filter has been applied to remove fluctuations with periods less than a few months. Negative values of the SOI and positive temperature anomalies correspond to El Niño episodes (e.g., the years 1982–3 and 1997–8), whilst a positive SOI and a negative temperature anomaly in the eastern equatorial Pacific are typical of La Niña periods (e.g., the years 1988–9, 1995–6). (Source: http://www.cpc.ncep .noaa.gov/data/indices/)

dry conditions over Indonesia and northern Australia during El Niño years (Ropelewski and Halpert 1987). ENSO is also associated with perturbations outside the tropical Pacific as it produces anomalies of the atmospheric circulation across nearly the whole world (see Figure 5.6). For instance, El Niño years tend to be much drier and warmer in Mozambique, whilst part of the south-western United States tends to be wetter. Important characteristics of these remote effects are associated with the classical Pacific North American pattern (PNA) which is triggered by the tropical anomalies. The PNA pattern is a dominant mode of variability over the North Pacific and North America characterised by a low SLP over the North Pacific off the coast of America, a high SLP over the high latitudes of North America and a low over the south-eastern United States (Wallace and Gutzler 1981) (see Figure 5.6). The equivalent for the southern hemisphere is the Pacific South American pattern (PSA pattern) (Mo and Paegle 2001). Because of these global **teleconnections**, ENSO is probably the internal mode of variability which has the greatest impact on human activities. As a consequence, forecasting its development a few months in advance is an intense area of research (see Section 6.2.2).

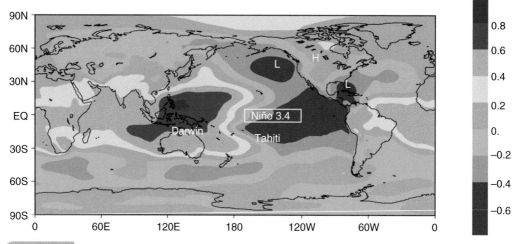

**Fig. 5.6** Correlation between the SST in the Eastern Tropical Pacific (Niño 3.4 index) and sea-level pressure in January. The approximate locations of the Tahiti and Darwin stations, used to define the SOI, of the Niño 3.4 box and of the low (L) and high (H) pressure system associated with the PNA pattern are indicated. (Source: Courtesy of W. Lefebvre using (http://climexp.knmi.nl/); Oldenborgh et al. 2005, data from NCEP-NCAR **reanalysis**. Reproduced with permission.)

## 5.2.2 North Atlantic Oscillation

The large-scale atmospheric circulation in the North Atlantic at middle latitudes is characterised by **westerlies** driven by the SLP difference between the Azores high and the Icelandic low (see Figure 1.6). As described in Section 5.2.1 for circulation in the tropical Pacific, there are irregular changes in the intensity and location of the maxima of these westerlies. These changes are associated with a North-South oscillation of the pressure, and thus of the atmospheric mass (Figure 5.7), known as the 'North Atlantic Oscillation' (NAO) (see, e.g., Wanner et al. 2001; Hurrell et al. 2003). The intensity of this mode of variability is measured by the normalised SLP difference between meteorological stations in the Azores and Iceland:

$$\mathrm{NAO}_{\mathrm{index}} = \frac{\mathrm{SLP}'_{\mathrm{Azores}}}{\sigma_{\mathrm{Azores}}} - \frac{\mathrm{SLP}'_{\mathrm{Iceland}}}{\sigma_{\mathrm{Iceland}}} \tag{5.2}$$

where $\mathrm{SLP}'_{\mathrm{Azores}}$ and $\mathrm{SLP}'_{\mathrm{Iceland}}$ are the SLP **anomalies** in the Azores and Iceland, and $\sigma_{\mathrm{Azores}}$ and $\sigma_{\mathrm{Iceland}}$ are the standard deviations of these anomalies, respectively. Because of the longer time series available, the station in the Azores is sometimes replaced by one in Portugal. When the NAO index is high, the westerlies are stronger than average, whilst they are weaker than the mean when the NAO index is negative.

The NAO can be observed in all months, but its amplitude is greater in winter when the atmosphere is more dynamically active, and its pattern varies between the seasons. When the NAO index is positive in winter, the strong westerly winds

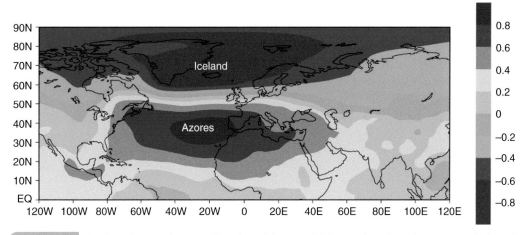

**Fig. 5.7**    Correlation between the winter NAO index and the winter SLP (averaged over December, January and February). The locations of the Azores and Iceland stations, used to define the NAO index, are indicated. (Source: Courtesy of W. Lefebvre using (http://climexp.knmi.nl/); Oldenburg et al. 2005, data from NCEP-NCAR reanalysis. Reproduced with permission.)

transport warm and moist oceanic air towards Europe. This leads to warming and increased precipitation at middle and high latitudes in Europe as well as in large parts of northern Asia, the Greenland Sea and the Barents Sea (Figure 5.8). In the Barents Sea, the warming is associated with a decrease in the extent of sea ice extent.

By contrast, the anomalous circulation when the NAO index is high brings cold air to the Labrador Sea, inducing cooling (see Figure 5.8) and an increase in the extent of sea ice there. Further southward, the stronger flow around the subtropical high leads to a drop in temperature over Turkey and North Africa and a rise in the eastern United States.

An SST tripole is associated with a positive NAO index over the Atlantic Ocean: the temperature anomaly is positive around 30 to 40°N, whilst it is negative north and south of this latitude band (see Figure 5.8). The dominant cause of this pattern appears to be air-sea interactions. Indeed, the SSTs tend to be lower in areas where the wind speed is higher, leading to higher evaporation rates and heat losses from the ocean to the atmosphere.

In contrast to the ENSO, which is a coupled ocean–atmosphere mode, the NAO appears to be mainly an intrinsic mode of variability of the atmosphere. It has been found in many types of atmospheric models, whether or not they are coupled to an oceanic layer. The mechanisms governing the existence of the NAO are related to interactions between the mean and transient circulations (in particular, transient cyclones and anti-cyclones). However, its amplitude is also influenced, for instance, by changes in SST and by the external forcing (see Section 6.2.7).

An equivalent in the Pacific to the NAO is the North Pacific Oscillation (NPO). Although interesting in their own right, the NAO and NPO are sometimes considered to be a regional manifestation of a larger-scale oscillation of the pressure

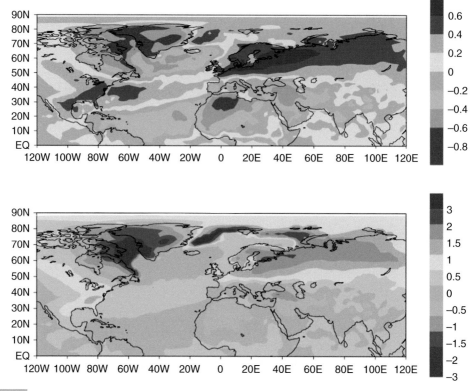

**Fig. 5.8**   Correlation (top) and regression (in °C, bottom) between the winter NAO index and the winter surface air temperature (averaged over December, January and February). (Source: Courtesy of W. Lefebvre using (http://climexp.knmi.nl/); Oldenborgh et al. 2005, data from NCEP-NCAR reanalysis. Reproduced with permission.)

between subtropical areas and high latitudes. Since this nearly hemispherical mode shows a high degree of **zonal** symmetry, it is referred to as the 'Northern Annular Mode' (NAM) but is also sometimes called the 'Arctic Oscillation'. Like the NAO (with which it is highly correlated), the NAM is associated with changes in the position and intensity of the westerlies at middle latitudes.

## 5.2.3  Southern Annular Mode

The equivalent of the NAM in the southern hemisphere is the SAM (Thompson and Wallace 2000). Various definitions of the 'SAM index' have been proposed: a convenient one is the normalised difference in the **zonal** mean SLP between 40°S ($P^*_{40°S}$) and 70°S ($P^*_{70°S}$) (Nan and Li 2003):

$$\text{SAM index} = P^*_{40°S} - P^*_{70°S} \qquad (5.3)$$

As expected, the SLP pattern associated with SAM is a nearly annular pattern with a large low-pressure anomaly centred on the South Pole and a ring of

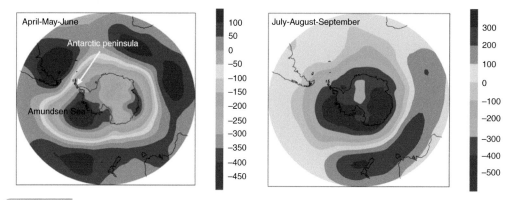

**Fig. 5.9**    Regression between the atmospheric surface pressure and the SAM index for the period 1980–99 (in Pa)
for (left) the averages in April, May and June and (right) July, August and September. (Source: Data from
NCEP-NCAR reanalyses; Kalnay et al. 1996. Figure courtesy of W. Lefebvre. Reproduced with permission.)

high-pressure anomalies at the middle latitudes (Figure 5.9). By geostrophy, this
leads to an important zonal wind anomaly in a broad band around 55°S with
stronger westerlies when the SAM index is high and a southward shift of those
westerlies.

Because of the southward shift of the **storm track**, a high SAM index is associated with higher precipitation at high latitudes and lower precipitation at middle
latitudes. The stronger westerlies above the Southern Ocean also increase the
insulation of Antarctica. As a result, there is less heat exchange between the tropics and the poles, leading to a cooling of Antarctica and the surrounding seas.
However, the Antarctic Peninsula warms because of a western wind anomaly
which brings maritime air onto the peninsula (Figure 5.10). Indeed, the ocean
surrounding the Antarctic Peninsula is in general warmer than the peninsula
itself, and stronger westerly winds mean more heat transport onto the peninsula.
The interaction between the circulation and the topography of the peninsula also
contributes strongly to the warming in some regions. Over the ocean, the stronger westerly winds tend to generate stronger eastward currents. Furthermore, the
divergence of the currents at the ocean surface around 60°S is enhanced because
of a larger wind-induced **Ekman transport**. This results in a stronger oceanic
**upwelling** there.

Most of the effects of the SAM can be explained by its annular form and the
related changes in zonal winds. However, the departures from this annular pattern
have large consequences for sea ice because they are associated with meridional
exchanges and thus large heat transport. In particular, a low-pressure anomaly is generally found in the Amundsen Sea during high SAM index years (see
Figure 5.9). This induces southerly wind anomalies in the Ross Sea (Pacific sector
of the Southern Ocean) and thus lower temperatures and a larger extent of sea
ice there (see Figure 5.10). However, because of the stronger northerly winds, the
area around the Antarctic Peninsula is warmer when the ice index is high, and the
ice concentration is lower there.

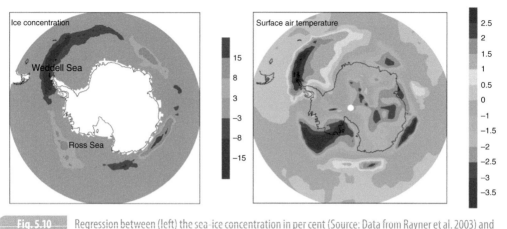

**Fig. 5.10** Regression between (left) the sea-ice concentration in per cent (Source: Data from Rayner et al. 2003) and (right) the surface air temperature in °C (Source: Data from Kalnay et al. 1996) and the SAM index averaged over July, August and September for the period 1980–99. (Source: Courtesy of W. Lefebvre. Reproduced with permission.)

### 5.2.4  Atlantic Multi-Decadal Oscillation and Pacific Decadal Oscillation

The SST in the Atlantic is characterised by pronounced decadal and multi-decadal variations, with a warm phase roughly from 1925 to 1965 and a cold phases from 1905 to 1925 and from 1965 to 1990. This can be interpreted as the combination of a signal associated with global warming (see Section 5.6.3) and that of more regional processes. To focus on the latter, the 'Atlantic Multi-Decadal Oscillation (AMO) index' is generally defined on the basis of the temperature changes over the Atlantic Ocean between 0 and 60°N in which the contribution from larger-scale changes has been removed. This has been done by subtracting the linear trend over the period analysed (Enfield et al. 2001) or the average of SST over nearly the whole globe (region 60°S–60°N) (Trenberth and Shea 2006), the latter method having been selected here.

Because the AMO is a mode characterised by dominant variations at lower frequencies than the ENSO or the NAO, the instrumental data cover only one full positive and two negative phases. The connections between the AMO and other climate phenomena thus are not always clear, but current analyses suggest some robust links in particular with Sahel droughts and hurricane frequency in the North Atlantic. Furthermore, because of the short time series available and the lack of sub-surface observations, the mechanisms responsible for the AMO are still debated. It has been proposed to associate a warm phase of the AMO with a positive anomaly of the **meridional overturning circulation** in the Atlantic which would bring more heat northward. In addition, other interpretations invoke a large contribution of the anthropogenic aerosol forcing over the Atlantic Ocean and natural forcings in the observed variations of the AMO (see, e.g., Delworth and Mann 2000; Booth et al. 2012).

In contrast to the AMO, which displays relatively homogeneous temperature variations over the whole North Atlantic, the PDO is characterised by

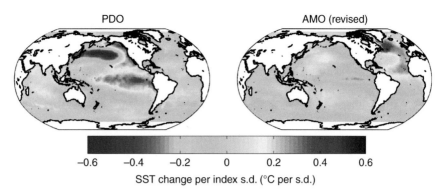

**Fig. 5.11**   Regression between PDO and AMO indices with SST. (Source: Hartmann et al. 2013.)

changes in one sign in the tropical Pacific and off the North American coast and the opposite sign off Japan and in the central Pacific at middle latitudes (see Figure 5.11) (Mantua et al. 1997). The PDO spatial pattern thus has similarities to the one associated with the ENSO. The 'PDO index' is defined as the first **principal component** of the anomaly of SST in the North Pacific between 20 and 70°N in which the global mean temperature has been subtracted for the same reasons as mentioned earlier for the AMO. It was in a negative phase between 1950 and 1970 and mainly in a positive phase from 1920 to 1950 and between 1970 and 2000. The goal of the principal-components analysis is to extract the mode which explains the largest fraction of the variance of a field. This definition thus is based on changes in the whole North Pacific without the need to specify dominant centres of action. Note that similar definitions based on principal components exist for other modes too, leading to characteristics similar to those presented in this chapter.

## 5.3   Reconstructing Past Climates

### 5.3.1  Records of Past Climate Changes

The longest instrumental observations of surface temperature in Europe cover a few centuries. Systematic measurements start in the mid-nineteenth century in many continental regions, but the data coverage is not that good everywhere. Over some oceanic areas, in Africa and Antarctica, data generally are available since the second half of the twentieth century at best (see Figure 5.41). Other surface observations of precipitation or atmospheric pressure, for instance, as well as of the vertical profile of atmospheric and oceanic properties, have been made routinely for many decades. Furthermore, the launch of satellites has strongly increased the quantity of observations performed each day. Nevertheless, not all the variables have been measured at all locations over a sufficiently long period. Consequently, adequate methods are needed to ensure the compatibility between records obtained with different techniques and to fill in the gaps. This can be done using statistical methods or via data assimilation which uses physical constraints to reconstruct as accurately as possible the full state of the system from incomplete observations and imperfect models (see Section 3.6.2).

The modern observations provide the most complete and precise representation of the climate system. They are essential to understand system dynamics, but they cover only a small fraction of the variations observed since Earth's formation, as already discussed in Section 3.5.2.2. To study climates which were much warmer or colder than the present one and, more generally, the variability at centennial and longer **time scales**, we have to rely on the indirect signal recorded in various archives. These are mainly natural archives, but historical documents also provide climate-related information, such as the dates of freezing of lakes or the dates of harvests, descriptions of meteorological conditions, and so on, which can be used to describe the climate of the last centuries and more in some cases. Reconstructing and analysing those changes before the systematic availability of instrumental observations are the focus of paleoclimatology.

The general principle is that some environment variables have an influence on a physical, chemical or biological characteristic of the recording system which can be stored in an archive via a sensor (Figure 5.12). Observations then are made on this archive to derive information on past variations of the climate system. For instance, the air temperature (environment variable) influences the growth of trees in a region (the forest is the sensor; the wood is the archive); growth is subsequently measured through the size of the tree rings or the wood density (paleoclimate observation). In this context, the wood density can be referred to as a proxy for air temperature and, in short, as a '**proxy**' or 'indirect record'. Note that several sensors can carry complementary environment signals in the same archive if they are sensitive to different variables or to the conditions during different seasons.

In some cases, only qualitative information can be obtained because of the nature of the mechanisms involved in the link between the observed quantity and the environmental conditions. For instance, it is possible to state that the formation of some type of rocks requires strong evaporation but without being able to

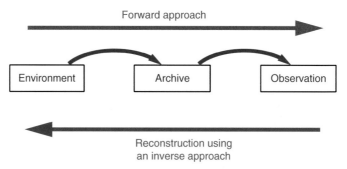

**Fig. 5.12**    Schematic representation of the link between environmental conditions (described here generally by climate variables) and the observations obtained from natural archives. The forward approach directly follows the processes occurring in reality in which climate conditions imprint a signal on the archive, which may be observed subsequently, whilst the aim of the inverse approach is to reconstruct the climate variation which is consistent with the indirect observations. (Source: Based on the framework applied by Evans et al. 2013.)

estimate quantitatively the changes in temperature or precipitation. Observations may imply that sea ice was present during some parts of the year because such conditions are needed for the growth of a particular algae present in the archive but without being able to determine the number of months of sea-ice coverage during the year.

Nevertheless, a quantitative estimate of the reconstructed climate variable, with error bars, is always looked for because it is the best way to test hypotheses on the mechanisms primarily explaining past changes and to evaluate model performance. This can be described mathematically by stating that the observed variable $V_{obs}$ depends on local environmental conditions such as temperature $T$ and precipitation $P$ at the time the archive was formed as well as of non-climatic factors $NC$; thus,

$$V_{obs} = F\left(T, P, NC, ...\right) \qquad (5.4)$$

where $F$, termed the 'response function', can be a relatively simple function of the dependent variables but more generally involves many complex mechanisms occurring in the physical, chemical and/or biological sensor which leaves the signal observed in the archive. Determining $F$ thus requires precise mechanistic understanding of the system and its response to environmental changes.

If we simply follow the approach behind Eq. (5.4), we first need estimates of past climate conditions to obtain $V_{obs}$, knowing $F$ and assuming that the influence of non-climatic factors can be neglected. In practice, this can be achieved only by model-data comparison or data assimilation, Eq. (5.4) providing the equivalent of the observation operator **H** included in Eq. (3.54). In this framework, the simulated temperature and precipitation are used in Eq. (5.4) to obtain an estimate of $V_{obs}$ which can be compared directly with the observation to test the validity of the model results on the basis of the agreement (or not) between the two values (see Sections 3.5.2 and 3.6.2). This methodology, which is known as 'forward modelling', has been applied successfully to different proxies such as tree-ring width and stable **isotopes** in corals (see, e.g., Evans et al. 2013 and references therein). It has many advantages, as the comparison is really made on the same variables in the model world and in observations, taking into account all the steps required to make the link between the climate state and the paleoclimate observation and the associated uncertainties.

However, in many cases, the goal is not to test model performance but to reconstruct past climate conditions from the paleoclimate observations only. It is then necessary to 'solve' Eq. (5.4) to estimate the temperature or another climate variable knowing $V_{obs}$. This corresponds to an inverse method because the goal is to determine from the paleoclimate observation the climatic state which is at the origin of the observed signal.

This inversion problem is generally not well conditioned because a sensor is often influenced by many different climate variables. The same observation then can correspond to several different environment conditions, and determining exactly which one really occurred in the past may become impossible. If we stay with the example involving trees, a low tree growth measured by a small ring width can be due to colder or dryer than normal conditions if both conditions negatively influence the metabolism of the tree in the region investigated. Several options are available to try to disentangle this problem. Observations can be made in regions where it is a reasonable approximation to consider that the quantity measured is a function of one variable only. In cold and wet regions at high latitudes, tree growth is generally much more influenced by temperature than by precipitation: enough water is available, whilst a cold summer will strongly limit growth and a warm summer will favour it. A small ring width thus can be reasonably attributed to low temperatures during the corresponding year. Another method is to select different proxies which have different sensitivities to the variables and combine the information in what is called 'multi-proxy reconstructions'. The tree-ring width may be sensitive to both temperature and humidity, but if another proxy record shows that the conditions were not drier than normal, the small tree-ring width must have been due to cold conditions.

The function which allows reconstructing the climate variable from proxy records is usually referred to as a 'transfer function' or an 'empirical function'. We have noted it in Eq. (5.5) as $G_t^{-1}$ to illustrate the inverse approach, but it is not necessarily mathematically the inverse of $F$ in Eq. (5.4), in particular, as several observed variables may be needed to reconstruct one climatic variable:

$$\hat{T} = G_t^{-1}\left(V_{obs1}, V_{obs2}, ...\right) \tag{5.5}$$

where $\hat{T}$ is the estimate of the real temperature $T$.

The simplest format for the transfer or empirical function is a linear relationship which could be obtained, for instance, via a classical **regression** between proxy records and direct observations of the climatic variable. This step, called the 'calibration', can be done using results obtained in the laboratory under controlled conditions or by comparing the proxy records with instrumental data for a period of common overlap. A validation of the relationship then is required using independent data (data not used in the calibration phase). This also provides estimates of the reconstruction uncertainty based on predictions of observations not entering into the calibration.

Other methods do not try to formally estimate $G_t^{-1}$. Firstly, they can be based on inversion of the forward models mentioned earlier. Alternatively, it is possible to reconstruct past changes by comparing the characteristics recorded in the archive at different times with modern characteristics observed at various locations. This approach assumes that the environmental conditions which prevail at the time of archive formation are similar to those observed presently in the region where the characteristics are the closest, taking into account some potential biases, for instance, due to different atmospheric $CO_2$ concentration. In other words, the goal is to look for the best analogue to past conditions in the modern observations. These similarity methods have been applied successfully in the reconstructions based on the relative abundance of various faunal and floral species in marine sediments and in pollen records (Guiot and de Vernal 2007).

**Table 5.1** Some examples of natural archives, the proxy records which are extracted from them and the variables which can be reconstructed, focussing on the ones which will be discussed later in this chapter

| Archive | Observation (proxy record) | Reconstructed variable |
|---|---|---|
| Wood | Tree-ring width<br>Wood density<br>Stable isotopes | Surface temperature, soil moisture |
| Ice core | Gas content in air bubbles<br>Stable isotopes<br>Cosmogenic isotopes<br>Concentrations of various chemical species (aerosols) | Atmospheric composition<br>Temperature<br>Solar and volcanic forcing |
| Marine sediment core | Stable isotopes<br>Faunal and floral abundance<br>Composition, organic compounds | Ocean temperature, salinity, sea-ice cover<br>Continental ice volume |
| Lake sediment core | Stable isotopes, mineralogy,<br>Faunal and floral abundance<br>Abundance of pollen species | Temperature and humidity<br>Lake level, lake pH<br>Precipitation |
| Coral colony | Stable isotopes | Ocean temperature, salinity |
| Speleothems in caves | Stable isotopes | Precipitation |

*Note*: For a more comprehensive list, see, for instance, the annex of Cronin (2010).

The first step is to analyse the pollen composition in an archive, for instance, a sediment core drilled in a lake. The second step is to determine which region presently displays the pollen assemblage which is the closest to the one observed in the archive. The reconstructed past temperature and/or precipitation then is the modern value in this region because similar climate conditions are assumed to lead to a similar distribution of vegetation and thus of pollen assemblages. This method is powerful because it deals easily with many different variables and is well adapted to non-linear relationships between the observed variables and the reconstructed variables. However, it is not always possible to find good analogues of past conditions in the modern world, which is a limitation of the applications of such methods.

Many records are available from different types of archives (Table 5.1). They can be used to reconstruct the time evolution of forcing agents or of various climate variables. Individual items of paleoclimate data from specific archives may have unique geographical distributions, recoverable time windows and time scales of variation, time resolution and chronological control. For instance, ice cores are only available in polar or mountainous regions where trees generally are not found. In general, high-resolution proxies, such as tree-ring width, lead to reconstruction of inter-annual climate variations but cover only few centuries to millennia because it is hard to obtain older records in archives. Proxies from marine cores generally provide information on longer time scales but with low time resolution, averaging the information over decades to centuries. The various proxies thus are complementary, each bringing a piece to a complex puzzle, and should be used in combination. We will discuss only some of these proxies this chapter, when needed, and present more specifically some isotope records in Section 5.3.3 because they are used repeatedly in subsequent sections. A more general overview can be found in paleoclimatological textbooks such as those of Cronin (2010) and Bradley (2014).

As described in Eq. (5.4), the observed variables are influenced by the local climate but also by many other environmental conditions as well as by processes specific to the archive itself. For instance, the growth rate of a tree depends on its age, on the possible outbreak of pests, on the availability of nutrients in the soil, on competition with its neighbours for light, and so on, as well as on climate variations. The selection of sites and some processing of the observed signal reduce the influence of the non-climatic effects, but with limited information to control for them, it is difficult to know that they are completely removed from the processed data sets. Reconstruction methods also have limitations, as they generally assume, for instance, that the link between the climate and the proxy record remains stable over time, whilst factors not evident within the calibration period may have modified this link in the past. It is thus important to keep in mind the hypotheses underlying the various reconstruction methods. Reconstructions derived from proxy records are always based on an interpretation of the way the paleoclimate observation is related to the environment; this interpretation may be well adapted to some applications but less so to others.

## 5.3.2 Dating Methods

In parallel to estimation of the climatic variables themselves, it is necessary to date the signal precisely in order to be able to determine whether two events recorded at different locations occurred simultaneously or one lead the other, to measure the duration of a transition and to analyse the potential synchronicity with some known forcing variations. This is essential to understanding the mechanisms ruling the observed changes and to study the way the climate system responds to perturbations.

Different dating methods are applied routinely, but many of them rely more or less directly on the sequential formation of the archive, in which new material accumulates on top of older material in the case of sediments and ice cores (forming layers or strata) or in the radial (or outward) direction in trees, speleothems, and corals. Additionally, some techniques provide a local estimate of the time elapsed since formation of the archive, that is, its 'age', such as, for instance, radiometric methods based of the radioactive decay of certain elements stored in the archive. The ages are generally measured in years 'before present' (B.P.), with the 'present' corresponding to 1950 A.D.

### 5.3.2.1 Radiometric Dating

The chemical elements are all characterised by the number of protons they contain, but they can have various numbers of neutrons, forming different **isotopes**. Because they have the same number of protons, the isotopes of one element share the same dominant chemical properties but have a different mass. Most isotopes are stable, but some are unstable, meaning that they decay spontaneously to form another element.

The basis of radiometric methods is the known rates of decay of unstable isotopes into other unstable or stable isotopes. By measuring isotopic abundances, we can then infer time since closure of the isotopic system. Consider the simple case of the decay of a radioactive isotope (or radioisotope) to a stable isotope, and assume that the amount of a radioactive isotope in the archive is known at an initial time $t = 0$. This is in general related to the equilibration within a large reservoir such as the ocean or the atmosphere, whose composition is supposed to be estimated with sufficient precision. After its formation, the archive becomes a closed system. The concentration of the radioisotope then declines (and the concentration of the product of the decay increases) at a rate which is proportional to the number of atoms of the radioisotope. The constant of proportionality is the decay constant. Solving the associated ordinary differential equation gives a negative exponential function

$$N = N_0 e^{-\lambda_R t} \tag{5.6}$$

where $N_0$ is the initial concentration of the radioisotope at time $t = 0$, and $N$ is the concentration at time $t$. $\lambda_R$ is the decay constant of the radioisotope, which is related to its half-life time $T_{1/2}$, that is, the time needed to reduce by a factor of 2 the quantity of radioactive material. Thus,

$$T_{1/2} = \frac{\ln 2}{\lambda_R} \tag{5.7}$$

If we measure $N$ in a sample and compare it to the initial value $N_0$ at the time of system closure, we can obtain from Eq. (5.6) the time since the closure. Finally, this leads to estimates of the time elapsed since incorporation of this sample into the archive and its age.

A commonly used radiometric dating technique is based on carbon-14 ($^{14}$C), an unstable isotope of carbon containing eight neutrons and six protons, whilst the most abundant carbon isotope, carbon-12 ($^{12}$C), which is stable, has six protons and six neutrons. Another stable isotope is carbon-13 ($^{13}$C), which has six protons and seven neutrons (see Section 5.3.3). The $^{14}$C isotopes are formed in the upper atmosphere from $^{14}$N isotopes (seven neutrons and seven protons) as a result of the action of **cosmic rays** and are subsequently transported to the lower atmosphere. This is the reason why $^{14}$C is called a '**cosmogenic** isotope'.

As living plants exchange carbon with the atmosphere via photosynthesis and respiration, their concentration of $^{14}$C is in equilibrium with the $^{14}$C concentration of the atmosphere. This is valid for animals too because they ingest carbon coming from plants whose $^{14}C$ concentration is in equilibrium with that of the atmosphere. These exchanges stop when the plant or animal dies, which thus corresponds to the time of closure $t = 0$ in Eq. (5.6). No $^{14}C$ is further incorporated in the archive, and its concentration starts to decay, providing an efficient way to derive the age of any sample which contains biogenic carbon. The time scale over which this radiocarbon dating can be applied is limited, however, because the half-life of $^{14}C$ is 5,730 years. This means that after 57,300 years (ten half-lives), less than 0.1% of the original material remains in the archive. The concentration thus becomes harder to measure with precision, not to mention the effect of any contamination with modern carbon. The same principles govern other decompositions, which can be applied on longer periods, allowing, for instance, use of the decay of potassium-40 ($^{40}K$) in argon-40 ($^{40}Ar$) (potassium-argon method) and of uranium in thorium, the latter being very useful for dating corals.

Radiometric methods are an essential tool in paleoclimatology, but some of the hypotheses on which the theory is based are not strictly met in most applications. Firstly, the initial composition of the atmospheric or oceanic reservoirs varies with time and is not precisely known. For $^{14}C$, the production rate depends on the quantity of cosmic rays that enter the atmosphere. This quantity varies with the intensity of Earth's magnetic field, which diverts the trajectory of the cosmic rays. Because the intensity of the magnetic field is not constant over time owing to the intrinsic behaviour of Earth and interactions with the solar magnetic field, which is itself variable, the initial atmospheric composition at the formation of an archive can be uncertain. The oceanic $^{14}C$ concentration is also different from the atmospheric concentration as a result of delays in the $CO_2$ equilibrium between the atmosphere and the ocean (see Section 2.3.2) and further mixing in the ocean reservoir. Indeed, the water in the deep ocean has not been in contact with the atmosphere for centuries to millennia (see Section 1.3.2.). Thus, its $^{14}C$ concentration has declined since it originally equilibrated with the atmosphere, making it apparently 'older' in terms of $^{14}C$ if we apply Eq. (5.5) using the atmospheric value for $N_0$. This has two consequences. Firstly, for oceanic records, it is necessary to determine the concentration of the reservoir when the archive was formed. This 'reservoir age' correction is typically several centuries. Secondly, any modification of the oceanic circulation will bring more or less old carbon (i.e., water with a low concentration of $^{14}C$) to the surface with an impact on atmosphere-ocean exchanges and thus on the atmospheric concentration.

In addition, differentiation of isotopic compositions of natural materials occurs by physical and chemical processes and biological activity (see Section 5.3.3), and the hypothesis of a closed system after formation of the archive is not totally valid. Because of all these factors, the age derived from Eq. (5.6) is not the true age of the sample but is referred to as the 'radiocarbon age'. Corrections thus have been developed to report radiocarbon ages as calendar dates, in particular, on the basis of comparison of the radiocarbon age with absolute values derived from other techniques in samples where different dating approaches are available (see, e.g., Reimer et al. 2013).

### 5.3.2.2  Methods Based on Stratigraphy: Annual-Layer Counting

The most direct dating method is probably to count the annual layers since the formation of the archive, if such layers can be identified visually or from the composition of the material deposited. Tree rings provide a classic example because the properties of wood formed in different seasons are different, leading to incremental growth of varying density. In the case where those density variations, associated with cell addition, enlargement and senescence, are closely associated with the active growing season and whose pattern may be replicated across multiple samples within a region, annually resolved tree-ring series may be developed. This is the basis of 'dendrochronology'. For a living tree, the outermost rings correspond to present-day growing seasons, and the age of a layer may be estimated by the number of replicated and validated annual increments between the point of interest and this known age. The method may be extended using dead or fossil wood, comparing the sequences of tree-ring width variations with a particular or characteristic signature (which may be related to a climatic signal; see Section 5.3.1) to identify overlap in time and thereby extend the continuous time series further into the past.

Such annual layering is also observed in ice cores and in some lake sediments (Figure 5.13), where, for instance, winter sediments can be mainly of mineral origin, whilst the summer material contains more biogenic elements because of the higher biological activity during summer. If these layers are not perturbed by organisms or destroyed by oxidation of the biogenic material, they may be used to establish time relative to that of the most recent deposition.

Additional chronological control may be obtained using radiometric dating techniques (see earlier) or identification of a signal event, such as the deposition of a volcanic ash layer, which is well dated by other means and/or within other

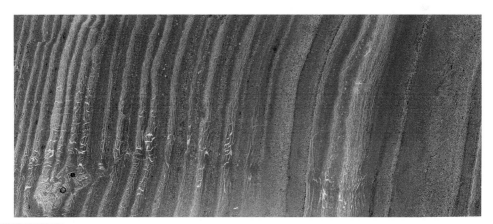

**Fig. 5.13**  A 5-cm-long section from the lake sediment of Cape Bounty, East Lake, Nunavut, Canada. Each annual layer, termed a 'varve', consists in a layer of coarse silts appearing in brown-green with small little bright dots and a layer of clay appearing as horizontal wavy white and dark olive layers. The silt layer is deposited in spring when snow melts, whilst the clay layer, which provides the criteria for counting a year of sedimentation, is deposited in winter when the lake is ice covered. (Source: Courtesy of François Lapointe, INRS-ETE, Quebec, Canada.)

archives. On longer time scales, the identification of reversals of the Earth's magnetic field in sedimentary rocks also provides information on the age of the material because those characteristic magnetic reversals are observed globally, and in some cases, the transitions may be dated via radiometric means. This is an example of the principle of 'stratigraphic correlation', which is an additional chronological constraint permitting the synchronisation of various records.

The age can be interpolated between the points that have been precisely dated in order to have an age model for a material not dated directly. This can be done using relatively simple methods or by applying models representing the processes leading to formation of the archive. Age models may consider, among other effects, variable sedimentation rates as well as the changes in layer thickness with depth because of the weight of the column itself or as a consequence of flow in case of ice cores (see Figure 3.15).

### 5.3.3   An Important Example: Reconstructions Based on Isotopes

Many reconstructions are based on oxygen isotopic composition observations because oxygen is found throughout the climate system. In particular, oxygen is present in water ($H_2O$) and thus in the ice that accumulates in ice sheets and glaciers, as well as in the calcium carbonate ($CaCO_3$) that forms the shells of small marine organisms found in marine sediments. The most abundant oxygen isotope is by far oxygen-16 ($^{16}O$; more than 99%), but $^{17}O$ and $^{18}O$ are also stable, the latter being measured more systematically in paleoclimatic records because of its higher abundance than $^{17}O$. The associated signal is commonly described using the delta value $\delta^{18}O$ (in ‰), which is the ratio of $^{18}O$ and $^{16}O$ in the sample compared to a standard:

$$\delta^{18}O = \left[ \frac{\left(^{18}O/^{16}O\right)_{sample}}{\left(^{18}O/^{16}O\right)_{standard}} - 1 \right] \times 1000 \tag{5.8}$$

where the factor 1000 is included to avoid dealing with small numbers.

When water evaporates, a higher fraction of the lighter $^{16}O$ is transferred from the ocean to the atmosphere compared to the heavier $^{18}O$. This is referred to as an **'isotopic fractionation'** process. For evaporation, it is associated with a decrease in $\delta^{18}O$ of about 10‰ near the equator and a bit more at high latitudes because the fractionation is inversely proportional to temperature. Since sea water at the sea surface has a $\delta^{18}O$ value close to 0, the $\delta^{18}O$ of atmospheric vapour is negative (Figure 5.14). In turn, precipitation is relatively enriched in heavy isotopes compared to the water vapour from which it condenses. As the air mass makes its journey away from its source and loses moisture through precipitation, the remaining water becomes increasingly depleted of the heavy isotope. For this reason, the water which falls in polar regions and accumulates in ice sheets is characterised by a very low $^{18}O$ relative abundance. For present-day conditions, the value of $\delta^{18}O$ in Greenland is around −30‰, and it is around −50‰ in Antarctica. The growth of ice sheets thus is associated with a larger storage of $^{16}O$ than $^{18}O$ and a global

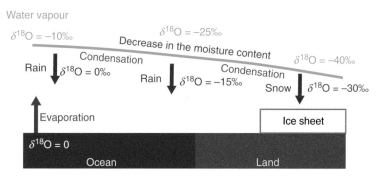

**Fig. 5.14**  Schematic illustration of the changes in isotopic composition of water vapour and precipitation. Depletion of the heavy isotopes in water vapour after initial evaporation is due to condensation as the air cools and to the subsequent removal of liquid water by the rain. The numbers are just indicative. Such a process is referred to as 'Raleigh distillation', but more complex representations of the system are available (see, e.g., Ciais and Jouzel 1994).

decrease in the amount of $^{16}O$ available in the other reservoirs, in particular, the ocean. This can be used to reconstruct past changes in ice-sheet volume, a high $\delta^{18}O$ value in the ocean being associated with a large volume of continental ice (see Section 5.5.2).

More generally, the ratio between $^{18}O$ and $^{16}O$ in precipitation is strongly influenced by the temperature history of the water from which it precipitated. This provides a useful way to estimate temperature changes, sometimes called the 'isotopic thermometer'. It can be illustrated by the very good correlation found presently between the $\delta^{18}O$ of precipitation in various regions and the local temperature. Nevertheless, this correlation is a function of climate conditions and may change with time, complicating interpretation of past records.

The strong link between the $\delta^{18}O$ value of precipitation and temperature is the result of various processes, one being the temperature dependence of the fractionation itself. Furthermore, the condensation occurring in the atmosphere before precipitation is also strongly conditioned by temperature (see Section 1.2). Nevertheless, the $\delta^{18}O$ of precipitation is not just a function of the local temperature or of the temperature at which the condensation occurred. It is influenced by the cycle of precipitation and re-evaporation, the trajectory of the air and thus the distance from the location where the water first evaporated (called the 'source'). The amount of precipitation also modifies the isotopic composition because the first rain would include more heavy isotopes, but if during an event all the water included in the air mass falls, the rain itself will include the heavy isotopes in a later stage. In some regions, a first-order estimate of temperature changes can be deduced from the isotopic ratio using a simple linear relationship, but all these processes (and additional ones not mentioned here) should be taken into account to have a precise quantitative interpretation of past isotopic changes. This can achieved using complex forward models, following the conceptual picture described in Figure 5.12, which are now coupled to an Earth system model and are able to track the transport of several isotopes and the fractionation associated with the various exchanges (see, e.g., Schmidt et al. 2005). This has allowed us to better understand the signals recorded in the archives and to create more precise model-data comparisons, as discussed in Section 5.3.1.

In the ocean, temperature influences the $^{18}O/^{16}O$ isotopic fractionation between sea water and carbonate. This permits estimation of past temperatures from the isotopic composition of various carbonate archives, if the composition of the water from which the carbonate was formed is known. Since the $\delta^{18}O$ in sea water depends of the ice volume, the signal recorded in the carbonate present in ocean cores is a mixture of temperature and ice-volume influences which is not straight-forward to disentangle. On time scales of hundreds of millions of years, the exchanges with the deep Earth also contribute to the variations in $\delta^{18}O$, making interpretation of the recorded fluctuations quite uncertain, as the mechanisms involved are still not well known.

Oxygen is one example among a wide range of elements. Hydrogen isotopes in ice are also widely used to reconstruct past climates, often in conjunction with oxygen (see, e.g., Masson-Delmotte et al. 2008). As for oxygen, $\delta D$ measures the relative abundance of deuterium (D or $^2H$) compared to the dominant form of hydrogen $^1H$.

During photosynthesis, $^{12}C$ is taken preferentially to $^{13}C$ because it is lighter. This implies that organic matter has a lower content of $^{13}C$ than the atmosphere or oceanic waters. The isotopic composition is also commonly measured by the delta value, that is, $\delta^{13}C$:

$$\delta^{13}C = \left[ \frac{\left(^{13}C/^{12}C\right)_{sample}}{\left(^{13}C/^{12}C\right)_{standard}} - 1 \right] \times 1000 \tag{5.9}$$

The past values of $\delta^{13}C$ can be measured directly in the $CO_2$ contained in the bubbles trapped in ice cores. In the ocean, the $\delta^{13}C$ values are recorded in carbon-ate sediments. This provides essential information on the functioning of the car-bon cycle (see Sections 5.4.2, 5.5.3 and 5.7.2).

## 5.4   Climate since the Earth's Formation

### 5.4.1 Precambrian Climate

Some indirect estimates provide information on the Earth's climate during the first billions years of its history (Figure 5.15), but they are scarce. For instance, a trace of glacial sediments during a particular period suggests glaciation, at least on a regional scale. Specific conditions are required for the formation of various rock types, providing additional indications of past climate changes which could complement the analysis of the **isotopic** composition of some deposits. However, the uncertainties are very large. Some qualitative conclusions appear robust, but interpretation of the climate reconstructions should be considered with caution.

| Eon | Era | Period | Epoch | Date (million years BP) |
|---|---|---|---|---|
| Phanerozoic | Cenozoic | Quaternary | Holocene | 0.012–0 |
| | | | Pleistocene | 2.6–0.012 |
| | | Neogene | Pliocene | 5.3–2.6 |
| | | | Miocene | 2.3–5.3 |
| | | Paleogene | Oligocene | 34–23 |
| | | | Eocene | 56–34 |
| | | | Paleocene | 66–56 |
| | Mesozoic | Cretaceous | | 145–66 |
| | | Jurassic | | 201–145 |
| | | Triassic | | 252–201 |
| | Paleozoic | Permian | | 299–252 |
| | | Carboniferous | | 359–299 |
| | | Devonian | | 419–359 |
| | | Silurian | | 443–419 |
| | | Ordovician | | 485–443 |
| | | Cambrian | | 541–485 |
| Precambrian | Proterozoic | | | 2500–541 |
| | Archean | | | 4000–2500 |
| | Hadean | | | 4600–4000 |

**Fig. 5.15**  A simplified geological time scale. Please note that the vertical times scale in the figure is highly non-linear. For a more detailed chart, see the website of the International Commission on Stratigraphy: http://www.stratigraphy.org/.

When the Earth was formed about 4.6 billion years ago, the solar irradiance was about 30% lower than at present. If the conditions at that time (e.g., **albedo**, composition of the atmosphere, distance between the Earth and the Sun, etc.) had been the same as they are now, a simple calculation using the models described in Section 2.1.2 would lead to an average surface temperature which is more than 20°C below today's. During the first hundreds of millions of years of Earth's existence, the continual bombardment by small **planetesimals** and meteorites and heating from the Earth's interior likely warmed the surface. Nevertheless, in such conditions, the Earth would have been frozen during a large part of its history. This contrasts with geological evidence for a liquid ocean at least 3.8 billion years ago and thus relatively mild temperatures. The apparent discrepancy is called the 'faint young Sun paradox' (see, e.g., Feulner 2012).

Resolution of this paradox is thought to be found in a much stronger greenhouse effect. A lower Earth albedo also may have played a role. At that time, the atmosphere was indeed very different from today, with a much higher $CO_2$ concentration (maybe reaching more than 10%, i.e., more than 250 times the present-day value) and nearly no oxygen. In the absence of oxygen, the methane was not quickly oxidised, as in the present atmosphere [see Eq. (2.53)]. The concentration of methane thus also was much higher than at present. This strongly affected the radiative balance of the Earth and allowed the presence of temperatures well above the freezing point of water despite the smaller amount of radiation coming from the Sun.

With time, the atmospheric composition has been modified. In particular, the formation of continents and increase in weathering at the surface (potentially due to the warming associated with the increase in the TSI; see Section 4.3.4) lead to a decrease in atmospheric $CO_2$ concentration. The burial of organic matter also may have contributed to this decrease (see Section 5.4.2). Furthermore, photosynthetic organisms started liberating oxygen about 2.7 billion years ago. This oxygen was first used to oxidise the minerals exposed to the atmosphere. Subsequently, the oxygen accumulated in the atmosphere, leading to a large increase in atmospheric oxygen concentration around 2.3 billion years ago as well as the formation of an **ozone** layer in the stratosphere (Catling and Claire 2005; Lyons et al. 2014). Because of these higher oxygen concentrations, the oxidation of methane became more efficient, and its concentration decreased markedly. Since the rise in the amount of oxygen coincided with a glaciation, it has been suggested that the reduced methane concentration was responsible for the cooling. However, the relative importance of this element is still unknown.

Other glaciations occurred during the Precambrian. Several well-documented glaciations took place between 550 and 750 million years ago. Some of them, in particular, the Marinoan glaciation around 635 million years ago, were apparently so severe that the whole Earth might have been totally covered by ice during some periods. This is the 'snowball Earth hypothesis' (Hoffman et al. 1998), which is, however, debated because some tropical oceanic regions may have remained ice free.

During that period, the atmospheric $CO_2$ concentration is assumed to have been relatively low because of the large **weathering** rate associated with dislocation of the Rodinia **supercontinent** around 800 million years ago (Donnadieu et al. 2004). After some initial cooling, probably during a time when the orbital configuration favoured the accumulation of snow over high-latitude continents (see Section 5.5), the ice-albedo feedback (see Section 4.2.2) may have been strong enough to generate an additional temperature decrease, leading to a progression of ice towards the Equator and eventually covering the whole Earth.

A key problem associated with the snowball Earth hypothesis is to explain the deglaciation. Indeed, it has been argued that such a totally frozen Earth may have been a stable equilibrium state of the climate system in which the Earth would have remained permanently if it had occurred at one time in the past. To solve this problem, it was proposed that during the snowball phase, volcanoes continued to outgas $CO_2$ into the atmosphere. Because the Earth was covered with ice, no weathering of continental rocks would have compensated for the $CO_2$ input, and the atmospheric $CO_2$ concentration would have increased greatly. Furthermore, volcanic ash and dust might have modified the albedo of the ice in this very cold and thus dry environment with very little precipitation. This may have led eventually to a melting of the ice in the tropics and a deglaciation of the Earth, thanks to the ice-albedo feedback. Finally, because the $CO_2$ concentration was still high after this deglaciation, adjustment of the carbon cycle to such perturbations being slow, the snowball Earth may have been followed by very warm conditions.

## 5.4.2 Phanerozoic Climate

As discussed in preceding sections, the carbon cycle is controlled mainly on time scales of millions of years by the interactions between rocks and the surface reservoir, exchanges within the surface reservoir being much faster (i.e., ocean, atmosphere, biosphere; see Section 2.3.4). This long-term carbon cycle, which determines the atmospheric concentration of $CO_2$ ([$CO_2$]), can be represented in a very simplified way by

$$\frac{\partial[CO_2]}{\partial t} = \text{volc}(t) - \left[\text{weath}(t) + \text{org}(t)\right] \tag{5.10}$$

where 'volc' describes the influence of the out-gassing of $CO_2$ associated with **metamorphism** during **subduction**, volcanic eruptions and at mid-oceanic ridges; 'weath' measures the combined role of silicate **weathering** and calcium carbonate sedimentation in the ocean, which removes carbon from the atmosphere and the ocean; and 'org' is associated with long-term burial of organic matter and the potential release of this carbon if it is brought back to the surface and oxidised. Actually, on a million-year time scale, the atmospheric concentration of $CO_2$ is very close to equilibrium with the conditions prevailing at that time, and the sum of the three terms of the right-hand side of Eq. (5.10) is close to 0.

Perturbations of this balance have been responsible for variations in the atmospheric $CO_2$ concentration and the climate for billions of years. Unfortunately, information about the various processes is not precise enough to estimate their magnitude for the whole Precambrian. The situation is a bit better for the Phanerozoic eon (the last 542 million years). Nevertheless, the uncertainties are still large, and it is necessary to use various approaches to improve the robustness of the estimates.

The most uncertain term is probably the out-gassing of $CO_2$ from the solid Earth. It has been argued that when tectonic activity is intense, high production rates of the oceanic crust at the mid-ocean ridges result in more buoyant oceanic plates that push sea water upwards. This results in flooding of the low-lying parts of the continents. Since such high tectonic activity may be related to large subduction rates and more frequent/stronger volcanic eruptions, it has been suggested that reconstructions of sea levels can be used to derive the time evolution of the out-gassing of $CO_2$. However, no clear consensus on the validity of this approximation has been reached yet. Additionally, the total release of $CO_2$ has been estimated for known periods of strong volcanic activity, such as the one that lead to the formation of the Deccan traps in India about 65 million years ago (see Section 5.4.3).

The burial of organic matter can be derived from the **isotopic** composition of the carbon in sea water. As organic matter is depleted in $^{13}C$ compared to the atmosphere or the ocean (see Section 5.3.3), a larger organic transfer to sediments is associated with a decrease in the relative amount of $^{12}C$ and thus to an increase in $\delta^{13}C$ in the atmosphere. Based on such measurements, it has been possible to determine that burial was particularly high during the transition from the Carboniferous to the Permian period, around 300 million years ago, a period characterised by a relatively large production of fossil-fuel source rocks and a relatively low atmospheric $CO_2$ concentration.

The weathering rate depends on many processes. It is more active in wet and hot environments. This implies that its magnitude tends to be higher when the continents are located close to the Equator. Different types of rocks are more or less sensitive to weathering, which thus depends on their relative presence at the surface. The slope of the terrain and thus mountain uplift affect mechanical erosion and then weathering by exposing new material to alteration. Finally, vegetation modifies the characteristics of the soils and the efficiency of weathering. This leads to interactions between the evolution of species and in particular the colonisation of land surfaces and the global carbon cycle.

Based on this information, it is possible to build models of the long-term carbon cycle (see, e.g., Berner 2006; Donnadieu et al. 2006), but they still have large uncertainties, and some of the hypotheses they use are contentious. These models can be very complex because they can also include, in addition to the carbon cycle, the cycles of other elements such as oxygen or sulphur and have to be coupled with estimates of past climate changes. For instance, Figure 5.16 illustrates the influence of **equilibrium climate sensitivity** on the simulated $CO_2$ concentration in one of these models. The equilibrium climate sensitivity affects the stabilizing feedback between the temperature increase due to higher $CO_2$ concentration and the intensity of weathering, which tends to lower the $CO_2$ concentration (see Section 4.3.4). With a low equilibrium climate sensitivity, this feedback is weak. $CO_2$ has only a moderate influence on the climate. As a consequence, huge variations in $CO_2$ are required to produce a modification in temperature and precipitation which is large enough to induce a change in the weathering which can compensate for any perturbation of the carbon cycle. By contrast, with high equilibrium climate sensitivities, the feedback is strong and restrains the amplitude of the changes in $CO_2$. For the model presented in Figure 5.16, the best agreement with reconstructions of the atmospheric $CO_2$ concentration based on various **proxy** records is obtained for a value of climate sensitivity of around 3°C. This is in the middle of the range provided by global climate models for present-day conditions (see Section 4.1.3), suggesting a relative stability of this number over long time scales.

The relatively good match between simulated and reconstructed $CO_2$ concentrations provides some confidence in the proposed interpretation of the dominant factor influencing the long-term carbon cycle. Furthermore, the periods of low $CO_2$ concentration generally correspond well with recorded glaciation, for example, during the end of the Carboniferous and beginning of the Permian, around 300 million years ago, in agreement with the hypothesis of a long-term relationship between $CO_2$ and climate. This also underlines the strong potential role of plate tectonics in past climate changes because it modifies the production of oceanic crust and the out-gassing of $CO_2$ as well as mountain uplift and the climates of continents, which both control the weathering. This influence on the climate of the continents is due to the changes in their latitudinal position, as already briefly mentioned in Section 5.4.1, and in the way continents are located relative to each

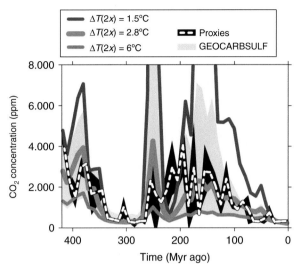

**Fig. 5.16** Comparison of the $CO_2$ concentration calculated by the GEOCARBSULF model for varying climate sensitivities [noted as $\Delta T(2x)$ in the figure] with an independent $CO_2$ record based on different proxies. All curves are displayed in ten-million-year time steps. The proxy error envelope (black) represents one standard deviation of each time step. The GEOCARBSULF error envelope (yellow) is based on a combined sensitivity analysis of different factors used in the model. (Source: Royer et al. 2007. Reprinted by permission from Macmillan Publishers, Ltd.: *Nature*; copyright 2007.)

other. In particular, when all the continents are grouped together into **supercontinents**, such as Pangaea, around 200 million years ago, and Pannotia, 600 million years ago, the interior of the continents tended to be very dry because of the long distance to the sea. This lead to an extension of desert there. Consequently, both the dry conditions and the reduced vegetation cover contributed to a reduction in the weathering (Donnadieu et al. 2006), leading to a complex and rich interplay among tectonics, climate and vegetation.

### 5.4.3 Cenozoic Climate

Over the last 65 million years, the $CO_2$ concentration has gradually decreased from more than 1000 ppm during the Paleocene and the beginning of the Eocene epochs to less than 300 ppm during the Pleistocene. This long-term decrease has been tentatively attributed to a reduction of volcanic emissions, which were probably larger during the Paleocene and Eocene epochs but which have diminished since then, and to the mountain uplift. In particular, the formation of the Himalaya over the Cenozoic induced steep slopes, large erosions and thus an increase in the rate of weathering of silicate rocks. Furthermore, this led to a larger transport of sediment and a more efficient burial of organic matter in the sea, contributing to the lower atmospheric $CO_2$ concentration [see Eq. (5.10)]. Finally, the uplift of the Himalaya and the Tibetan Plateau also strongly modified the monsoon circulation and thus precipitation in these regions.

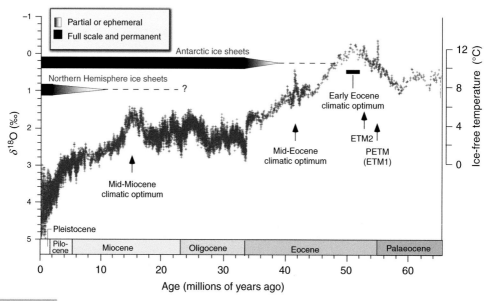

**Fig. 5.17** Evolution of the global climate over the past 65 million years based on deep-sea oxygen isotope measurements in the shells of benthic foraminifera (i.e., foraminifera living at the bottom of the ocean) which record deep-sea temperature. The $\delta^{18}O$ temperature scale, on the right axis, is only valid for an ice-free ocean because oceanic $\delta^{18}O$ is also influenced by the volume of ice sheets. This therefore applies only to the time preceding the onset of large-scale glaciation in Antarctica (about 35 million years ago; see inset in the upper left corner). (Source: Zachos et al. 2008. Reprinted by permission from Macmillan Publishers, Ltd: *Nature*; copyright 2008.)

The decline in $CO_2$ concentration is associated with a general cooling trend that has been deduced from high-resolution $\delta^{18}O$ records (Figure 5.17). This shift from the warm conditions of the early Eocene climatic optimum between 52 and 50 million years ago to the colder conditions of the last million years is often referred to as a 'transition from a greenhouse climate to an icehouse'. The latter is associated with the presence of large permanent ice sheets over Antarctica (starting around 35 million years ago) and over Greenland (starting around 3 million years ago).

In conjunction with the decrease in atmospheric $CO_2$ concentration, the climate was affected by significant changes in oceanic circulation over the Cenozoic (which also may have induced modifications in the carbon cycle). About 50 million years ago, the locations of the continents were quite close to their present-day locations (Figure 5.18), but a relatively large seaway was present between North and South America, whilst Antarctica was still connected to South America. The uplift of Panama and closure of the Central America seaway likely modified the circulation in the Atlantic Ocean, possibly influencing the glaciation over Greenland. More importantly, the opening, deepening and widening of the Drake passage (between South America and Antarctica) and the Tasmanian passage (between Australia and Antarctica) allowed the formation of an intense Antarctic circumpolar current which isolates Antarctica from the influence of the milder middle latitudes. This amplified the cooling there and contributed to the

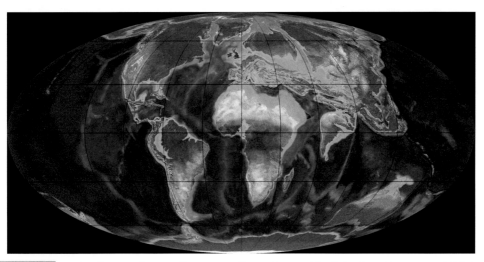

**Fig. 5.18**   Land configuration about 50 million years ago. (Source: Courtesy of Ron Blakey (https://www2.nau.edu/rcb7/mollglobe.html). Copyright 2014 by Ron Blakey, Colorado, Plateau Geosystems, Inc. Reproduced with permission.)

formation of an ice sheet over Antarctica, although this had likely a secondary role compared to the $CO_2$ decline (DeConto and Pollard 2003). These few examples illustrate again the strength of the driving force associated with changes in boundary conditions due to plate tectonics.

In addition to the low-frequency changes just described, relatively brief events are also recorded in the geological archives. One of the most spectacular was the large meteorite impact that occurred 65 million years ago at the boundary between the Cretaceous and Tertiary periods (called the 'K-T boundary'). This cataclysm has been hypothesised to have caused the extinction of many plant and animal species, including the dinosaurs. It certainly had a large climatic impact, inducing a cooling of several degrees over a few years to decades, but the exact magnitude of the cooling is uncertain, and the long-term climatic influence of this event is not clear. During the same period, volcanic activity that lead to the formation of the Deccan traps released to the atmosphere large amount of $CO_2$, likely inducing a warming of a few degrees. After a few hundred thousand years, this was compensated for by the high weathering rate of the newly exposed basaltic rocks resulting from the lava produced by the eruptions. Over a longer term, the initial release was even likely over-compensated for by the weathering, leading to a cooling.

The warming during the Paleocene-Eocene thermal maximum (PETM; 55 million years ago) also had a clear impact on life on Earth (see, e.g., McInerney and Wing 2011). During this event, which lasted about 200,000 years, the global temperature increased by more than 5°C in about 10,000 years (Figure 5.19). Similar but smaller-amplitude events also occurred during the same period (see Figure 5.17). The changes during the PETM are abrupt compared to the geological time scale, which deals with variations taking millions of years, but are slower than the rapid events during the last glacial period (see Section 5.5.4) or the recent warming (see Section 5.7). The PETM is characterised

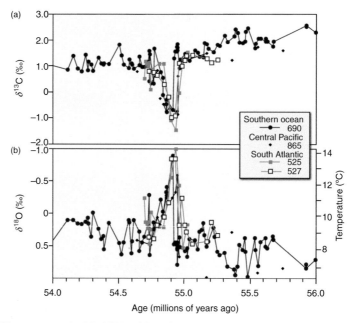

**Fig. 5.19** Marine stable-isotope records of the PETM and the carbon isotope excursion. The carbon isotope (a) and oxygen isotope (b) records are based on benthic foraminiferal records. Panel b also shows inferred temperatures. Ocean drilling site locations are indicated in the keys. (Source: Zachos et al. 2008, adapted from Zachos et al. 2001 and reprinted with permission from AAAS. Reprinted by permission from Macmillan Publishers, Ltd: *Nature*; copyright 2008.)

by a massive injection of several thousands of petagrams of carbon into the atmosphere-ocean system. This has been recorded based on a large decrease in the $\delta^{13}C$ measured in sediments [referred to as the 'carbon isotopic excursion' (CIE)] as well as by a significant acidification of oceanic water leading to carbonate dissolution. The source of these massive inputs of carbon remains uncertain. The most likely candidates are the release of the methane stored in the marine sediments and permafrost (see Section 2.3.5), the methane in marine sediment being the source the most often cited. Indeed, large stocks of methane were available in those reservoirs, and they were characterised by a very negative $\delta^{13}C$ because of their biological origin (see Section 5.3.3). A large fraction of this methane would have been oxidised quickly, increasing the atmospheric $CO_2$ concentration. A second critical issue is the cause of an initial destabilisation of the methane reservoir. This may be due to a temperature increase which triggered the event, possibly associated with a particular orbital configuration which strongly warmed the high latitudes (DeConto et al. 2010). The release of methane then provided a feedback which amplified the initial perturbation (see Section 4.3.1.4) and prolonged the duration of the event because it takes a long time to warm up oceanic sediments and destabilise them (Zeebe 2013). Despite these uncertainties, the PETM provides another clear example of carbon release associated with a warming of the climate.

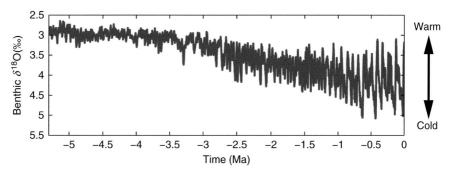

**Fig. 5.20** Benthic $\delta^{18}O$, which measures global ice volume and deep-ocean temperature, over the last 5.3 million years, as given by the average of globally distributed records. (Source: Data from Lisiecki and Raymo 2005.)

Closer to the present, large climate fluctuations have occurred over the last 5 million years. This is not clear on the scale of Figure 5.18, but a higher-resolution plot shows fluctuations with a dominant period of 100,000 years over the last million years and 41,000 years before that (Figure 5.20). These periodicities are very likely related to variations in the **insolation**, as discussed later.

## 5.5 The Last Million Years: Glacial-Interglacial Cycles

### 5.5.1 Variations in Astronomical Parameters and Insolation

At the top of the atmosphere, the **insolation** at a particular time and location on the Earth's surface is a function of the Sun–Earth distance and the cosine of the solar zenith distance [Eq. (2.19)]. These two variables can be computed from the time of day, the latitude, and the characteristics of the Earth's orbit. In climatology, the Earth's motion is determined by three **astronomical parameters** (Figures 5.21 and 5.22): the **obliquity** $\varepsilon_{obl}$, measuring the tilt of the ecliptic compared to the celestial equator (see Figure 2.8); the **eccentricity** $ecc$ of the Earth's orbit around the sun; and the **climatic precession**, which is related to the Earth–Sun distance at the summer solstice. Depending on the convention, the climatic precession is defined as $ecc\sin\tilde{\omega}$, where $\tilde{\omega}$ is the true longitude of the **perihelion** measured from the moving vernal equinox ($\tilde{\omega} = \pi + \text{PERH}$ in Figure 2.8), or as $ecc\sin\omega$, where $\omega$ is the true longitude of the perihelion measured from the moving autumn equinox ($\omega = \text{PERH} = \tilde{\omega} - \pi$ in Figure 2.9). In the following, we will use $\omega$ except in Eq. (5.11) to follow the notation applied in the original study.

Because of the influence of the Sun, the other planets in the solar system and the Moon, the astronomical parameters vary with time. In particular, the Earth is not a perfect sphere (the distance from the surface to the Earth's centre is larger at the equator than at the poles). Consequently, it is submitted to a torque induced by the gravitational forces due to the Sun and the Moon. This torque is largely responsible for the variations in obliquity and plays an important role in the

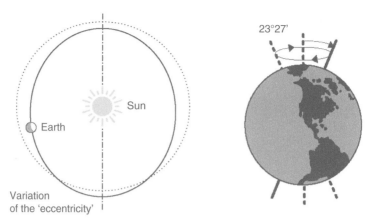

**Fig. 5.21**    Schematic representation of the changes in the eccentricity *ecc* and the obliquity $\varepsilon_{obl}$ of the Earth's orbit. (Source: Berger 2001. Reproduced with permission.)

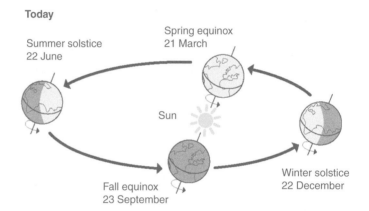

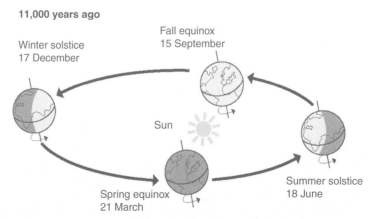

**Fig. 5.22**    Because of the climatic precession, the Earth was closest to the Sun during the boreal summer 11 ka (ka = 1000 years) ago, whilst it is closest to the Sun during the present boreal winter. (Source: Berger 2001. Reproduced with permission.)

changes in the climatic precession. The eccentricity is particularly influenced by the largest planets of the solar system (Jupiter and Saturn), which also have an impact on the climatic precession.

The way these parameters varied over time can be calculated from the equations representing the perturbations of the Earth–Sun system resulting from the presence of other celestial bodies and the fact that the Earth is not a perfect sphere. The solution then can be expressed as the sum of the various terms (Berger 1978):

$$ecc = ecc_0 + \sum_i E_i \cos\left(\lambda_i t + \phi_i\right)$$

$$\varepsilon_{\text{obl}} = \varepsilon_{\text{obl},0} + \sum_i A_i \cos\left(\gamma_i t + \xi_i\right) \qquad (5.11)$$

$$ecc \sin \tilde{\omega} = \sum_i P_i \sin\left(\alpha_i t + \eta_i\right)$$

The values of the independent parameters $ecc_0$ and $e_{\text{obl},0}$; of the amplitudes $E_i$, $A_i$ and $P_i$; of the frequencies $\lambda_i$, $\gamma_i$ and $\alpha_i$; and of the phases $\phi_i$, $\xi_i$ and $\eta_i$ are provided in Berger (1978) and updated in Berger and Loutre (1991). Equations (5.11) can be used to clearly illustrate that the astronomical parameters vary with characteristic periods (Figure 5.23). The dominant periods for the eccentricity are 404, 95, 123 and 100 ka (1 ka = 1000 years). For the climatic precession, the dominant periods are 24, 22 and 19 ka and for the obliquity 41, 40 and 54 ka. To completely determine the Earth's orbit, it is also necessary to specify the length of the major axis of the ellipse. However, taking it as a constant is a very good approximation, at least for the last 250 million years.

The eccentricity of the Earth's orbit has varied over the last million years between nearly zero, corresponding to a nearly circular orbit, to 0.054 (see Figure 5.23). Using Eq. (2.26), it can be shown that the annual mean energy received by the Earth is inversely proportional to $\sqrt{(1 - ecc^2)}$. As expected, this value is independent of the obliquity because of the integration over all latitudes and is independent of $\omega$ because of integration over a whole year. The annual mean energy received by the Earth therefore is at its smallest when the Earth's orbit is circular and increases with eccentricity. However, as the variations in eccentricity are relatively small (see Figure 5.23), there are only minor differences in the annual mean radiation received by the Earth. The maximum relative variation is 0.15% ($1.5 \times 10^{-3} = 1 - 1/\sqrt{1 - 0.054^2}$), corresponding to about 0.5 W m$^{-2}$ ($0.5 = 1.5 \times 10^{-3} \times 340$ W m$^{-2}$).

Over the last million years, the obliquity varied between 22 and 24.5° (see Figure 5.23). This corresponds to maximum changes in daily mean insolation at the poles of up to 50 W m$^{-2}$ (Figure 5.24). Obliquity also has an influence on the local annual mean insolation. With a large obliquity, the insolation is much higher in the polar regions in summer, whilst it is still zero in winter during the polar night, increasing the annual mean insolation by a few watts per square metre. By contrast, the annual mean insolation decreases (but to a lesser extent) at the equator.

Finally, the position of the seasons relative to the perihelion (i.e., the precession) also has an influence on insolation. If Earth is closer to the Sun during the boreal

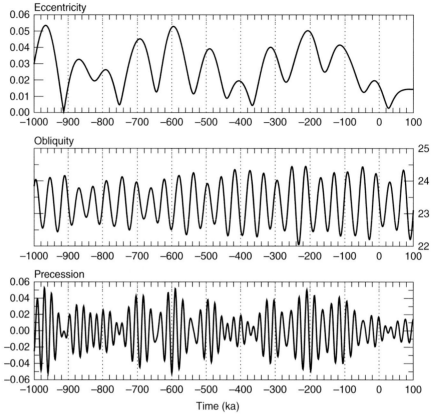

**Fig. 5.23** Long-term variations in eccentricity, climatic precession and obliquity (in degrees) for the last million years and the next 100,000 years (zero corresponds to 1950 A.D.). The maximum value of the climatic precession corresponds to the boreal winter (December) solstice at perihelion. (Sources: Computed from Berger 1978. Figure courtesy of Marie-France Loutre. Reproduced with permission.)

summer and further away during the boreal winter, the summer in the northern hemisphere will be particularly warm, and the winter will tend to be cold. However, if the Earth is closer to the Sun during the boreal winter, the seasonal contrast will be smaller in the northern hemisphere. This effect is particularly marked if the eccentricity is large. If the eccentricity is nearly zero, the distance between the Earth and the Sun is nearly constant, implying no impact of the changes in the position of the seasons relative to the perihelion. The climatic precession varies roughly between −0.05 and 0.05. This produces changes in **insolation** which can be greater than 20 W m$^{-2}$ at all the latitudes (see Figure 5.24). As a consequence, the climatic precession potentially induces larger variations in daily insolation than the variations associated with obliquity at low and middle latitudes.

## 5.5.2 The Astronomical Theory of Paleoclimates

The information recorded in terrestrial sediments and marine (see Figure 5.20) and ice cores (Figure 5.25) documents the alternation between long glacial

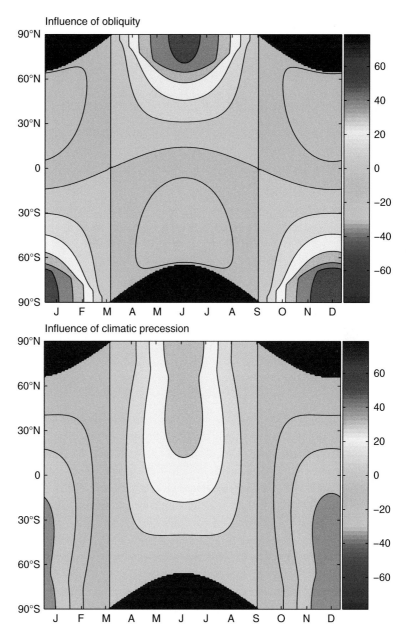

**Fig. 5.24** Changes in insolation (in W m$^{-2}$) caused by (top) an increase in the obliquity from 22.0 to 24.5° with $ecc = 0.016724$, $\omega = \text{PERH} = 102.04$ (i.e., present-day value) and (bottom) by a decrease of the climatic precession from its maximum value (boreal winter at perihelion) to its minimum value (boreal summer at perihelion) with $ecc = 0.016724$ and $\varepsilon_{obl} = 23.446°$ (i.e., present-day value). Contour interval is 10 W m$^{-2}$. The brown areas correspond to a zone with zero insolation. Time of the year is measured in terms of true longitude $\lambda_t$. It is assumed that $\lambda_t = -80°$ corresponds to 1 January and that one month corresponds to 30° in true longitude. (Sources: Data from Berger 1978. Figure courtesy of Marie-France Loutre. Reproduced with permission.)

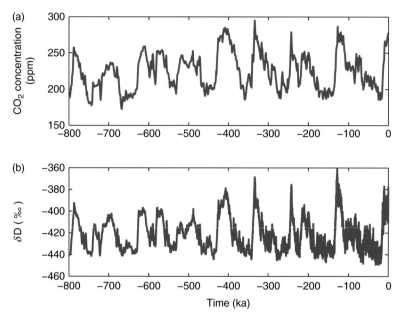

**Fig. 5.25**   (a) Variations in the atmospheric concentrations of $CO_2$ [in ppm, stack of seven different ice cores, as in Masson-Delmotte (2013)] and (b) in deuterium in the Antarctica Dome C (EDC) ice core [$\delta$D in ‰; Jouzel et al. (2007)]. Deuterium is a **proxy** for local temperature (see Figure 5.27).

periods (or ice ages) and relatively brief **interglacials** over the past million years. We are currently living in the latest of the interglacials, the Holocene. All previous interglacials display relatively low ice volumes, by definition, but they have different characteristics associated with different astronomical configurations and greenhouse gas concentrations (Yin and Berger 2012). The interglacials between 450,000 and 800,000 years ago appear in some records to be a bit less warm than the most recent ones, in particular, in Antarctica (see Figure 5.25). The transition between these two regimes around 450,000 years B.P. has been called the 'Mid-Brunhes event'. Whether this event was induced by a shift in some environmental factors or simply was associated with a large variance between the interglacials is still debated. The temperature during the last interglacial (LIG) was likely higher around 125,000 years ago than in pre-industrial times, with a larger temperature increase at high latitudes. The LIG also was characterised by a lower volume of the ice sheets and a sea level which was higher by more than 5 m compared to the Holocene (Masson-Delmotte et al. 2013).

The best known glacial period is the latest one. It peaked around 21 ka B.P. at the so-called last glacial maximum (LGM). At that time, the ice sheets covered most of the high latitudes in Europe and North America, with ice sheets as far south as 40°N. Because of the accumulation of water in the form of ice over the continents, the sea level was lower by around 120 m, exposing new land to the surface. For instance, there was a land bridge between America and Asia across the present-day Bering Strait and another between continental Europe and Britain. The **permafrost** and the **tundra** stretched much further south than at present,

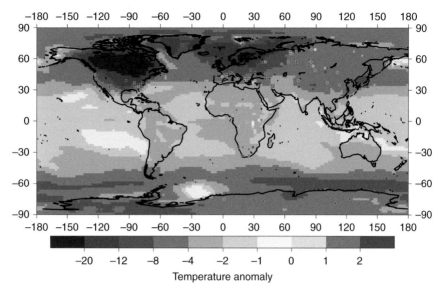

Fig. 5.26 Reconstruction of the difference in surface air temperature (°C) between the last glacial maximum and pre-industrial times based on a combination of model results and proxy data. The local reconstruction based on the proxy time series is given as coloured dots. The land-sea distribution used corresponds to the present day. (Source: Annan and Hargreaves 2013.)

whilst the rain forest was less extensive. Tropical regions were about 2°C cooler than now (Figure 5.26). The cooling was greater at higher latitudes, in particular, over the ice sheets in northern America and northern Europe, as expected. The sea ice also extended much further equator-ward. Overall, the global mean temperature is estimated to have been about 4°C lower than in pre-industrial times, but the uncertainty is on the order of 1°C at least (Masson-Delmotte et al. 2013).

The astronomical theory of paleoclimates assumes that alternations in the glacial and interglacial periods are controlled mainly by the changes in the astronomical parameters with time. The summer insolation at high northern latitudes, which is strongly influenced by obliquity and precession (see Figure 5.24), appears to be of critical importance. A classical argument is based on updates of the initial suggestion of Milankovitch (see Berger 1988); this is the reason why the astronomical theory of paleoclimate is sometimes referred to as the 'Milankovitch theory'. If summer insolation at high northern latitudes is too low, the summer is relatively cold, and only a fraction of the snow that had fallen over land at the higher latitudes during winter melts. As a consequence, snow accumulates from year to year, and after thousands of years, the large ice sheets characterizing the glacial periods build up. Conversely, if summer insolation is high, all the snow on land melts during the relatively warm summer, and no ice sheet can form. Furthermore, because of the summer warming, snow melting over existing ice sheet can exceed winter accumulation, leading to an ice-sheet shrinking and a deglaciation. Via this feedback (and others; see Section 4.2), the effect of relatively small changes in insolation can be amplified, leading to the large variations observed in the glacial-interglacial cycles. This role of the continental ice justifies

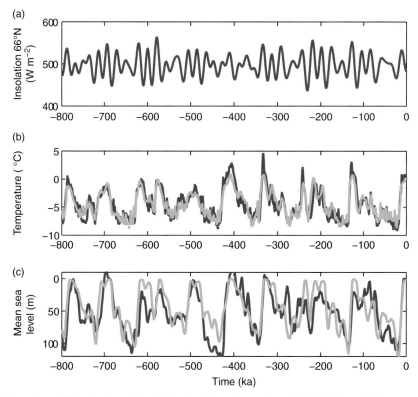

**Fig. 5.27**    (a) Insolation at 66°N on the June solstice (in W m$^{-2}$, red) according to Berger (1978). (b) Anomaly of Antarctic temperature reconstructed from the deuterium record displayed in Figure 5.25 (blue) and in the simulation of Ganopolski and Calov (2011) (green). (c) Sea level reconstructed by Elderfield et al. (2012) (blue) and deduced from the change in continental ice volume simulated in Ganopolski and Calov (2011) (green). Positive values correspond to a larger ice volume and thus a lower sea level than presently.

the classical focus on insolation at high northern latitudes. Indeed, it is the region where most of the land masses which can host an ice sheet are presently located.

One of the most convincing argument in favour of the control of the timing of glacial-interglacial cycles by changes in astronomical parameters comes from the fact that the dominant frequencies of the those parameters are also found in many proxy records covering the Quaternary (Hays et al. 1976) as well as the more distant past. Another important argument comes from paleoclimate modelling. Climate models driven by changes in orbital parameters and by the observed evolution of greenhouse gases over the last 800,000 years reproduced quite well the estimated past ice-volume variations (Figure 5.27). If the changes in astronomical parameters are not taken into account, it is not possible to simulate adequately the pace of glacial-interglacial cycles (see, e.g., Berger and Loutre 2003).

However, the link between climate change and insolation is far from being simple and linear (see, e.g., Crucifix 2012b; Paillard 2015). In particular, the correspondence between summer insolation at high latitudes and ice volume is not clear at first sight (see, e.g., Figure 5.27). It seems that ice sheets can grow when summer insolation is below a particular threshold (see, e.g., the low value around

120–115 ka B.P., when the last glaciation started). However, because of the powerful feedbacks, in particular, related to the altitude and the albedo of the ice sheets (see Section 4.2.2), the insolation has to be much stronger to induce a deglaciation. To have a full picture of the changes, we also must take into account that the insolation changes in different ways at every location and during every season. Analysing the sole summer insolation at high latitudes can provide only a simplified view of the forcing and the potential response of the climate system.

The most intriguing point is the predominance of strong glacial-interglacial cycles with a period of around 100 ka over the last million years, whilst this period is nearly absent from the insolation curves. The eccentricity does exhibit some dominant periods around 100 ka, but this is associated with very small changes in insolation (Berger et al. 2005). Furthermore, until about 1 million years ago, the ice volume mainly varied with a period of 40 ka (see Figure 5.20), and 40 ka is the dominant period of the obliquity. Thus, the presence of this time scale in the records is in accordance with the strong influence of obliquity on insolation at high latitude (see Figure 5.24). It has even been speculated that recent deglaciations actually may occur every second or third obliquity cycle, giving an apparent periodicity between 80 and 120 ka – that is, close to the proposed value of 100 ka (Huybers and Wunsch 2005). The exact value of the dominant period in the records is indeed hard to estimate precisely because of the small number of cycles available.

The 100-ka cycle also may be related to the periodicity of a mode of variability associated with the internal dynamics of ice sheets or of the carbon cycle or both (see, e.g., Ganopolski and Calov 2011; Abe-Ouchi et al. 2013; Paillard 2015). The timing of this mode then would be controlled by the small changes resulting from the eccentricity cycle. In this framework, the shift between the dominant 40- and 100-ka periods about 1 million years ago would be related to the long-term cooling. Crossing some thresholds, such as the maximum volume the ice sheets can reach in a colder world, would have allowed those modes to operate fully, whilst it was not possible in warmer conditions. Alternatively, the long-term cooling may have directly modified the insolation value required at high latitudes to induce ice-sheet formation or melting, and then the frequency (40 or 100 ka) at which orbital configuration leading to such values is occurring (see, e.g., Berger et al. 1999). Nevertheless, explaining in detail and in a convincing way the mechanisms at the origin of the 100-ka period is still a challenge.

### 5.5.3  Glacial-Interglacial Variations in the Atmospheric $CO_2$ Concentration

Greenhouse gas concentrations have varied nearly synchronously with temperature and ice volume over the last 800 ka at least (see Figure 5.25), with the difference between recent interglacial and glacial periods reaching about 90 ppm for carbon dioxide and 350 ppb for methane. This corresponds to a radiative forcing of more than 2 W m$^{-2}$ for $CO_2$ only and thus to a strong contributor to the temperature decrease during glacial periods. Using a climate sensitivity between 2 and 4°C, such a 2 W m$^{-2}$ forcing would lead to a cooling ranging between 1.1 and 2.2°C, that is, a significant fraction of the difference between glacial and interglacial periods given in Section 5.5.2. This means that whilst the carbon cycle was stabilizing the climate on geological time scales through the link between temperature changes and weathering (see Section 5.4.2), it contributed actively to the climate variations over the last million years.

A full quantitative explanation of the causes of these glacial-interglacial variations in greenhouse gases is still lacking. A large part of the changes in methane concentration is likely related to modifications in the size and characteristics of wetland areas during glacial times.

The land biosphere cannot be responsible for the decrease in the $CO_2$ concentration during glacial periods. Because of advance of the ice sheets, the land area available for vegetation growth declined significantly. Furthermore, the lower temperatures induced less evaporation over the oceans and less precipitation over land. The fraction of dry areas and desert, which only store a small amount of carbon compared to forest, thus was larger. Furthermore, the lower $CO_2$ concentration decreased the efficiency of photosynthesis (negative $CO_2$ fertilisation effect; see Section 4.3.1). All these factors lead to a lower carbon storage over land which was not compensated for by the growth of terrestrial vegetation on the new land area made available by the lower sea level. This lower carbon storage on land released carbon characterised by a low $\delta^{13}C$ value because of its origin in the terrestrial biosphere (see Section 5.3.3), which let a negative $\delta^{13}C$ signal in ocean sediments during glacial time. Overall, these changes in the land biosphere during glacial periods would have contributed to an increase in the atmospheric $CO_2$ concentration estimated to be around 25 ppm (Ciais et al. 2013).

The cause of the observed decline in $CO_2$ concentration therefore must lie in the ocean, the geological processes being too slow to account for the observed changes. Because of the accumulation of freshwater with nearly zero **dissolved inorganic carbon** (DIC) and **alkalinity** in the ice sheets, the salinity, DIC and alkalinity of the ocean increases. This leads to an increase in the $p^{CO_2}$ in the ocean. However, this is outweighed by the greater solubility of $CO_2$ in the ocean owing to cooling. The net effect is a small decrease in the atmospheric concentration of $CO_2$ of about 10 ppm, which is still insufficient to explain all the observed decrease between interglacial and glacial periods.

An additional decrease therefore must be related to changes in the ocean circulation and/or the soft-tissue and carbonate pumps. All these factors have a large influence on the distribution of DIC and alkalinity in the ocean and thus on the ocean-atmosphere $CO_2$ exchanges (see Section 2.3.2). Most hypotheses emphasise the role of the Southern Ocean. A strong argument in favour of this is the very similar evolution of atmospheric $CO_2$ concentration and Antarctic temperatures. At present, there is a strong **upwelling** of deep water, rich in nutrients and DIC, in that area. Biological activity is insufficient to fix the excess carbon, and some of the carbon coming from the deep ocean is transferred to the atmosphere. If in glacial periods the connection between surface and deep waters or the biological production changed, this would have had a considerable influence on the concentration of atmospheric $CO_2$.

The upwelling in the Southern Ocean might have been reduced at the LGM because of a northward shift of the westerlies and the divergence associated with the wind-driven **Ekman transport**, but this still needs to be confirmed. Changes in sea ice also may have affected ocean stratification and thus the exchanges between the surface and deep layers. Such an increase in stratification also has been proposed for other regions. Overall, the potential impact of changes in oceanic circulation thus is relatively large but also uncertain. The best estimate on a global scale provided by Ciais et al. (2013) gives a value of a decrease of 25 ppm between glacial and interglacial conditions.

The weaker hydrological cycle during cold periods and the associated increase in the Earth's surface covered by dry areas probably lead to greater dust transport towards the Southern Ocean. This dust brought a large amount of iron to the Southern Ocean. Consequently, biological production might have been higher during glacial times because this micro-nutrient strongly limits primary production for present-day conditions in the area, contributing to the observed atmospheric $CO_2$ concentration decrease. It also has been suggested that the supply of iron to the Southern Ocean by dust has induced a large-scale shift in the ecosystem from phytoplankton producing calcium carbonate towards species which do not form $CaCO_3$. This would have decreased the intensity of the carbonate pump, so inducing a decrease in the atmospheric $CO_2$ concentration. The total effect of the additional iron supply is estimated to be around 10 ppm but with values strongly differing between the various studies (Ciais et al. 2013).

Additional explanations also have been suggested, but it seems that, on its own, no mechanism alone can account for the 90 ppm change. It is likely that the ones just discussed contributed significantly. However, the exact importance of the various explanatory factors and their interactions is still unknown.

### 5.5.4  Millennial-Scale Variability during Glacial Periods

In addition to the low-frequency variations of glacial-interglacial cycles, more rapid changes have been observed during the past million years. The best documented ones are associated with the millennium-scale variability which took place during the last glacial period, but similar variations have been reported for the preceding glacial period too. This millennium-scale variability is particularly clear in Greenland ice cores (Figure 5.28), where it is characterised by a rapid transition from cold conditions, referred to as the 'stadial phase' or 'Greenland stadial', to the warmer inter-stadial phase (see, e.g., Wolff et al. 2010). This was followed by a more gradual temperature decrease and, finally, by a faster cooling back to cold stadial conditions. The full cycle is referred to as the 'Dansgaard-Oeschger cycle' and the warming phases as 'Dansgaard-Oeschger events' (Dansgaard et al. 1993). For an easier reference, the warming events have been numbered, the most recent ones having the smallest numbers (see Figure 5.28).

The warming phases are quite spectacular, with a temperature increase in Greenland of more than 10°C in a few decades at most during some of the events. Several definitions of '**abrupt changes**' have been proposed. They could be described as forced transitions to a new state 'at a rate determined by the climate system itself and faster than the cause' (NRC 2002). A recent definition focusses more on the time scale of the transition than on the time scale of the

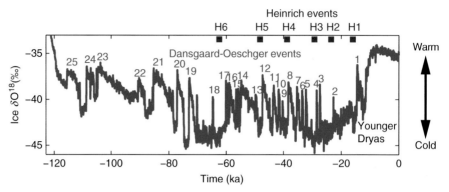

**Fig. 5.28** Time series of δ¹⁸O measurements obtained in the framework of the North Greenland Ice Core Project (NGRIP. (Source: North Greenland Ice Core Project Members 2004.)

cause, arguing that an abrupt change is 'a large-scale change in the climate system that takes place over a few decades or less, persists (or is anticipated to persist) for at least a few decades, and causes substantial disruptions in human and natural systems' (CCSP 2008). This definition is the one used in the fifth assessment report of the Intergovernmental Panel on Climate Change (IPCC) (Collins et al. 2013). Other approaches are centred on the switch between two qualitatively different states which may occur if the system is forced to cross a critical point, sometimes referred to as a 'tipping point'. In this case, the speed of the transition appears less important than the potential existence of this critical point (Lenton et al. 2008). All these definitions share common features, and the warming phase during the Dansgaard-Oeschger events in Greenland can be considered to be an abrupt event, whatever the definition selected.

Another emblematic characteristic of the millennial variability during glacial periods is the presence of thick layers of debris in the sediments of the North Atlantic. Six of them have been reported for the last glacial period (Figure 5.29) and are referred to as 'Heinrich events' (Heinrich 1988; Hemming 2004). Because of the size of the debris in those layers, they could not have been transported by ocean flow. They must have their origin in the material eroded on land by the ice sheets. This material is included in the icebergs when they are calved and is transported over long distances in the ocean before it is released as the icebergs melt. The quantity of this ice-rafted debris (IRD) is so large during Heinrich events that a huge number of icebergs must have been released from the ice sheets, in particular, from the region of the Hudson Strait, leading to sea-level rise of several meters at least (Siddall et al. 2008; Yokoyama and Esat 2011).

The millennial variability is not restricted to Greenland and the North Atlantic. Each Dansgaard-Oeschger cycle has its counterpart in Antarctica. Although the changes are less abrupt than in Greenland, each warm phase of the Dansgaard-Oeschger cycle is associated with a cooling there. The warm phases of the cycle also correspond to modifications in the precipitation regime in the tropical Atlantic which have been interpreted as a northward shift of the **intertropical convergence zone**. Furthermore, among other examples, variations in the strength

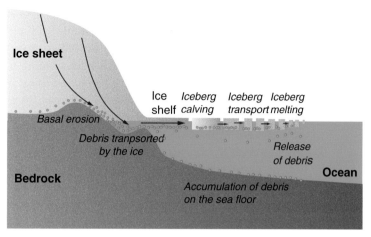

Schematic representation of the massive iceberg release leading to the sediment deposits characteristics of Heinrich events.

of Asian monsoons have been convincingly related to the Dansgaard-Oeschger cycles, illustrating a global pattern of millennial-scale variability.

Despite the multiple evidence collected over the last decades, the mechanisms ruling this millennial variability are not yet fully understood (Clement and Peterson 2008). One of the difficulty is to obtain a precise enough dating of the various records in order to determine the synchronicity between events reconstructed at various locations and the potential leads or lags between them. This is, however, an essential step in determining the processes that potentially could explain the recorded signal.

In contrast to glacial-interglacial cycles, astronomical forcing is not the origin of the changes at millennial time scales, the dominant variability in this forcing being at a lower frequency (see Section 5.5.1). Besides, modifications of the oceanic circulation in the Atlantic could explain many of the observed characteristics of the Dansgaard-Oeschger cycles. A stronger **meridional overturning circulation** in the North Atlantic would lead to larger northward oceanic heat transport (see Section 1.3.2), which could be responsible for the observed warming in Greenland, whilst colder condition there would be associated with a weaker oceanic circulation. These changes at high latitudes likely would have been amplified by various feedbacks, in particular, associated with sea-ice changes. A stronger northward heat transport also would cool the southern hemisphere. This north-south connection, referred to as the 'bipolar seesaw' (Stocker and Johnsen 2003), is perfectly consistent with the good correspondence of the events between Greenland and Antarctic records and provides one of the most convincing arguments in favour of the dominant contribution of the meridional overturning circulation in the observed millennial variability. Furthermore, changes in the magnitude of this circulation are associated with precipitation patterns in the tropics similar to those reconstructed for Dansgaard-Oeschger cycles.

The next question, then, is the cause of the fluctuations in the intensity of the meridional overturning circulation. A standard hypothesis finds their origin in perturbations of the freshwater balance in the North Atlantic. In particular, climate models driven by changes in freshwater input there are able to reproduce some of the key characteristics of the observed changes (see, e.g., Ganopolski and Rahmstorf 2001; Menviel et al. 2014). According to this view, the warm phases of the cycle are associated with periods of lower freshwater inflow from the northern hemisphere ice sheets which increases the salinity in the Atlantic and strengthens the circulation (see Section 4.2.5). This lower freshwater inflow to the ocean corresponds to a positive mass balance of the ice sheets and thus ice-sheet growth. By contrast, a decrease in the volume of the ice sheet, caused by melting or greater iceberg calving, induces a stronger freshwater flux to the North Atlantic which weakens the circulation and leads to colder conditions. In the framework of this hypothesis, the Heinrich events could be considered to be an extreme case where the massive discharge of icebergs leads to a total (or nearly total) collapse of the meridional circulation in the North Atlantic. Finally, the storage of water in ice sheets or the release of icebergs at the origin of the perturbation in the North Atlantic might be related to the internal dynamics of the various ice sheets present in glacial conditions. Additionally, some feed-backs between the ocean circulation and the ice-sheet dynamics also may play a role. A reduced meridional oceanic circulation is associated in models with lower heat losses at the oceanic surface and a sub-surface warming. The temperature of the water in contact with ice shelves then may increase, favouring subsequent melting of the ice (see Sections 3.3.6 and 6.3.3).

Despite this attractive general concept, the amount of freshwater required to induce significant changes in the oceanic circulation during glacial times is still unknown. The response of various models to freshwater forcing is qualitatively similar but displays clear quantitative differences, precluding a robust estimate of the sensitivity of the real system. Furthermore, it has not been possible up to now for many abrupt events to derive from observations a clear, robust link with a freshwater perturbation. Consequently, other hypotheses have been proposed to explain millennial-scale variability during glacial periods. For instance, the varia-tions in the topography as an ice sheet grows or melts have a clear influence on the wind patterns in the northern hemisphere which may affect the meridional over-turning circulation in the Atlantic and lead to abrupt events (see, e.g., Zhang et al. 2014). Additionally, the nature of the teleconnections between the tropics and the higher latitudes shapes a large part of the characteristics of the millennial vari-ability, but the origin of the links and the mechanisms involved deserve additional investigation.

## 5.6  The Last Deglaciation and the Holocene

### 5.6.1  The Last Deglaciation

The last deglaciation started about 19,000 years ago and ended about 11,000 years ago (Figure 5.30). As the glacial period, it was characterised by strong

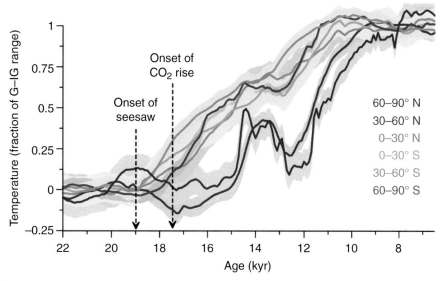

**Fig. 5.30**  Time series of temperatures averaged over different latitude bands reconstructed from a compilation of various proxies. The values are given in relative units corresponding to the ratio between the warming reconstructed at a particular time compared to the difference between the full glacial and interglacial warming in the corresponding latitude band. (Source: Shakun et al. 2012. Reprinted by permission from Macmillan Publishers, Ltd.: *Nature*; copyright 2012.)

millennium-scale variability. In the northern hemisphere, the warm period which corresponds to the Dansgaard-Oeschger event 1 (see Figure 5.28) is referred to as the 'Bølling-Allerød interval'. It peaked around 14,000 years ago and preceded a cold interval, during which the temperature came back to nearly full cold stadial conditions, known as the 'younger Dryas'.

The warming during the deglaciation appears faster in the southern hemisphere than in the northern hemisphere. This may seem to contradict the classical astronomical theory which assumes a melting of the ice sheets triggered by a larger summer insolation in the higher latitudes of the northern hemisphere (see Section 5.5.2). This apparent paradox can be resolved if we make the hypothesis that the initial melting of the ice in the northern hemisphere reduced the surface salinity in the North Atlantic. This decreased the intensity of the Atlantic **meridional over-turning circulation** and thus the heat transport towards the higher latitudes of the northern hemisphere. Consequently, the warming there was delayed compared to other regions. In the southern hemisphere, this reduction in the northward heat transport favoured a temperature increase because of the bipolar seesaw mechanism discussed in the preceding section. Furthermore, the intensity of the meridional overturning circulation was not constant during the whole deglaciation. When its magnitude was higher, the temperature rose faster in the northern hemisphere, as likely during the Bølling-Allerød interval, whilst temperatures at higher southern latitudes increased less quickly or even decreased.

Another key element of the deglaciation is the increase in $CO_2$ concentration which is synchronous with the temperature rise in Antarctica. Indeed, the timing

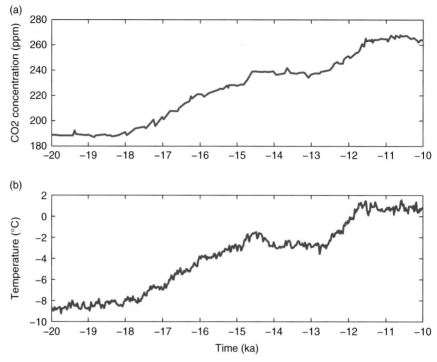

**Fig. 5.31** Times series of (a) $CO_2$ concentration measured in the EDC ice core and (b) Antarctic temperatures estimated from a composite of five Antarctic ice core records during the deglaciation. (Source: Data from Parrenin et al. 2013.)

differences between the developments of these two variables is within the uncertainty of the current dating of ice cores (Figure 5.31). This link between $CO_2$ concentration and Antarctic temperatures advocates again for a strong contribution of the Southern Ocean to the modifications in the carbon exchanges between the ocean and the atmosphere which caused the observed changes in atmospheric $CO_2$ concentration (as discussed in Section 5.5.3).

According to this hypothesis presented, the deglaciation was initiated by an increase in northern high-latitude insolation. Combined with changes in the oceanic circulation, this lead to a temperature increase in the Southern Ocean and to a rise in atmospheric $CO_2$ concentration. In turn, the $CO_2$ provided a global forcing that significantly enhanced warming on a global scale, leading to the end of the glacial period. This scenario appears to be consistent with the temperature reconstruction in various latitude bands, with indirect estimates of the intensity of the deep ocean circulation during the deglaciation and with model results for this period (Shakun et al. 2012). Nevertheless, this still needs to be confirmed in the future by additional observations and modelling work and cannot explain alone the full temporal and spatial complexity of the deglaciation. For instance, the ice sheets may have had some mechanical instabilities during their melting. Local insolation in the higher latitudes of the southern hemisphere likely played a role in the warming there. The melting of the Antarctic ice sheet also had an influence on the oceanic circulation in the Southern Ocean, thus affecting the local temperatures and the carbon cycle, with a likely global impact.

## 5.6.2 The Current Interglacial

Compared to the glacial period and the deglaciation, the climate of the Holocene, whose start has been set at 11,700 years B.P., appears to be relatively stable in Greenland (see Figure 5.28). Some fluctuations have been observed, but their amplitude is much smaller than those seen in glacial periods, likely because of the absence of big ice sheets over North America and Europe. Nevertheless, the vestiges of the ice sheets formed during the previous glacial period still influenced the climate of the early Holocene, in particular, over western Canada, where the final remains of the Laurentide ice sheet disappeared around 6000 years ago. Locally, the presence of an ice sheet induced a cooling because of the higher albedo and higher elevation, as well as changes in atmospheric circulation imposed by the presence of this ice mass. Furthermore, the freshwater flux to the North Atlantic resulting from ice-sheet melting was responsible for a long-term reduction in the **meridional overturning circulation** as well as some abrupt events. A particular example occurred around 8.2 ka B.P. (the so-called 8.2 event) which was likely caused by the rapid discharge of a lake fed by melt-water which was dammed by the ice. When enough ice melted, the dam broke, and the water of the lake was transported to the Atlantic in a few months, leading to perturbation of the oceanic circulation and a cooling of the North Atlantic and surrounding regions (see, e.g., Alley et al. 1997; Barber et al. 1999).

The maximum of summer **insolation** at high latitudes over the Holocene was reached at the beginning of the interglacial, mainly because of the influence of **precession** because the Earth was closer to the Sun in summer at that time (see Figure 5.22). In particular, the summer insolation at the North Pole was up to 50 W m$^{-2}$ higher 10,000 years ago than now (Figure 5.32). As a consequence, the summer temperature in the northern hemisphere was relatively high during the early Holocene. However, because of the remaining influence of the ice sheets, the maximum temperature was delayed by several millennia in some regions compared to the maximum insolation, and the timing of the maximum thus depends strongly on the location (see, e.g., Renssen et al. 2012). Nevertheless, if we ignore the last 150 years, for which different forcings were in action (see Section 5.7), the highest summer temperatures of the Holocene in the middle and high latitudes of the northern hemisphere generally have been found to occur between 9 and 6 ka B.P. Consequently, this period is often referred to as the 'Holocene thermal optimum' or 'Holocene climatic optimum'.

Because of the lower thermal inertia of the land surfaces (see Section 2.1.5.1), the summer warming during the Holocene thermal optimum was larger over the continents than over the ocean. Since the **monsoons** are driven by the temperature contrasts between land and sea, this led to an intensification of the summer monsoons in the northern hemisphere, as illustrated by stronger precipitation over northern Africa, northern India and eastern China (Figure 5.33). This

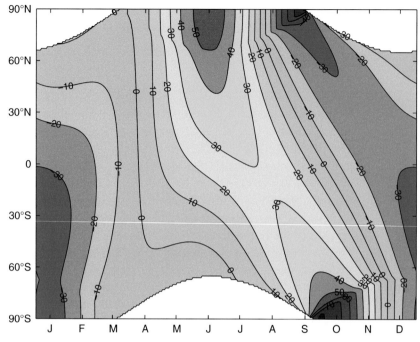

**Fig. 5.32**    Deviations from present-day values at 10 ka B.P. of the 24-h mean solar irradiance (daily insolation) for **calendar months** (in W m$^{-2}$). (Sources: Data from Berger 1978. Figure courtesy of Marie-France Loutre. Reproduced with permission.)

signal is recorded by various proxies and is in good agreement with model results (see, e.g., Kutzbach and Otto-Bliesner 1982). The models generally tend, however, to underestimate the magnitude of the changes reconstructed from data. This is due to the fact that although the processes at the origin of monsoon intensity changes are well understood, several feedbacks are less well quantified. For instance, the dynamics of the ocean mixed layer and ocean transport had an influence on SST, which reinforced the monsoon circulation in some regions (see, e.g., Zhao and Harrison 2012). Furthermore, the increase in precipitation over North Africa led to a Sahara which was largely covered by savannah and lakes during the early Holocene. This period, referred to as the 'African humid period' or simply the 'Green Sahara', clearly contrasts with the dry desert state observed today or at the LGM. The vegetation cover reduces the surface albedo, inducing a higher absorption of solar radiations (see Section 4.3.2). Consequently, it amplifies the local warming and thus potentially the land-sea contrast and the monsoon circulation. The temperature changes associated with this **biogeophysical feedback** was partly compensated for by the cooling produced by the larger evaporation from the surface covered by vegetation, whilst this increased evaporation may have contributed to more local precipitation too (see Section 4.2.4).

The decrease in northern hemisphere **insolation** during the Holocene is associated with a long-term summer cooling, a reduction in the intensity of the monsoon

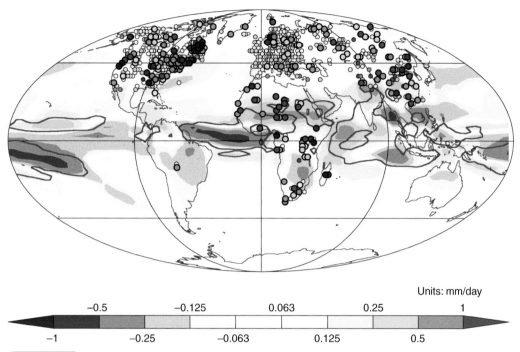

Units: mm/day

| −0.5 | −0.125 | 0.063 | 0.25 | 1 |

| −1 | −0.25 | −0.063 | 0.125 | 0.5 |

**Fig. 5.33**   Difference in precipitation (mm/day) between the middle Holocene (6,000 years B.P.) and pre-industrial times for the ensemble mean of Paleoclimate Modelling Inter-Comparison Project Phase 2 (PMIP2) simulations. The dots represent the local reconstructions given in Bartlein et al. (2011). The red line represents the location where the mean root-mean-square difference between two PMIP simulations reached 0.5 mm/day, providing an idea of model spread. (Source: Reproduced with permission Braconnot et al. 2011.)

and then a desertification of the Sahara region. Although the interpretation of some of the proxies is still debated, the vegetation shift appears to have been quite rapid, in some parts of Africa at least, compared to the relatively slow changes in astronomical forcing. Indeed, some reconstructions indicate major changes in less than one millennium. The timing of the transition seems region dependent but occurred generally around 5000 years B.P. It has been suggested that the positive biogeophysical feedbacks mentioned earlier may be responsible for the rapid transition to a desert state in some regions when the precipitation rate goes below a threshold under which vegetation growth cannot be sustained (Claussen 2009). Alternatively, the non-linear response of the vegetation itself when precipitation changes or the response of the monsoon circulation to surface temperature variations also has been invoked as a possible cause of such rapid transitions.

### 5.6.3 The Past 2000 Years

#### 5.6.3.1 Temperature Changes at Hemispheric Scale

The last millennium is certainly the period for which we have the largest number of **proxy** records, coming from various archives: tree rings, lake and marine sediments, ice cores, corals, historical documents and so on. The various

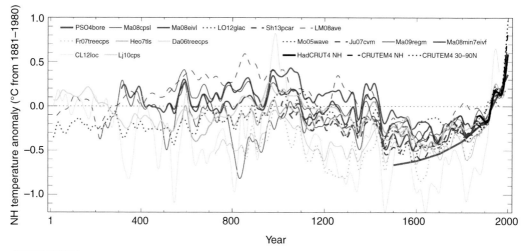

Reconstructed northern hemisphere temperatures over the last 2000 years. Individual reconstructions are grouped by colour according to their spatial representation (red: land only, all latitudes; orange: land only, extra-tropical latitudes; light blue: land and sea, extra-tropical latitudes; dark blue: land and sea, all latitudes). Instrumental temperatures are shown in black. All series represent **anomalies** (°C) from the 1881–1980 mean and have been smoothed with a fifty-year filter. See the source for a full reference to all the reconstructions. (Source: Masson-Delmotte et al. 2013.)

reconstructions of global mean temperature show common features (Figure 5.34), with relatively mild conditions roughly between 950 and 1250 A.D. (the so-called medieval climate anomaly), followed by a colder period roughly between 1450 and 1850 A.D. (the so-called little ice age) (see, e.g., Mann et al. 2009). This cooling trend during the past millennium follows the longer-term temperature decrease described in the preceding section (see also Liu et al. 2014), making the little ice age one of the coldest period of the middle to late Holocene, if not the coldest. Moreover, the reconstructions display a large warming since 1850.

Nevertheless, the range of available reconstructions at hemispheric scale is still large. This is due to the low number of records in some regions and the uncertainty of interpretation of individual proxy time series and the techniques applied to obtain estimates of large-scale changes from those series (Jones et al. 2009). Because of the decreasing quality of the reconstructions as we go back in time, comparisons between the late twentieth century and the preceding 2000 years are also uncertain. However, according to the latest IPCC report, it is likely that the last thirty years were the warmest thirty-year period of the last 1400 years in the northern hemisphere (Masson-Delmotte et al. 2013).

At inter-annual to millennial timescales, the variations in insolation at the top of the atmosphere due to changes in **astronomical parameters** are relatively weak but still have an influence on long-term trends. Additionally, changes in (TSI) and large volcanic eruptions (see Section 4.1.2.4) are key natural forcings over the last 2000 years. In contrast to the astronomical forcing, whose time development is very well known, we are still uncertain about the past variations in solar and volcanic forcings. Volcanic eruptions let sulphate deposits which are measured in

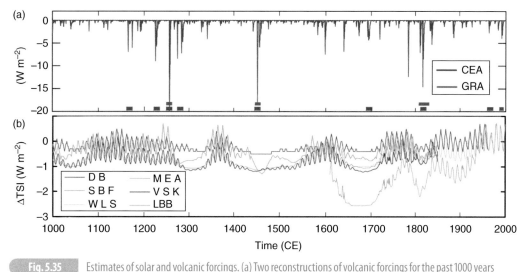

**Fig. 5.35** Estimates of solar and volcanic forcings. (a) Two reconstructions of volcanic forcings for the past 1000 years used for the Paleoclimate Modelling Inter-Comparison Project Phase III (PMIP3). (b) Reconstructed total solar irradiance (TSI) anomalies back to the year 1000. For the years prior to 1600, the eleven-year cycle has been added artificially to the original data with an amplitude proportional to the mean anomaly of the TSI. See the source for a full reference to all the reconstructions. (Source: Masson-Delmotte et al. 2013.)

ice cores. The information obtained then be used to derive indirectly the volume and characteristics of the aerosols released by past eruptions (Figure 5.35). This approach requires, however, strong assumptions on the locations and timing of the eruptions, as well as on the transport by the atmosphere to the points where the ice cores are collected. Significant uncertainties thus are present in estimates of the reconstructed volcanic forcing (Gao et al. 2008).

The standard way to extend estimates of the changes in TSI before direct satellite observations is based on the concentration of **cosmogenic** isotopes such as $^{10}$Be and $^{14}$C in ice cores. As explained in Section 5.3.3, when solar activity is low, the shielding of the Earth from energetic **cosmic rays** is weaker, and there is an increase in the production of these isotopes. However, the link between the concentration of cosmogenic isotopes in ice cores, solar activity and solar forcing is far from simple. The periods with strong or weak solar forcing are relatively robust between the various available TSI estimates, but the amplitude of the reconstructions can differ by a factor of 5. Most of the recent reconstructions suggest relatively small changes between the present and pre-industrial periods corresponding to an increase in TSI smaller than 0.1% since 1750 (see, e.g., Wang et al. 2005). Such a weak forcing has been applied in most PMIP3 experiments, but previous simulations generally were driven by a larger forcing characterised by a difference in TSI of 0.2 to 0.3% between 1750 and the late twentieth century. These two groups of simulations are separated in Figure 5.36 between 'weak solar variability' and 'strong solar variability'. Even larger TSI changes have been proposed, but they lead to model results which are not consistent with most of the reconstructions.

The last millennium is an ideal test case for climate models to compare natural and human-induced changes. When driven by astronomical, solar and volcanic

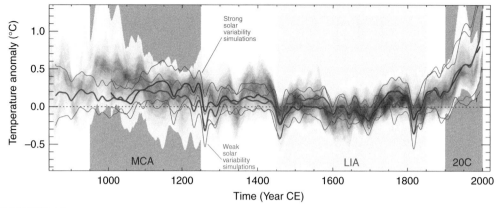

Comparison of simulated and reconstructed changes over the past millennium in the northern hemisphere. MCA stands for 'medieval climate anomaly', LIA for 'little ice age' and '20C' for the twentieth century. The range of reconstruction included in Figure 5.34 is given by the grey shading. The thick blue line is the ensemble mean of the simulations driven by a weak solar forcing, the thin blue lines representing the 90% range of the ensemble. Similarly, thick and thin red lines correspond to the mean and 90% range of the simulations driven by a stronger solar forcing. The temperatures are given as anomalies compared to the mean over 1450–1850. All the time series have been smoothed with a thirty-year filter. (Source: Masson-Delmotte et al. 2013.)

forcings, as well as by anthropogenic forcings (increases in greenhouse gas concentration and sulphate aerosol load and land-use changes; see Section 4.1.2), the simulated temperatures are within the range provided by the reconstructions (see Figure 5.36). This gives us additional confidence in the validity of the models. However, most models are in the lower range of the reconstructions for the medieval climate anomaly, in particular, for the period around 1000 C.E. Consequently, they indicate a cooling between the medieval climate anomaly and the little ice age which is smaller than the one provided by most reconstructions.

These simulations can be used to analyse the causes of the observed global mean. The influence of volcanic forcing is dominant in the years immediately following an eruption. This also explains to a large extent the cold periods at multi-decadal time scales during the little ice age, when several major eruptions occurred in a short period. The role of solar irradiance is less clear. For instance, the agreement between model results and temperature reconstructions is not improved significantly whenever a solar forcing with a low or high amplitude is selected to drive them. Solar forcing probably played a role in the cooling during some periods of the little ice age, but disentangling the contribution of solar forcing from that of volcanic forcing is difficult. Indeed, several major eruptions occurred by chance during times of low solar irradiance, both forcings thus acting in the same direction to cool the system.

### 5.6.3.2  Temperature Changes at Continent Scale

As observed for the northern hemisphere mean, the continental-scale reconstructions covering past millennia display a cooling trend that ended in the nineteenth century (Figure 5.37). Consequently, the medieval climate anomaly was relatively

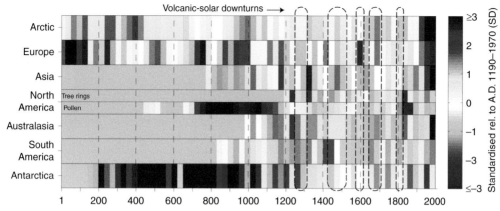

Volcanic-solar downturns ⟶

**Fig. 5.37**   Temperature reconstructions for seven continental-scale regions. The time series have been averaged over thirty-year periods and standardised to have a zero mean and a standard deviation equal to 1 over the period of overlap among records (A.D. 1190–1970). North America includes a shorter tree-ring-based and a longer pollen-based reconstruction. Dashed outlines enclose intervals of pronounced volcanic and solar negative forcing since A.D. 850. (Source: PAGES 2K 2013. Reprinted by permission from Macmillan Publishers, Ltd.: *Nature Geoscience*; copyright 2013.)

mild on most continents, whilst the little ice age was cold. The warm and cold periods were not synchronous between the different regions, however. The difference is particularly clear between the hemispheres, with warm conditions at the beginning of the second millennium delayed in Australasia and South America compared to Europe, Asia and the Arctic. A clear warming is seen at the end of the records in all the continental-scale reconstructions, except in Antarctica. Nevertheless, for several continents, the temperatures reached over the period 1970–2000 were not the highest of the last two millennia. This contrasts with the hemispheric mean and indicates that the last thirty years are perhaps not the warmest locally, but this is the period for which the warming appears to be the most homogeneous on a global scale.

At the hemispheric scale, the response to the **forcings** has played a clear role in the variations in temperature (see Figure 5.36), with the **internal variability** also having contributed to the observed changes (see Section 5.1.1). The ratio between forced and internal variability is not precisely known for the last millennium, but it is clear that the relative contribution of the internal variability is larger at regional than at hemispheric scale. The astronomical forcing is likely responsible for the cooling trend in some regions, such as the Arctic. Periods of large volcanic activity and low TSI can be identified as cold intervals in many areas. Nevertheless, the internal variability, for instance, related to standard modes of atmospheric or oceanic variability (see Section 5.2), can completely mask the influence of the forcing locally. As a consequence, the temperature in the first part of the second millennium was generally higher than in the period 1450–1850, but warm and cold periods occurred at different times in different locations.

This contribution of the internal variability, whose origin is found entirely in the chaotic nature of the climate system, can be illustrated by performing an

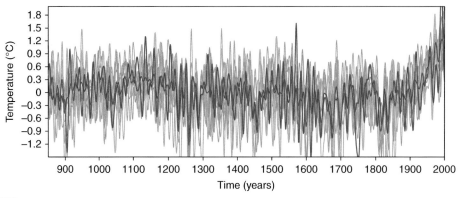

Fig. 5.38 Surface temperature anomaly (°C) in the Arctic (defined here as the area northward of 65°N) over the last millennium driven by both natural and anthropogenic forcings in an ensemble of ten simulations. The **ensemble members** differ only in the initial conditions. Eight simulations are in grey, whilst two are highlighted in blue and green for easy reference. The ensemble mean is in red. The reference period is A.D. 850–1850, and a decadal smoothing has been applied to the time series. (Source: Data from Crespin et al. 2013.)

ensemble of experiments using a climate model. For Figure 5.38, ten simulations have been run with exactly the same model and same forcing but slightly different initial conditions. Because of these small differences and the sensitivity of the climate system to small perturbations (see Section 5.1.1), the **ensemble of simulations** diverges quickly, and each simulation finally includes a different sample of the internal variability of the modelled system. Because this contribution of internal variability is nearly independent among all the simulations, averaging over them tends to filter out the internal variability and gives an estimate of the response to the forcing, which is identical in each experiment. If an infinite number of simulations were averaged, only the influence of the forcing would remain. This response to the forcing provides some common characteristics for all simulations. Nevertheless, the large differences between the forced response and each individual simulation indicate that internal variability can be a dominant source of climate changes before the twentieth century, even at continental scale, for multi-decadal variations.

It should be recalled here that the observed climate also corresponds to one realisation of the internal variability of the real system, among all the possibilities (if the model and the forcing were perfect, observations simply would correspond to an additional ensemble member in Figure 5.38). Consequently, the results of one simulation covering the past millennium cannot be compared to a reconstruction without taking into account that they both include a component resulting from internal variability, which has a priori no reason to be synchronous between the model and the observation. The role of internal variability also must be kept in mind when using the terms 'medieval climate anomaly' and 'little ice age'. It is often not simple to justify that a local warm or cold period is related to a large-scale phenomenon such as the medieval climate anomaly because simultaneous warm conditions in different regions may have different origins and occur at similar times only by chance.

Furthermore, the distinction between internal and forced variability is not as sharp as presented earlier. Some changes in atmospheric or oceanic circulation can be part of the response of the climate system to the forcing. For instance, a tendency towards a positive NAO index has been mentioned in the winter following a major volcanic eruption. The volcanic forcing may pace the variability of the AMO. The solar forcing also may influence the atmospheric circulation. It therefore is hard to disentangle such a potential response of the circulation to the forcing from the internal variability which would be present in the absence of any forcing change.

### 5.6.3.3  Hydrological Changes

The temperature reconstructions discussed in preceding sections represent just one manifestation of climate changes during past millennia. For instance, glacier lengths have displayed dramatic variations, with spectacular advances in Europe during the little ice age compared to the medieval climate anomaly. Large variability in drought and floods also has been reconstructed, putting the short number of events recorded by instruments over the last centuries into a well-needed longer perspective. The region whose hydrological changes are best documented is probably North America, where drought indexes have been reconstructed from tree-ring data for more than a millennium. These estimates display large variability, with, in particular, several multi-decadal dry intervals between 1000 and 1300 in the south-western United States (Figure 5.39). These so-called mega-droughts are more severe than what was observed during the last century and likely had a dramatic effect on local human populations (Cook et al. 2007). Their origin has been tentatively linked to a response to low temperatures in the Eastern Tropical Pacific corresponding to la Niña–like conditions (see Section 5.2.1), with potentially local feedbacks between precipitation changes, soil moisture and vegetation.

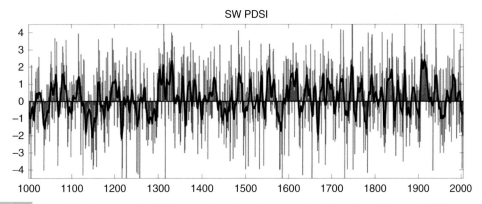

**Fig. 5.39**  Palmer drought severity index (PDSI) averaged over the south-western United States (32–40°N, 105–125°W). The PDSI is a normalised index that measures the severity of drought, taking into account the influence of changes in both precipitation and evapotranspiration. Dry conditions correspond to negative values and wet conditions to positive values. The original data (green and brown bars) are smoothed using a ten-year filter to obtain the dark black line. (Source: Cook, B. I., J. E. Smerdon, R. Seager and E. R. Cook (2014). Pan-continental droughts in North America over the last millennium. *Journal of Climate* 27, 383–97. © American Meteorological Society. Used with permission.)

## 5.7 The Last Century

### 5.7.1 Observed Changes

The global mean surface temperature averaged over 1986–2005 is about 0.6°C warmer than during the period 1850–1900 (Figure 5.40). A linear trend over the years 1901–2012 provides a global mean temperature increase of 0.89°C over that period (90% uncertainty interval 0.49–1.08°C). The warming rate, however, is not constant with large warming during the years 1910–1940 and 1970–2000 and more stable conditions during the years 1950–1970 and the last decade. Still, the first decade of the twenty-first century is the warmest since 1850, and ten of the warmest years on record have occurred since 1997 (Hartmann et al. 2013).

The warming is seen in nearly all regions, with a generally slower warming over ocean than over land. Some oceanic areas, for example, close to the southern tip of Greenland, even display a slight cooling over the period 1901–2012 (Figure 5.41). Furthermore, the signal is more uncertain in some regions such as Antarctica because of the lack of long series of observational records. More generally, the heat content of the climate system, mainly associated with the ocean warming at depth and in particular in the top 700 m, has increased by more than $250 \times 10^{21}$ J between 1971 and 2010 (Rhein et al. 2013). This is equivalent to a flux of about 0.4 W m$^{-2}$ averaged over the entire Earth for this period, showing the strong role of the ocean in storing heat and thus moderating the transient response to the forcing (see Section 4.1.5).

The warming is associated with clear modifications of the **cryosphere**. Most glaciers have retreated over the last decades, the snow cover in the northern hemisphere has declined and the Greenland ice sheet has lost mass over the last twenty years. The sea-ice extent in the Arctic has been reduced by about 4% per decade since 1979. The decrease in the extent of the sea ice is even larger in summer, at a rate of about 7% per decade, with the very large summer minima in 2007 and

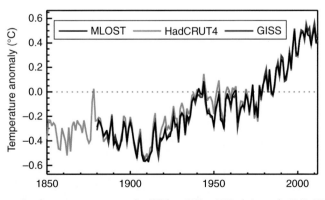

**Fig. 5.40**　Global mean annual surface temperature anomalies (°C) from 1850 to 2012 relative to the 1961–1990 mean from three different data sets: MLOST (Vose et al. 2012), HadCRUT4 (Morice et al. 2012) and GISS (Hansen et al. 2010). (Source: Hartmann et al. 2013.)

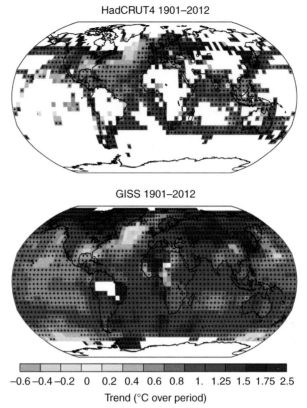

GISS 1901–2012

−0.6 −0.4 −0.2  0   0.2  0.4  0.6  0.8  1.  1.25 1.5 1.75 2.5
Trend (°C over period)

**Fig. 5.41**  Linear trend of annual temperatures between 1901 and 2012 in HadCRUT4 and GISS data sets (°C over the period). Areas in white have insufficient data to produce reliable trends. The main difference between the data sets is that some extrapolation to fill in data gaps is allowed in GISS but not in HadCRUT4. (Source: Hartmann et al. 2013.)

2012 being more than 30% below the 1981–2010 mean. By contrast, the sea ice in the Southern Ocean has increased slightly at a rate of about 1% per decade over the last thirty years (Vaughan et al. 2013). The response to anthropogenic forcings simulated by models is relatively weak at surface there partly because of the large thermal inertia of the Southern Ocean (see Section 6.2.4), and the internal variability is large. The observed increase thus might result from an increase due to internal processes which overwhelm the slight decrease associated with global warming. Nevertheless, the natural variability and response to various forcings at high southern latitudes are not well understood. Consequently, many uncertainties remain on the origin of the recent changes in sea-ice extent in the Southern Ocean.

## 5.7.2   Detection and Attribution of Recent Climate Changes

The anthropogenic origin of the rise in atmospheric $CO_2$ concentration since the nineteenth century is unequivocal. The increase in the mass of carbon in the atmosphere corresponds to a fraction only of the carbon released by human activities, the remaining part being stored in the ocean and the biosphere over land (see Section 2.3.1). Furthermore, fossil fuels have a strong isotopic signature. Because they have a biological origin, they have a low $^{13}C$ content and contain nearly no $^{14}C$ because they have not been in contact with the atmosphere for millions of years (see Section 5.3.3). Consequently, the massive burning of fossil fuels which began with the industrial revolution has led to a release of a larger fraction of $^{12}C$ compared to the pre-industrial atmospheric ratio of the three carbon isotopes. It therefore should be associated with a decrease in the relative amount of $^{13}C$ and $^{14}C$ in the atmosphere (the so-called Suess effect). Such a decrease has been seen consistently for $^{13}C$. For $^{14}C$, the decrease lasts until the mid-twentieth century. The nuclear weapons explosions at that time during tests injected into the atmosphere large amounts of $^{14}C$ which has induced a peak in its atmospheric concentration.

A more critical issue is to determine which part of the recent temperature changes are compatible with the natural variability of the system and which part cannot be explained without a contribution from anthropogenic forcing. Such an analysis can be based on observations and empirical models, trying to estimate the amplitude of the changes associated with known modes of variability such as ENSO or the AMO and the ones which simply could be related to the history of natural and anthropogenic forcings. However, this approach is sensitive to the underlying hypotheses, in particular, to the way in which the impact of the forcing is taken into account (see, e.g., Lean and Rind 2008; Imbers et al. 2013).

A second method is to compare different types of simulations performed with climate models with observations. When models are driven by both natural and anthropogenic forcings, observed temperature changes at global and continental scales are well in the simulated range (Figure 5.42). By contrast, when driven by natural forcing, climate models cannot reproduce the warming observed at a global scale and for all continents except Antarctica. Note that in both cases, the model range also includes the contribution of the internal variability present in any GCM simulation (see Section 5.1). This means that according to those simulations, observations are compatible with the hypothesis that anthropogenic forcing is needed to explain the recent temperature changes. Besides, they are not compatible with the alternative hypothesis stating that the changes in climate observed recently are in the range of **natural variability** on decadal to centennial time scales (i.e., internal variability plus response to natural forcings). Similarly, models can reproduce the large decreases in sea-ice extent in the Arctic only if anthropogenic forcings are included. In the southern hemisphere, the small increasing trend in sea-ice extent is consistent with the natural variability of the system according to current simulations and could be reproduced with or without including the anthropogenic forcings.

The validity of these conclusions depends on the ability of the models to reproduce the characteristics, and, in particular, the amplitude, of the natural variability. For instance, an underestimation of the magnitude of the natural variability may result in a wrong interpretation of the anthropogenic origin of some events, whilst they would indeed be compatible with natural variations. In turn, an over estimation of this magnitude would prevent the detection of an

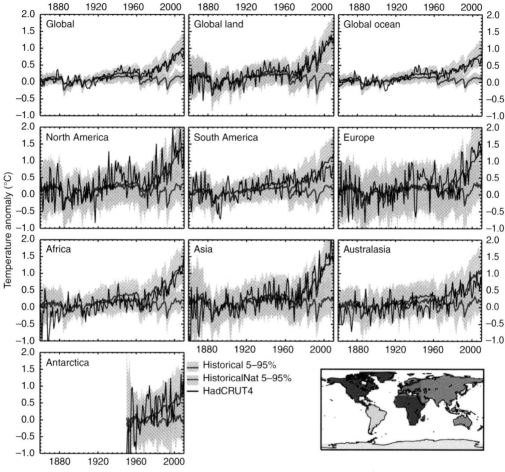

**Fig. 5.42** Comparison of observed continental- and global-scale changes in surface temperature (HadCRUT4, black) with results simulated by climate models. Multi-model means are thick lines, and the 5–95% ranges are shown by shading. Model results using natural and anthropogenic forcings are in red, whilst results using natural forcings only are in blue. The values are computed as anomalies with respect to the corresponding average for 1880–1919, except for Antarctica, where the average over 1950–2010 is used. The continental regions over which the averages are performed are shown at the bottom right. (Source: From Bindoff et al. 2013, adapted from Jones et al. 2013.)

anthropogenic signal which is already present in observations. For many simulated variables such as the global surface temperature, the agreement between model results and observations over the twentieth century and reconstructions for the more distant past is good (see, e.g., Section 5.6.3 for the last millennium). There is thus no reason to consider that models are biased in one way or the other. This is not the case for some other variables such as the sea-ice concentration in the southern hemisphere, for which models have trouble reproducing the mean state and variability of the observations (see Section 3.5.2.3). This reduces the confidence in the conclusion mentioned earlier that the recent change in sea-ice extent in the Southern Ocean can be explained by natural variability only.

The contributions of internal variability and the forced response can be more formally estimated through a detection and attribution method (Hasselmann 1993; Hegerl and Zwiers 2011). This method is based on the assumption that the observed changes $\mathbf{Y}(\mathbf{x}, t)$ could be represented by a linear combination of the response to $n$ different forcings $\mathbf{X}_i(\mathbf{x}, t)$ ($i$ varying from 1 to $n$) and internal variability $\mathbf{U}(\mathbf{x}, t)$, where $\mathbf{x}$ includes the spatial coordinates (latitude, longitude and potentially depth or altitude), and $t$ is the time; thus,

$$\mathbf{Y}(\mathbf{x},t) = \sum_{i=1}^{n} \beta_i \mathbf{X}_i(\mathbf{x},t) + \mathbf{U}(\mathbf{x},t) \qquad (5.12)$$

where $\mathbf{X}_i(\mathbf{x}, t)$ is called a 'fingerprint' and represents both the spatial and temporal characteristics of the response to forcing $i$. Note that in practise, this can correspond to a group of forcing agents rather than to a single agent. The $\beta_i$ are the scaling factors of those patterns. The fingerprints are generally determined from the results of simulations performed with climate models driven by the corresponding forcing(s) only. $\mathbf{Y}(\mathbf{x}, t)$ contains all the information provided by observed variations in different regions, but the same methodology can be applied on integrated variables such as the global mean temperature or the sea-ice extent in one hemisphere. In this case, the observed changes are only one time series, and the fingerprint is the temporal signal of the response to one particular forcing.

To illustrate the methodology using a simple, idealised situation, consider an observed time series $T(t)$ representing, for instance, the global mean temperature (Figure 5.43). There is good reasons to test the hypothesis that two forcings contribute to the observed changes. Using a physical model of the system, the responses to those forcings are estimated as $\mathrm{Resp}_1(t)$, characterised by a linear trend, and $\mathrm{Resp}_2(t)$, characterised by an oscillatory behaviour. Detection and attribution methods then look for the optimal parameters $\beta_1$ and $\beta_2$ to have the best match between $T(t)$ and $\beta_1 \mathrm{Resp}_1(t) + \beta_2 \mathrm{Resp}_2(t)$ to obtain

$$T(t) = \beta_1 \mathrm{Resp}_1(t) + \beta_2 \mathrm{Resp}_2(t) + u(t) \qquad (5.13)$$

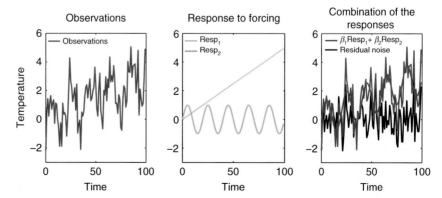

**Fig. 5.43** Simple illustration of the detection and attribution method. The observed time series $T(t)$ is represented as the linear combination of the response to a forcing 1, $\mathrm{Resp}_1(t)$, and a forcing 2, $\mathrm{Resp}_2(t)$, plus a residual $u(t)$ as in Eq. (5.13). In the chosen example, the coefficient of the linear combination are $\beta_1 = 0.6$ and $\beta_2 = 1.2$.

where $u(t)$ is a residual, usually representing high-frequency noise. It is important to insist here on detection and attribution being not only a statistical method. It is necessary to ensure that the selection of the forcing is performed on the basis of knowledge of the climate conditions analysed and that the estimate of the response is based on a physical understanding of the mechanism involved to take advantage of all the information available on the system.

Equation (5.12) has the format of a linear **regression**, and the coefficients $\beta_i$ can be determined by classical statistical methods in which $\mathbf{U}(\mathbf{x}, t)$ appears as the residual (see, e.g., Wilks 2011). The value of $\beta_i$ then allows determination of whether the contribution of a forcing to the observed changes can be separated from the internal variability of the system. If the scaling factor is significantly different from 0 (i.e., if its confidence interval does not include 0), the fingerprint is likely to have contributed to observed variations and thus is considered to be detected in the observations. Additionally, if 1 is in the confidence interval of $\beta_i$, the amplitude of the simulated response is consistent with the observations. Formally, it is then said that attribution has been achieved. Since the detection does not require this consistency, the methodology is more robust than the direct comparison of model results and observations presented earlier. Indeed, in this detection method, the central information is included in the spatio-temporal patterns $\mathbf{X}_i(\mathbf{x}, t)$. The uncertainties in the forcing time series or in the **climate sensitivity** of the system (see Section 4.1) strongly influence the magnitude of the response, whilst the impact on its shape is likely smaller. This bias on the magnitude of the changes can be compensated for by values of $\beta_i$ which are different from 1, whilst direct model-data comparison can be sensitive to such uncertainties. Besides, uncertainties on the patterns themselves potentially may have a larger impact in detection and attribution studies.

Equation (5.12) cannot be used directly in most cases. Firstly, the fingerprint estimates based on GCM results also contain noise resulting from internal variability because those simulations include both internal variability and the response to the forcing. This contribution of the internal variability can be reduced by using the average over an ensemble of simulations driven by the same forcing (see Figure 5.38). However, to take into account the role of this noise, $\mathbf{X}_i(\mathbf{x}, t)$ is replaced in Eq. (5.12) by $\mathbf{X}_i(\mathbf{x}, t) - \mathbf{U}_i(\mathbf{x}, t)$, where $\mathbf{U}_i(\mathbf{x}, t)$ is the estimate of internal variability still present in the fingerprint. The method then is close to a total least-squares regression (Allen and Stott 2003). Secondly, the signals associated with the response to different forcings are better detected if they display much contrasted patterns in space or time, as in the example shown in Figure 5.43. For instance, the solar forcing has a very distinct time history compared to the greenhouse gas forcing, which has increased steadily over the last century. Nevertheless, some filtering is still needed to maximise the signal-to-noise ratio and obtain useful results. Simple steps include, for instance, a temporal and spatial smoothing of the data and model fields, but more sophisticated procedures are often used. This leads to what is called 'optimal fingerprint methods' (for more details, see, e.g., Hegerl and Zwiers 2011).

Since the role of detection is to separate the contribution from a forcing and internal variability, good estimates of the characteristics of the latter are required, although it might appear less clear up to now in contrast to the direct comparison between models and observations presented in Figure 5.42. Firstly, the consistency of the residual $U(x, t)$ with the natural variability of the system has to be checked to ensure that the procedure worked properly and that some factors or additional forcings have not been overlooked in the procedure. Secondly, the uncertainty of the factors $\beta_i$ is derived from tests using different samples of internal variability added to observations or in the fingerprints. Finally, the information provided by estimates of the characteristics of natural variability of the system is often used to amplify the signal-to-noise ratio and bring a clearer conclusion. This confirms that an adequate estimate of the magnitude of the internal variability is important in detection and attribution too. Because observation time series are too short, this estimate is generally derived from model simulations, but the sensitivity of the results to an inflation in the magnitude of the internal variability is generally performed to test the robustness of the results.

When this method is applied to the twentieth century, it allows us to detect the contribution of the increase in greenhouse gas concentrations in the atmosphere in the recent warming with a scaling coefficient $\beta$ close to 1 (Figure 5.44). From the value of the scaling coefficient, the increase in greenhouse gas concentration has been estimated to lead to a warming of between 0.5 and 1.3°C over the years 1951–2010 compared to an observed global warming over this period of 0.6°C. Other anthropogenic forcings, including aerosols, likely induce a cooling, but a small warming could not be excluded, with estimates of the response being between –0.6 and +0.1°C. The contributions of natural forcing and natural variability appear to be smaller than 0.1°C over this period (Bindoff et al. 2013).

The contribution of greenhouse-gas forcing is dominant in explaining the long-term trend in temperature, but the situation is more complex on decadal or interannual time scales. Major volcanic eruptions occurring in 1963 (Agung), 1982

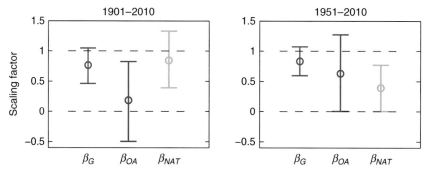

**Fig. 5.44**  Scaling coefficient in a detection and attribution study separating the temperature changes over the period 1901–2010 and 1951–2010 in a response to greenhouse gases ($\beta_G$), to other anthropogenic forcings ($\beta_{OA}$) and to natural forcings ($\beta_{NAT}$). The conclusions are based on the average of simulations performed with eight global climate models. The best estimates and 90% ranges are shown. (Source: Data from Jones et al. 2013.)

(El Chichon) and 1991 (Pinatubo) are associated with cooling of several tenths of degree on a global scale. A decrease in solar irradiance and a negative forcing due to volcanic eruptions probably contribute to explain the modest temperature increase during the last decade. Internal variability also plays a role in those variations on a decadal scale, potentially masking the signal associated with the forcing, as in the last decade, or reinforcing it, as in the first half of the twentieth century.

Many detection and attribution analyses have focussed on global surface temperatures because of the good signal-to-noise ratio. However, despite the larger contribution of internal variability (see Section 5.1.1, and compare the different panels of Figure 5.42, for instance), it also has been possible to detect the impact of human activities on temperature on regional scales on many continents, as well as on other variables such as the ice extent in the Arctic, precipitation and ocean salinity. This has led the IPCC to conclude, in its fifth assessment report, that 'It is extremely likely that human influence has been the dominant cause of the observed warming since the mid-20th century' (IPCC 2013) and that 'It is virtually certain that human influence has warmed the climate system' (Bindoff et al. 2013). 'Virtually certain' in this sentence means a likelihood higher than 99%, whilst 'extremely likely' means a likelihood higher than 95%. In the IPCC terminology, 'very likely' corresponds to the 90% level and 'likely' to a level higher than 66%. The conclusion about the whole climate system is stronger than for surface temperature because the changes observed in the different components are consistent with our understanding of the response to human activities. In the future, additional changes are expected, as discussed in Chapter 6.

## Review Exercises

1. The magnitude of the inter-annual changes associated with the internal variability of the climate system is much larger for global mean surface temperature than for the surface temperature averaged over a continent or a region.
   a. True
   b. False
2. The magnitude and timing of internal variations should be similar in a model and observations if the model is able to adequately represent the dynamics of the climate system.
   a. True
   b. False
3. The time scales of climate variations are set up by
   a. the time scales of the forcing.
   b. internal dynamics.
   c. the time scales of the forcing and internal dynamics.
4. In the equatorial Pacific, the Bjerknes feedback is a positive feedback which links the intensity of the easterlies with the zonal gradient in SST.
   a. True
   b. False

5. El Niño events are associated with
   a. a weaker than normal SLP gradient between the East and West equatorial Pacific.
   b. an oceanic warming close to the coast of Peru.
   c. less precipitation over northern Australia.

6. The positive phase of the North Atlantic Oscillation is associated with
   a. stronger westerlies over the North Atlantic.
   b. a cooling over northern Europe and Asia in winter.
   c. a cooling over the Labrador Sea.

7. Extracting information on past climate changes from observations made in natural archives must rely mainly on statistical techniques.
   a. True
   b. False

8. The $\delta^{18}O$ of the land ice in polar regions is related to temperature
   a. because isotopic fractionation during phase transition is a function of temperature.
   b. because the vapour condensation which leads to fractionation is itself a function of temperature.

9. Four billion years ago, the solar irradiance was about 25–30% lower than at present. The Earth was not totally covered by ice mainly because
   a. of the continual bombardment by small planetesimals and meteorites which released large amounts of energy.
   b. of a lower albedo of the surface at that time in the absence of vegetation.
   c. of a stronger greenhouse effect.

10. The oxygen liberated by photosynthesis has modified the composition of the atmosphere. This is assumed to have led to a surface cooling because
    a. of the direct radiative effect of oxygen.
    b. of more rapid oxidation of methane and thus a decrease in its concentration in the atmosphere.

11. The long-term evolution of the carbon cycle over the Phanerozoic, which is strongly coupled with climate, is mainly influenced by changes in
    a. out-gassing associated with metamorphism and subduction.
    b. the efficiency of the soft-tissue pump in the ocean.
    c. weathering and sedimentation.
    d. deep burial of organic matter.
    e. exchange of carbon between the atmosphere and the continental biosphere leading to storage in soils.

12. The temperature has decreased over the last 50 million years (since the early Eocene climatic optimum) because
    a. of the long-term decrease in solar irradiance.
    b. of modification of the astronomical parameters of the Earth and thus of the distribution of insolation at the surface.
    c. of a reduction in the $CO_2$ concentration of the atmosphere.

13. The eccentricity of the Earth's orbit around the Sun influences the annual mean insolation integrated over the whole Earth.
    a. True
    b. False

14. The obliquity, which is a measure of the tilt of the ecliptic compared to the celestial equator, strongly influences the amplitude of the seasonal cycle of insolation at high latitudes
    a. but has no effect on the annual mean insolation at every point on Earth.
    b. but has no effect on the annual mean insolation integrated over the whole Earth.

15. The climatic precession, which is related to the Earth–Sun distance at the summer solstice,
    a. plays a dominant role in variations in insolation at low and middle latitudes.
    b. has no impact on annual mean insolation at a particular point.
    c. has no effect on annual mean insolation integrated over the whole Earth.

16. In the astronomical theory of paleoclimates, the summer insolation at high latitudes is critical in explaining the glacial-interglacial cycles because
    a. it influences the strength of the westerlies and thus the inflow of moist air towards the continents.
    b. it influences the strength of the deep oceanic circulation.
    c. it influences the summer melting of snow over continents.

17. The cause of the decrease in atmospheric $CO_2$ concentration by about 90 ppm during the last glacial periods compared to interglacials is due to
    a. changes in the continental biosphere.
    b. changes in weathering over lands related to the drier conditions there.
    c. changes in the solubility of $CO_2$ in the ocean because of cooling.
    d. changes in the distribution of alkalinity and DIC in the ocean.

18. The climate recorded in Greenland ice cores was more stable during the Holocene than during the previous glacial period and the deglaciation.
    a. True
    b. False

19. The decrease in summer insolation in the northern hemisphere since the Early Holocene has led to
    a. a general cooling at high and middle northern latitudes in summer.
    b. an intensification of the summer monsoons.
    c. a desertification of the Sahara.

20. The dominant natural forcings on a global scale driving the climate of the last millennium before the industrial era are
    a. volcanic forcing (related to the injection of aerosols into the stratosphere).
    b. dust forcing (changes in dust concentration in the atmosphere).
    c. solar forcing (changes in total solar irradiance).
    d. astronomical forcing (changes in the astronomical parameters of the Earth).

21. Regional changes during the last millennium are only due to the response of the system to changes in external forcings.
    a. True
    b. False
22. The surface warming observed over the last 150 years
    a. can be explained by a response of the system to natural forcings and internal variability.
    b. can be explained by a combination of the response to anthropogenic forcing, natural forcing and internal variability.

# 6 Future Climate Changes

## OUTLINE

This chapter describes how estimates of future climate changes are derived and the main results which have been obtained up to now. Particular attention is paid to interpretation and the limitations of the forecasts at different time scales and on the most robust mechanisms which explain the changes. This allows underlining the links with previous chapters.

## 6.1 Scenarios

### 6.1.1 The Purpose of the Scenarios and Scenario Development

As discussed in Chapter 5, the changes in external **forcings** have driven major past climate variations. In order to "predict" the climate of the twenty-first century and beyond, it thus is necessary to estimate future changes in the forcings. This is achieved by the development of **scenarios** for the emission of **greenhouse** gases, **aerosols**, various pollutants in the atmosphere, land use and so on. These scenarios depend on many uncertain elements (as discussed later), and some of the uncertainties in the estimates of future climate changes are related to these factors (see Figure 6.10). This is the reason why, in the scientific literature, the term 'climate **projection**' is generally preferred for estimates of the changes during the twenty-first century and beyond to the term 'climate **prediction**' (see also Section 6.2.2). Climate projection emphasises the fact that the results depend on the scenarios chosen and the hypotheses employed in these scenarios. The scenarios are also used for analysing impact, adaptation and vulnerability, thus providing a consistent approach for socio-economic and climatic issues.

Various types of scenarios have been proposed in recent years and decades. In the fourth assessment report of the Intergovernmental Panel on Climate Change (IPCC), the climate projections were based on the Special Report on Emission Scenarios (SRES) scenarios (see Section 6.1.2), which cover the whole of the twenty-first century (Nakicenovic and Swart 2000). These scenarios were derived in a sequential form (Figure 6.1). Firstly, the main driving forces influencing the emissions from demographic, social and economic development have to be identified. This implies estimating population growth, future levels of economic

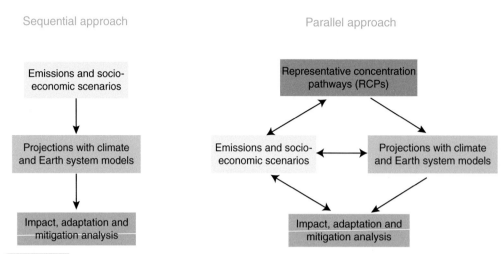

Sequential approach                                    Parallel approach

**Fig. 6.1**   Sequential and parallel approaches to the development of scenarios. In contrast to the sequential approach, the projections with climate models (driven by RCPs) and the choice of emission and socio-economic scenarios (corresponding to those RCPs) are performed concurrently in the parallel approach. (Source: Modified from Moss et al. 2007.)

activity, the way exchanges between different countries will be organised, the technology choices/opportunities of the countries and so on. On the basis of these estimates, some models produce scenarios for future emissions of greenhouse gases and aerosols and land-use changes. The concentrations of greenhouse gases and aerosols in the atmosphere are also derived for the models which do not include a representation of the carbon and/or aerosols cycle. In this framework, different combinations of demographic and socio-economic changes can lead to similar emission paths. For instance, large population growth combined with efficient technologies and renewable energy can lead to similar emissions to a smaller increase in the Earth's population with less efficient and more energy-demanding technologies.

For the fifth IPCC assessment report (IPCC AR5; Stocker et al. 2013), a slightly different approach is followed. Four representative concentration pathways (RCPs) (Moss et al. 2007, 2010) were selected, covering a wide range of future changes in **radiative forcing** (see Section 6.1.3). The concentrations of greenhouse gases corresponding to these four RCPs then were provided to the climate-modelling community, and emissions compatible with those concentrations were derived. In parallel, possible socio-economic scenarios were developed, providing different alternatives for the same RCP. If needed, the information provided by the climate-model projections can be used in the socio-economic scenarios to assess the impact of climate change on society. Such a parallel approach strengthens the collaboration between the different communities whilst ensuring that the climate-modelling groups only have to run a small set of well-contrasted scenarios with their models (which are very demanding of computer time). Another advantage of the RCP scenarios is that they include both more detailed short-term estimates (to about 2035) and stylised estimates to about 2300, in addition to the classical long-term estimates up to 2100 provided by the SRES scenarios.

Neither the SRES scenarios nor the RCPs made any attempt to provide a best guess or to assess the likelihood of the various scenarios. Many elements of the scenarios are too unpredictable for this to be feasible. As a consequence, all the scenarios should be considered as reasonably possible and equally probable.

### 6.1.2  Special Report on Emission Scenarios (SRES)

Among the infinite number of possible futures, four families have been proposed, comprising forty SRES scenarios covering a wide range of possibilities. Each family includes a so-called storyline, providing a coherent descriptive narrative of the choices made. The four families can be described very briefly as follows (for more details, see Nakicenovic and Swart 2000):

- A1 corresponds to very rapid economic growth, low population increase and the rapid introduction of efficient technologies. The A1 family assumes strong interactions between different countries and a reduction in regional differences in per capita income. In addition, the A1 family is separated into four groups related to technology choices, one group, for instance, being devoted to fossil-intensive energy production.
- A2 corresponds to a slow convergence between regions and a high population growth. Technological changes are more slowly implemented than in the other storylines, with more disparity between the regions.
- B1 corresponds to a low population growth and strong convergence between regions but with faster introduction of clean and resource-efficient technologies than A1.
- B2 corresponds to intermediate population and economic growth with less rapid introduction of new technologies than in the B1 and A1 storylines. It assumes an emphasis on local and regional solutions.

From these storylines, different research groups have proposed different scenarios. From the resulting options, four marker scenarios were selected, one to illustrate each storyline. Two additional scenarios were selected in the A1 family to illustrate alternative developments in energy systems. This resulted in six scenarios which have been used to perform climate projections.

It is important to remember that none of those storylines involves clear climate initiatives or climate-related regulations, although the policy choices described in the various scenarios would have a substantial impact on the emissions of greenhouse gases and aerosols. In all six illustrative scenarios, the emissions of $CO_2$ increase during the first decades of the twenty-first century (Figure 6.2). This trend continues up to 2100 in three scenarios, whilst the emissions peak between 2030 and 2050 and then decrease in scenarios A1T, B1 and A1B. Based on computations made by the teams which developed the scenarios, this induces an increase in atmospheric $CO_2$ concentration in 2100 of up to nearly 1000 ppm in scenario A1F and a bit less than 600 ppm in scenario B1. This last value

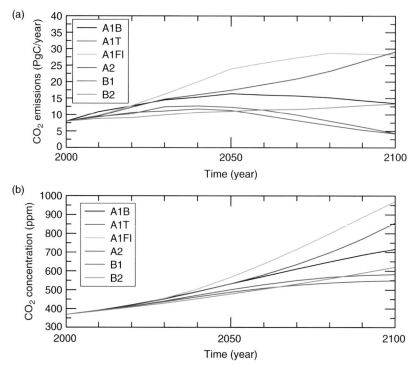

(a) Global emission (in PgCO₂ yr⁻¹) and (b) atmospheric concentration of CO₂ (in ppm) in the six illustrative SRES scenarios (A1B, A1T, A1FI, A2, B1 and B2).

roughly corresponds to a doubling of the $CO_2$ concentration compared to the pre-industrial level (~280 ppm; see Section 2.3.1).

SRES scenarios also provide estimates for future emissions and concentrations of other greenhouse gases (e.g., $N_2O$ and $CH_4$), as well as emissions of sulphur dioxide ($SO_2$), which leads to the production of sulphate **aerosols** in the atmosphere. In contrast to $CO_2$, $SO_2$ emissions reach their maximum in all the scenarios during the first half of the twenty-first century and then decrease (Figure 6.3), thanks to policies devoted to reducing air pollution. Because of the relatively short life of aerosols in the atmosphere (see Section 4.1.2.2), sulphate concentration changes in roughly the same way over time as emissions. As a consequence, the negative **radiative forcing** due to aerosols (see Figure 4.2) will decrease during a large part of the twenty-first century, whilst the positive forcing due to greenhouse gases will increase continuously in most scenarios.

## 6.1.3 Representative Concentration Pathways (RCPs)

The most extreme of the selected RCPs, RCP8.5, displays a continuous rise in **radiative forcing** during the twenty-first century, leading to a value of about 8.5 W m⁻² in 2100. RCP6.0 and RCP4.5 are characterised by a steady rise during the twenty-first century, up to a radiative forcing of about 6 and 4.5 W m⁻², respectively, and a stabilisation after 2100. Finally, in RCP2.6, the radiative

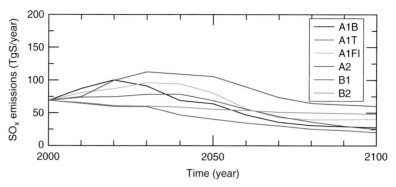

Fig. 6.3 Global emissions of sulphur oxides in six illustrative SRES scenarios (in TgS yr$^{-1}$).

forcing peaks before 2100 at about 3 W m$^{-2}$ and then declines. Emissions and atmospheric concentrations of CO$_2$ corresponding to these RCPs are shown in Figure 6.4. As expected, with CO$_2$ being the largest contributor to radiative forcing (see Figure 4.2), the time series of atmospheric CO$_2$ concentration have the same shape as the time series for radiative forcing. The emissions cover a wide range of possibilities with, for instance, CO$_2$ emissions of more than 25 PgC yr$^{-1}$ at the end of the twenty-first century in RCP8.5 (i.e., more than three times greater than in 2000). By contrast, the emissions are close to 0 after 2080 in RCP2.6. This implies strong **mitigation** actions and corresponds to values which are lower than in all the representative SRES scenarios discussed in Section 6.1.2 (see Figure 6.2 for comparison). Note that the estimates of the correspondence between emissions and concentrations for RCPs (and SRES scenarios) were derived using particular models and specific hypotheses. A climate model including a carbon-cycle model and thus its own representation of climate-carbon feedbacks, driven by the same emission or concentration scenario, will lead to different results than those plotted in Figure 6.4 (as discussed in Section 6.2.9).

The RCPs include estimates of emissions of a larger number of greenhouse gases and atmospheric pollutants than in the SRES scenarios (CH$_4$, N$_2$O, **chlorofluorocarbons**, SO$_2$, black carbon, etc.) as well as estimates of future changes in land use. For instance, in all the RCPs, the decrease in SO$_2$ emissions (Figure 6.5) is even larger than in the SRES scenarios (see Figure 6.3).

Furthermore, the RCPs have been extended to 2300 and even 2500 for studies of long-term climate change (Figure 6.6). Because of the very large uncertainties in the driving forces influencing the emissions, the long-term scenarios are kept as simple as possible and thus are highly idealised. Nevertheless, they provide a reasonable range for the possible changes, give time developments compatible with the RCPs over the twenty-first century and display a common framework in which the results of different models can be displayed. Among the various possible extrapolations, a forcing stabilisation path emerges for RCP4.5 and RCP6.0. For RCP2.6, the forcing is assumed to continue to decrease after 2100. For RCP8.5, the extension suggests an increase in forcing until at least 2200, although the emissions growth slows in the second part of the twenty-first century, leading to a nearly flat profile after 2100.

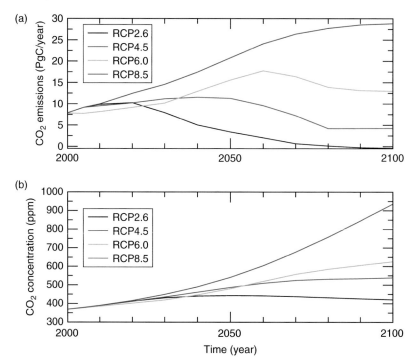

**Fig. 6.4** (a) Global emission (in PgC yr$^{-1}$) and (b) atmospheric concentration of CO$_2$ (in ppm) in four RCP scenarios.

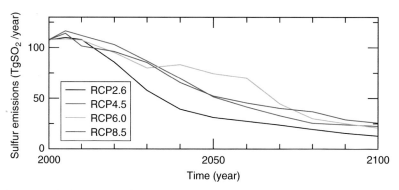

**Fig. 6.5** Global emissions of sulphur dioxide in four RCP scenarios (in TgSO$_2$ yr$^{-1}$).

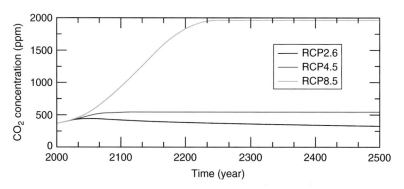

**Fig. 6.6** Global atmospheric concentration of CO$_2$ (in ppm) in three RCP scenarios.

## 6.2 Climate Changes over the Twenty-First Century

### 6.2.1 Model Ensembles

Most of the simulations discussed in the following sections are driven by the most recent SRES **scenarios** and have been performed in the framework of the fifth phase of the Coupled Model Inter-Comparison Project (CMIP5) (Taylor et al. 2012) (see also Section 3.5.2.2). CMIP5 was a major source of information for the Fifth Assessment Report of the IPCC (IPCC AR5) (see IPCC 2013) which is a basis for this chapter. All the simulations performed with the various models used the same experimental design. Note, however, that the ways in which the CMIP5 recommendations are applied in each of the models still may differ slightly between them because, for instance, they have components such as their soil model which require different inputs.

The mean and spread of this **multi-model ensemble** are classically used as a simple way to estimate the agreement between the models and the uncertainty of the response for a particular scenario. In an ideal case, the averaging of available simulations would strongly reduce the random errors present in each of the individual models and would give a more reliable estimate of the changes. This reasoning seems valid, for instance, for the mean state of the system because the **ensemble mean** is generally closer to observations than any of the models for present-day conditions (see Section 3.5.3). However, the interpretation should be performed with caution because the CMIP5 ensemble is an ensemble of opportunity based on existing models. It has not been specifically designed to cover the whole range of possible outcomes and may well underestimate this range. Furthermore, the models themselves are not all independent, as they may share common parameterisations, parts of the codes or even components and have similar spatial resolution. The models thus may have the same types of biases, leading to a strong agreement between them but low confidence in their results and no real improvement coming from their averages. Consequently, the uncertainty should be based on a physical understanding of the processes involved in the response, in addition to the spread of the ensemble.

Besides the systematic biases, the ensemble mean tends, by construction, to reduce the magnitude of the changes, in particular, if individual members of the ensemble have responses of different signs. It is standard to test the significance of the response of models to forcing by comparing the mean of the ensemble to the averaged **internal variability** of the system. The latter is estimated from the inter-annual variability of control simulations performed with the same models using a constant forcing (see, e.g., Figure 6.11). If the mean change simulated for a particular variable is larger than two standard deviations of its inter-annual variability, the influence of the forcing is considered to be dominant in the time development of this variable. If it is smaller than one standard deviation of the inter-annual variability, the internal variations largely would mask the forced response. Nevertheless, a mean signal smaller than one standard deviation of the inter-annual variability can be related to two different situations, and additional diagnostics are required to distinguish those two cases. Firstly, all the models can agree that the response to the forcing is weak. Second, all the models can simulate a large response, but because they strongly disagree on its sign, their mean is close to 0.

We also have to be careful when averaging the results of all the models because it may lead to a state which is not physically consistent and not representative of possible future changes. One clear example is the internal variability of the system. As it is uncorrelated between the **ensemble members**, averaging all the simulations strongly reduces their magnitude (see Figure 5.38). This, on the one hand, has the advantage of identifying more easily the common response to the forcing. On the other hand, the ensemble mean cannot be used directly to assess changes in the total variance of the system or in the probability of extreme events.

Instead of taking into account all the available simulations, it is tempting to select only the 'best' projections performed with the 'best' models or at least to give them a larger weight in the ensemble mean. This potentially would reduce the spread of the ensemble and in this way the uncertainty of the projections. In contrast with weather forecasts, for which the performance of the prediction system can be evaluated via a comparison of a large number of daily forecasts with observations, we could not estimate directly the **skill** of the projections themselves for the next decades or century. An alternative is to consider that the 'good' models are the ones which are closest to observations for recent or some past conditions.

There are a few cases where model selection appears relatively simple. If one model does not have the minimal representation of key mechanisms required for a meaningful climate projection for a particular variable or has a mean state that is so biased that projections are meaningless, then it can be rejected. A simple example would be a model which is simulating a totally ice-free Arctic in summer for present-day conditions, in disagreement with observations (see Section 1.4). Such a model obviously would be discarded from any analysis focussed on the probability of a future disappearance of sea ice in summer at high latitudes.

More generally, although some 'good' models perform better than others for many fields for present-day conditions, no model is better than others on all aspects. The criteria for model selection therefore must be based on a strong understanding of the processes which are essential to simulate the response of the variable of interest to the **forcing**. A link between the simulated state for present-day or any past condition and the future response to the forcing also should be demonstrated to justify a choice based on the agreement between model results and observations for a particular period. This is not straightforward for many of the variables because models with positive or negative biases in their mean state may display, for instance, the same range of future changes. If the magnitude of this bias is used as a selection criterion, there is a strong risk that the spread of the ensemble will be reduced for wrong reasons and that the projections will become over-confident. This topic of model selection and weighting is currently a subject of intense research (see, e.g., Knutti et al. 2010). Nevertheless, systematic and objective model selection or weighted averages based on model performance have been applied successfully up to now in a few specific studies only (see, e.g., Massonnet et al. 2012). This is the reason why such an approach will not be discussed in the following sections, which are based on the full ensemble of the available models, the ensemble mean being a simple average with the same weight for each model.

## 6.2.2 Decadal Predictions and Projections

Most of the studies devoted to future climate changes have analysed **projections** for the late twenty-first century and beyond. They are based on simulations initialised in the nineteenth century (the year 1850 in the CMIP5 protocol) or earlier, which are driven by observed changes in external forcing up to the present and then by some scenarios. This can be referred to as the solution to a 'boundary-condition problem' as it only deals with the response to the forcing – considered in this framework as a boundary condition. This procedure does not take any direct advantage of the observations collected over the last decades. The goal is to reproduce adequately the statistics of the late twentieth century and early twenty-first century climate (e.g., mean temperature over a decade, probability of extreme events, etc.). Besides, the results of the simulations are quite different from observations for any individual year in the past and provide only limited information for a specific year in the future because the **internal variability** of the modelled system displays a different timing compared to observations (see also Section 5.1.1).

Additional information thus may be gained by starting the experiments from a state which is close to observations, as is done, for instance, in weather forecasts. In particular, the phase of all the modes of internal variability should be similar initially in the model and in the observations. Because of the chaotic nature of the climate system, the unavoidable errors in the initial state (present because of a lack of data and instrumental errors, for instance) and in model physics grow with time. Nevertheless, we can expect that the uncertainties in the estimate of the future changes are reduced over a few years thanks to the initialisation. After some time, the range of simulated states becomes identical, as in projections (Figure 6.7). To make the distinction with the projections, the simulations initialised using observations are referred to as '**predictions**', 'climate forecasts' or, more specifically, 'decadal predictions', as it is often the time horizon of those experiments. By contrast, the projections are sometimes called 'uninitialised simulations' (see, e.g., Meehl et al. 2009, 2014). Predictions also include the response to the forcing; they are thus a mix of initial and boundary-condition problems, whilst weather forecasts can be considered to be 'initial-value problems' because of the dominant role of initial conditions in the quality of the forecast.

The difference between initial-value and boundary-condition problems can be illustrated by two simple examples. In the first case, by analysing satellite images for present-day conditions, we can estimate the direction of the winds and the position and velocity of the clouds in order to make a weather forecast for tomorrow. In the second case, we can state that the temperature at middle latitudes next winter will be, on average, lower than in summer because we know the changes in boundary conditions responsible for the seasonal cycle.

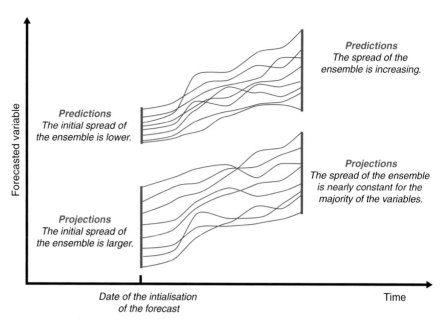

**Fig. 6.7**   Schematic representation of the difference between projections and predictions. Each blue line represents
a member of an ensemble of simulations performed with one particular model and using the same forcing
scenario. Projections are not initialised using observations. Although each simulation includes internal
variability, the ensemble can only provide an estimate of the response to the forcing represented here by
the long-term trend. Because of the initialisation which constrains the state of the system to be close to
observations, including the contribution of internal variability, the uncertainty of the predictions is smaller at
the beginning of the experiment but becomes similar to the uncertainty of the projections, generally after a few
months to a few years depending on the variable and the region investigated.

The growth rate of the errors is so large that there is no hope to have accurate
weather forecasts over two weeks. The predictions thus would never be able to
provide any precise estimate of the temperature or precipitation during a particular day two years in advance. Nevertheless, some general information on future
climate may be obtained, for instance, an estimate of the global mean surface
temperature or the probability of having a mean winter temperature in a region
which is higher than over a reference period. This distinction between the type of
information expected from the predictions or projections and weather forecasts
is already present in the preceding example, where it was stated that knowing the
processes responsible for the seasonal cycle, we can compare mean conditions in
summer and winter but not guess the exact temperature at a specific time.

More generally, a part of the variations not attributed directly to the forcing
will be intrinsically unpredictable. Unfortunately, the fraction of the variance
related to internal variability which is predictable from an adequate estimate of
initial conditions is still largely unknown. Model simulations suggest that the initialisation has an impact over a few years for surface temperature, with some maxima up to about a decade over the oceans at middle to high latitudes (Figure 6.8),
but observations are too short to assess this precisely. The **predictability** is lower

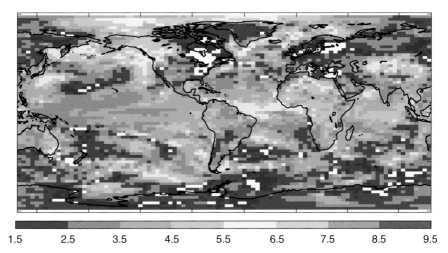

1.5        2.5        3.5        4.5        5.5        6.5        7.5        8.5        9.5

**Fig. 6.8**   The number of years during which the difference between the surface temperatures obtained in initialised and uninitialised simulations is significant at the 90% level. The figure represents the average over simulations started in 2011 and 2012 performed with ten climate models. (Source: Smith et al. 2013. Used with kind permission from Springer Science and Business Media.)

for precipitation, which is more variable, both in space and time, than surface temperature.

Several processes can be responsible for the predictability. Firstly, inertia, in particular, the inertia of the ocean, may lead to the persistence of existing **anomalies**. For instance, if initialised from a colder than average state, the temperature of the system may remain low during some time, leading to additional prediction skill compared to uninitialised simulations. Secondly, the time evolution of some modes of climate variability might be estimated knowing the current state of the system. In this framework, the oceanic circulation in the North Atlantic has received a lot of attention. Some of its components are predictable several years in advance in many models because of the long-term influence of the ocean temperature and salinity anomalies present in the initial conditions. Better estimates of the temperature of the ocean surface and of the adjacent continents then might be obtained from this information on the state of the North Atlantic circulation.

Although mechanisms have been proposed to justify predictability at inter-annual to decadal time scales and models suggest that initialisation influences the quality of predictions, achieving an actual skill in real conditions given the available observations requires tackling several challenges. Firstly, the number of observations in the ocean, which is a dominant source of predictability, has increased dramatically in recent years. Nevertheless, there are still some regions where coverage is insufficient to precisely estimate the initial state of the system, in particular, at depth. Secondly, only some variables, often at specific locations only, are observed, whilst the models need initial conditions for all their **state variables** at all the grid points. Estimating the full state of the system from such an incomplete set of observation is the subject of data assimilation (see Section 3.6.2). The field is mature for the atmosphere, thanks to the experience gained in weather forecasting, but new developments are still needed when dealing with the full climate system. Thirdly, generating an ensemble of simulations in such a way that it provides a reliable estimate of future changes and of the uncertainty of this estimate is challenging. If the range of the ensemble is too narrow, the prediction will tend to be over-confident, underestimating the uncertainty. By contrast, if the ensemble is too wide, the forecast may conclude that the predictability is low, whilst it may be potentially higher. Finally, the model biases must be taken into account. This is sometimes done directly during initialisation or through correction of the climate forecast, as explained in Section 3.6.1.

Predictions on a decadal time scale are a new domain of research, as illustrated by the very small number of studies published before 2007. It is, however, a growing field both because of the rise of important scientific questions and because of the high relevance of the decadal time scale for policy and decision makers. The Fifth Assessment Report of the IPCC has already reported the skill of decadal predictions compared to observations for global mean surface temperature, as well as for the surface temperature in many regions (Kirtman et al. 2013). The initialisation is mainly responsible for this skill over the first five years of the simulation. Later, the skill is mainly due to the response to the forcing. We should expect significant additional developments in the near future to determine more precisely the regions where the predictability is high on an inter-annual to decadal time scale compared to the perhaps large part of the world where the system is largely unpredictable on this time scale. The methodologies applied for decadal predictions, as well as how to interpret those projections, also should improve in coming years.

### 6.2.3   Changes in Global Mean Surface Temperature

In response to the specified forcing changes, a surface warming during the twenty-first century is obtained for all the RCP scenarios, but the magnitude of the increase is strongly different, showing the potential impact of **mitigation** policies. For scenario RCP2.6, the global annual mean temperature rise relative to pre-industrial values remains below 2°C during the whole of the twenty-first century for the average of the results of the models included in CMIP5. This is not the case for the other scenarios, with, on average, a warming for 2081–2100 relative to 1850–1900 of 2.4°C for RCP4.5, 2.8°C for RCP6.0 and 4.3°C for RCP8.5 (Figure 6.9). After 2100, the global temperature rise is still large for RCP8.5, whilst temperature nearly stabilises in RCP 4.5 and slightly decreases during the twenty-second century for scenario RCP2.6.

Figure 6.9 illustrates two sources of uncertainty in climate projections. The first is related to the **scenario**, as discussed earlier. The second source is model uncertainty, with different models displaying different responses to the same forcing. This is indicated in Figure 6.9 by the range of the results of all the models. Additional uncertainty is due to the **internal variability** of the system, that is, the natural fluctuations which would occur even in the absence of any change in **radiative forcing** and not predictable in the long term (see Section 6.2.2).

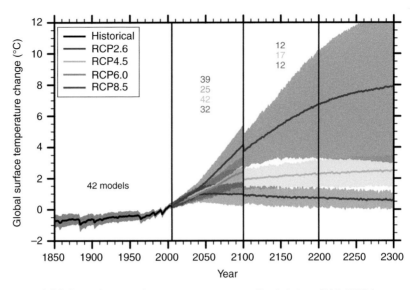

**Fig. 6.9** Time series of global annual mean surface air temperature **anomalies** (relative to 1986–2005) from an ensemble of model simulations performed in the framework of CMIP5. Projections are shown for each RCP for the multi-model mean (solid lines) and the 5–95% range (±1.64 standard deviations) across the distribution of individual models (shading). Discontinuities at 2100 are due to different numbers of models performing the extension runs beyond the twenty-first century and have no physical meaning. Numbers in the figure indicate the number of different models contributing to the different time periods. (Source: Collins et al. 2013.)

The relative importance of the three sources of uncertainty can be estimated for projections over different time periods (also known as 'lead times'). For estimates of the temperature over the next decade, the influence of the uncertainty about future emissions of greenhouse gases is small. This is consistent with Figure 6.9, where the curves for all the RCP scenarios lie close to each other until 2030–2040. On a global scale (Figure 6.10a), the relative importance of scenario uncertainty increases with time and is dominant in projections for the end of the twenty-first century. The internal variability plays a role for a few decades only, the internal fluctuations in global mean temperatures over decades and centuries having a much smaller magnitude than the changes expected by 2100. Model uncertainty is dominant for projections up to forty years ahead, but its relative contribution then decreases, although it is still significant in 2100. When analysing temperature changes over a smaller region such as Europe (Figure 6.10b), each source of uncertainty has more or less the same time evolution as discussed for the Earth as a whole. Additional uncertainty may be present on the forcing itself because of a weak knowledge of the potential contribution of aerosols and land-use changes on a regional scale. The clearest differences, however, are the larger contributions of modelling uncertainty and the internal variability in Figure 6.10b, as internal fluctuations have a higher amplitude on a regional scale than on a global scale (see Section 5.1.1).

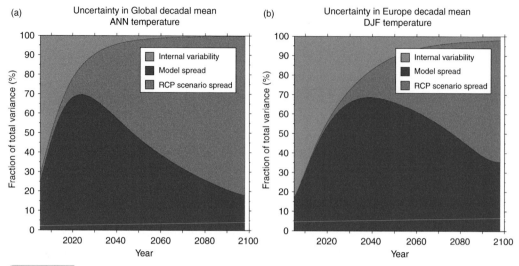

**Fig. 6.10** The fraction of total variance in decadal mean surface air temperature projections explained by the three components of total uncertainty is shown for (a) a global average of annual mean temperature and (b) winter (December-January-February) mean in Europe. Green regions represent scenario uncertainty, blue regions represent model uncertainty and orange regions represent the internal variability component. (Sources: Kirtman et al. 2013, based on Hawkins and Sutton 2009.)

## 6.2.4 Spatial Distribution of Surface Temperature Changes

The increase in global mean temperature by 2100 is associated with a warming in all regions for the multi-model average (Figure 6.11). The spatial distribution is similar in all the scenarios, with firstly a larger temperature change over land than over the ocean. This land-sea contrast, commonly defined as the warming over land divided by the warming over sea, is estimated to be around 1.5 for all the scenarios (Collins et al. 2013). The large thermal inertia of the ocean and the associated heat uptake play a role in some regions (see Section 2.1.5.1). This is the case at high latitudes, where the warming is particularly low because of the deep mixed layer and the contact with colder deep water which has not been exposed recently to surface warming. Nevertheless, a global land-sea warming contrast of similar magnitude is also found at equilibrium in response to a perturbation at a time when thermal inertia no longer plays a role.

The different moisture availabilities over ocean and land are a dominant cause of this land-sea contrast at equilibrium (Figure 6.12). Over the ocean, a warming will lead to higher evaporation and thus higher latent heat loss, which can mitigate the temperature change there, whilst evaporation will be limited over land as a function of the soil moisture content (see Sections 2.2.2 and 4.2.4). Furthermore, the increase in humidity resulting from the warming (see Section 6.2.5) induces a decrease in the saturated **lapse rate** (see Section 1.2.1). Since the lapse rate is closer to the saturated lapse rate in more humid environments than in a drier ones, the decrease in the lapse rate tends to be higher above the ocean than over land. If we assume that the temperature increase is relatively spatially homogeneous above

RCP2.6                                        RCP8.5
Change in average surface temperature (1986–2005 to 2081–2100)

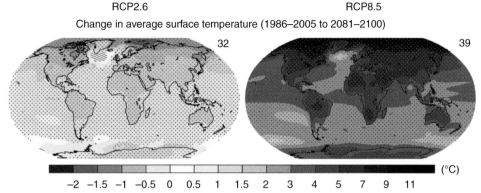

**Fig. 6.11**   Multi-model mean of surface temperature change for the RCP2.6 and RCP8.5 scenarios in 2081–2100 relative to 1986–2005. The number of CMIP5 models to calculate the multi-model mean is indicated in the upper right corner of each panel. Hatching indicates regions where the multi-model mean change is less than one standard deviation of internal variability. Stippling indicates regions where the multi-model mean change is greater than two standard deviations of internal variability and where at least 90% of models agree on the sign of the change. (Source: IPCC 2013.)

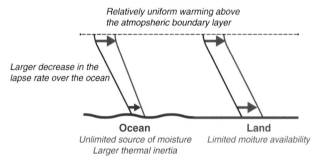

**Fig. 6.12**   Schematic representation of mechanisms influencing the land-sea contrast on global and regional spatial scales. (Source: Modified from Joshi et al. 2013.)

the **atmospheric boundary layer** because of large-scale transport and mixing, this will lead to a larger warming over land (Joshi et al. 2013). The difference in moisture availability between land and ocean also can be responsible for reduction in the low-level cloud cover in some land regions, further amplifying the warming.

Secondly, the changes simulated for the Arctic are also much larger than at middle latitudes. This so-called polar amplification, which is measured by the ratio between the mean warming at high northern latitudes (67.5–90°N) and the global average warming, is generally a bit higher than 2 in models for changes in 2081–2100 compared to 1986–2005 (Collins et al. 2013). The polar amplification is seen in both hemispheres in many past periods, but for the twenty-first century, it is mainly an Arctic phenomenon because of the large thermal inertia of the Southern Ocean. As the models reach equilibrium, the warming rate becomes higher there compared to other regions, leading to a strong polar amplification in the southern hemisphere too.

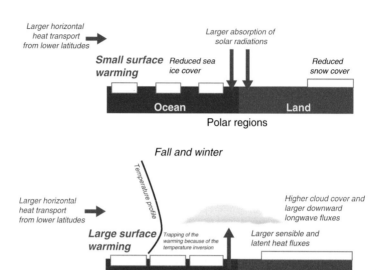

Spring and summer

Fall and winter

**Fig. 6.13**    Some processes potentially playing a role in the polar amplification.

Several processes take part in the polar amplification (Figure 6.13), but the relative role of each is still relatively uncertain (Serreze and Barry 2011). Major contributors are the snow-and-ice–albedo (or temperature-albedo) feedback (see Section 4.2.2) and the larger exchanges between the ocean and the atmosphere associated with the future retreat of sea ice and snow cover. The snow-and-ice–albedo feedback is mainly active in summer and spring. The additional solar radiation absorbed because of the decrease in the surface albedo induces ice melting or warms the ocean. However, this has only a limited effect on the temperature at this time of the year, which stays close to the freezing point in most of the Arctic (see Figure 6.13). After this period characterised by small surface temperature changes, the polar amplification peaks in autumn and winter as the heat stored in summer is released to the atmosphere at the ocean interface and through the thinner ice, leading to a strong surface air warming. During this season, the Arctic atmosphere displays a temperature inversion in the low levels owing to the large radiative cooling of the surface and the atmospheric heat inflow in the middle atmosphere. The resulting strong **stratification** reduces the vertical exchanges, traps the warming close to the surface, and is associated with a positive **lapse-rate feedback** if we follow the decomposition proposed in Section 4.2.

Additionally, the atmosphere has a larger water-holding capacity in a warmer world (see Section 6.2.5). The poleward horizontal transport of latent heat thus is expected to increase. This can be compensated for in part by the decrease in sensible heat transport resulting from the lower temperature difference between low and high latitudes associated with the larger warming there. Modified characteristics of the water masses in the Arctic and North Atlantic, as well as changes in the wind stress at the surface of the ocean, also can have an impact on the northward oceanic heat transport. Furthermore, higher temperatures and humidity in the Arctic may lead to an increase in the cloud cover which both reduces incoming solar radiation and increases the downward longwave fluxes. Among those effects, that of the low clouds on the longwave fluxes is the most important for the surface temperature changes in many models, contributing to the polar amplification.

## 6.2.5 Spatial Distribution of Precipitation Changes

A direct consequence of the higher temperatures is an increase in the **saturation vapour pressure**, as described by the **Clausius-Clapeyron equation**. Following this law, the water content of the atmosphere should increase by about 7% per degree Celsius. This is very close to the results obtained in climate models and means that the **relative humidity** remains constant at the first order. When refining this point, it appears that the relative humidity slightly decreases over most land areas. The temperature increase is larger over land than over the ocean (see Section 6.2.4). As a consequence, if humid air is transported from the ocean to the land, its saturation vapour pressure will increase because of the warming. In the absence of additional moisture sources, its relative humidity will decrease.

In parallel to this change in humidity, precipitation is strongly influenced by the energy budget of the climate system because the latent heat released in the atmosphere when water vapour condenses before precipitating balances to a large extent the radiative deficit of the atmosphere (see Section 2.1.6). As a consequence, evaporation and precipitation are strongly linked to changes in the radiative fluxes in the atmosphere. This leads in models to an increase in precipitation in a warmer world at a rate of about 1–3% per degree Celsius, which is much smaller than the increase in humidity.

The current uncertainty about precipitation changes is higher than for temperature and is dominated by model spread, illustrating a low agreement between the results of different models driven by the same scenario (Figure. 6.14). By contrast, the relative contribution of forcing uncertainty is much more modest. Spatially, the pattern of precipitation changes during the twenty-first century projected by climate models is more complex than that of temperature (Figure 6.15). There are also fewer regions where the changes are significant, in many cases because of the disagreement between model results.

Some features, however, are robust. First, the large-scale structure can be interpreted as an amplification of existing differences in precipitation minus evaporation $(P-E)$, often referred to as the 'wet-get-wetter' and 'dry-get-dryer response' (Held and Soden 2006). This can be explained by a purely thermodynamical reasoning. If atmospheric humidity increases as a consequence of the higher temperature and the atmospheric circulation remains constant, moisture transport will increase. The regions in which there is presently a net moisture inflow then will receive more water, leading to an increase in precipitation, whilst regions with a net moisture deficit will have an even higher deficit (Figure 6.16b). This justifies why models predict an increase in precipitation at high latitudes and close

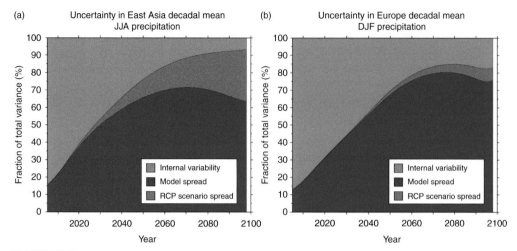

**Fig. 6.14**  The fraction of total variance in decadal mean projections of precipitation changes explained by the three components of total uncertainty is shown for (a) summer (June, July and August) mean in East Asia and (b) winter (December, January and February) mean in Europe. Green regions represent scenario uncertainty, blue regions represent model uncertainty and orange regions represent the internal variability component. (Sources: Kirtman et al. 2013, based on Hawkins and Sutton 2009.)

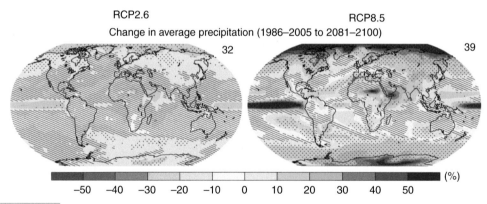

**Fig. 6.15**  Multi-model mean of average per cent change in annual mean precipitation for the RCP2.6 and RCP8.5 scenarios in 2081–2100 relative to 1986–2005. The number of CMIP5 models to calculate the multi-model mean is indicated in the upper right corner of each panel. Hatching indicates regions where the multi-model mean is less than one standard deviation of internal variability. Stippling indicates regions where the multi-model mean is greater than two standard deviations of internal variability and where at least 90% of models agree on the sign of change. (Source: IPCC 2013.)

to the equator, where precipitation exceeds evaporation in present-day conditions (see Section 2.2). By contrast, precipitation is predicted to decrease over many areas which are already dry in the subtropics. The higher atmospheric humidity also explains why the **monsoon** precipitation is expected to increase during the twenty-first century, although the monsoon circulation itself may decrease in some regions.

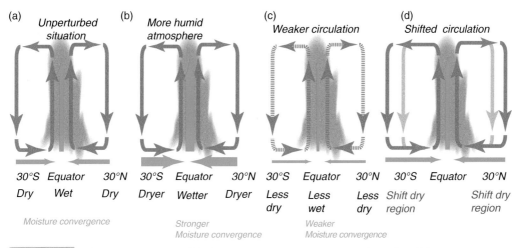

(a) *Unperturbed situation*

(b) *More humid atmosphere*

(c) *Weaker circulation*

(d) *Shifted circulation*

| 30°S | Equator | 30°N | 30°S | Equator | 30°N | 30°S | Equator | 30°N | 30°S | Equator | 30°N |
|---|---|---|---|---|---|---|---|---|---|---|---|
| *Dry* | *Wet* | *Dry* | *Dryer* | *Wetter* | *Dryer* | *Less dry* | *Less wet* | *Less dry* | *Shift dry region* | | *Shift dry region* |

*Moisture convergence*     *Stronger Moisture convergence*     *Weaker Moisture convergence*

**Fig. 6.16**   Schematic representation of the changes in precipitation associated with the Hadley cell due to an increase in specific humidity, a reduction in the strength of the overturning circulation and a shift in the location of the subsidence.

This reasoning is more valid over the oceans than over land, where the changes in water availability can play a dominant role. Furthermore, in addition to this thermodynamical component, dynamical changes associated with modification of the atmospheric circulation also must be taken into account. In tropical regions, the Hadley circulation, which is associated with upward transport and strong precipitation at the equator and **subsidence** of dry air in subtropical regions, is expected to weaken in the future. This would partly compensate for the effect of the thermodynamical component by reducing precipitation close to the equator and increasing it in the subtropics (Figure 6.16c). A similar reasoning also can be applied to the decrease in intensity of the Walker cell, simulated by models in response to the warming. Furthermore, the Hadley cells are also projected to expand, inducing a northward shift of the subsidence (Figure 6.16d), reinforcing the drying in some regions around the Mediterranean Sea, for instance.

Over the tropical oceans, the precipitation tends to increase in regions which are warming more than the mean over all the tropics. This feature, which seems robust in many models, has been referred to as the 'warmer-get-wetter' pattern and has been related to the influence of the local sea surface temperature on the vertical stability of the atmosphere, on vertical motion and thus on precipitation (see, e.g., Huang et al. 2013).

## 6.2.6  Changes in the Ocean and Sea Ice

The warming simulated at high latitudes is associated with year-long decreases in the extent and the thickness of sea ice in both hemispheres (Figure 6.17). The differences between the projections provided by the various models are quite large, and so are the uncertainties, in particular, in the Southern Ocean, where models also have trouble reproducing present-day conditions (see Section 3.5.2.3). There

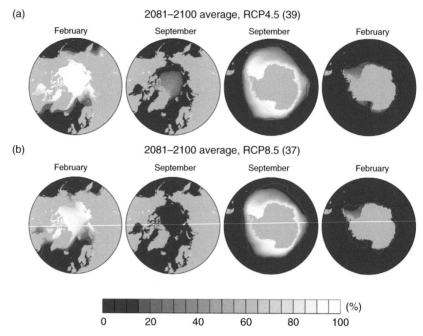

(a)                                    2081–2100 average, RCP4.5 (39)

(b)                                    2081–2100 average, RCP8.5 (37)

**Fig. 6.17**   February and September CMIP5 multi-model mean sea-ice concentrations (%) in the northern and southern hemispheres for the period 2081–2100 under (a) RCP4.5 and (b) RCP8.5. The pink lines show the observed 15% sea-ice concentration limits averaged over 1986–2005 (Comiso and Nishio 2008). The number of models available for each RCP is given in parentheses. (Source: Collins et al. 2013.)

is, however, an agreement on the conclusion that the projected decrease will be more pronounced in summer in the Arctic. As a consequence, although some ice still will be present in winter, many simulations forecast a totally ice-free Arctic in summer before the end of the twenty-first century, the best estimates being between 2040 and 2060 for scenario RCP8.5.

Ocean circulation is also projected to change during the twenty-first century. Because of the warming and the increase in precipitation at high latitudes (see Sections 6.2.4 and 6.2.5), the density of the water at the surface will tend to decrease, increasing the stratification in many regions. In the North Atlantic, this would imply less sinking of dense water and a weaker southward transport of dense water. As a consequence, the northward transport of warm surface water also will decrease, with potential implications for the heat budget of the North Atlantic and the surrounding regions. In particular, the changes in ocean currents may be responsible for the small amount of warming predicted for the North Atlantic (see Figure 6.11).

The intensity of the **thermohaline circulation** is generally measured by the maximum of the Atlantic **meridional overturning circulation** (MOC) in the North Atlantic, although the two concepts are slightly different (see Sections 1.3.2 and 4.2.5). The scatter of the results for the MOC from the different GCMs is very large, both for present-day and future conditions (Figure 6.18), but they nearly all simulate a decrease during the twenty-first century. The mean value depends

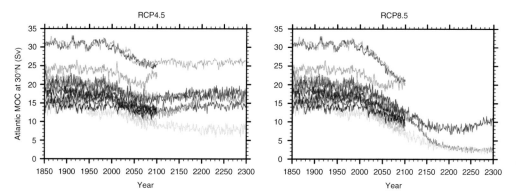

The changes in the Atlantic **meridional overturning circulation** (MOC) at 30°N in simulations with several coupled climate models (each in a different colour) from 1850 to 2300 in the RCP4.5 and RCP8.5 scenarios (in Sv; $1\,\text{Sv} = 10^6\,\text{m}^3\,\text{s}^{-1}$). (Source: Collins et al. 2013.)

on the scenario and is about 35% in RCP8.5 (Collins et al. 2013). When the forcing stabilises as in RCP4.5, the magnitude of the MOC tends to increase back, and none of the model simulates a complete collapse of the circulation for the twenty-first century (see also Section 6.3.3).

## 6.2.7  Changes in Modes of Variability

The main modes of variability explain a significant part of the present climate variations on a regional scale (see Section 5.2). This will continue in the future but with potential changes in the mean and variance of their index and/or in their spatial characteristics and **teleconnections**. Such modifications in the modes of variability, however, should not be mixed up with a trend in the mean state of the system. For instance, a general warming of the North Atlantic as a result of the thermodynamical response to a radiative perturbation should not be associated with a shift to a more positive phase of the Atlantic Multi-Decadal Oscillation (AMO) (see Section 5.2.4), although the latter is also characterised by a warming of this region.

Currently, many uncertainties remain in terms of the future evolution of these modes of variability (see, e.g., Collins et al. 2010; Christensen et al. 2013). The changes in the index of El Niño–Southern Oscillation (ENSO) and AMO are expected to be weak compared to the natural variability of the system, although the probability of an intense El Niño event may increase because of the warming in the east equatorial Pacific (see, e.g., Cai et al. 2014). Moreover, both the NAO index (see Section 5.2.2) and the Southern Annular Mode (SAM) index (see Section 5.2.3) become higher in model projections for the twenty-first century than for the twentieth century on average over December to January, that is, in winter in the northern hemisphere and in summer in the southern hemisphere (Figure 6.19). The changes in the NAO index are weak, whilst a stronger slope is seen for the SAM index during the last decades of the twentieth century before stabilisation in scenario RCP4.5. This summer trend is due to the combined influence of the greenhouse gas forcing and ozone depletion in the stratosphere (with the role of ozone

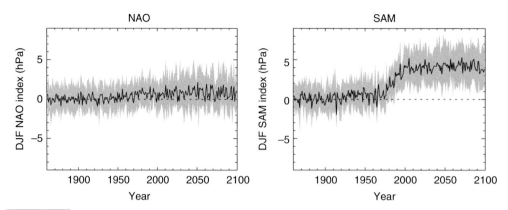

Multi-model ensemble simulations of December-to-February (DJF) mean NAO and SAM indices for historical and RCP4.5 scenarios using thirty-nine models. The ensemble mean is in black, and the inter-quartile range is in grey shading. The NAO index is defined here as the difference of regional averages of sea-level pressure: (90°W–60°E, 20°N–55°N) minus (90°W–60°E, 55°N–90N). The SAM index is defined as the difference in zonal mean sea-level pressure at 40°S and 65°S. All indices have been centred to have zero time mean from 1861–1900. (Source: Christensen et al. 2013.)

dominating) which affects the meridional temperature gradients at different heights and then the strength and position of the westerly winds. Over the twenty-first century, because of the projected recovery of ozone, the two effects tend to compensate each other, and the simulated SAM index does not increase further.

## 6.2.8 Changes in Climate Extremes

Changes in annual and seasonal mean temperature, precipitation or sea-ice extent are important elements of the projected climate for the twenty-first century. However, many other characteristics of the atmospheric state, such as wind intensity or cloud amount, are also expected to change. A particularly sensitive point is the modification in the probability of extreme events (e.g., heavy rains and heat waves) in a warmer climate. Such extreme events are difficult to model, and the available time series are often too short to detect the contribution of anthropogenic forcing on the observed changes. However, some simple arguments suggest that even a small modification in the mean temperature greatly increases, for example, the probability of experiencing a temperature above a particular threshold and thus of an increase in the number of very hot days (Figure 6.20). A rise in average temperature also decreases the probability of the temperature falling below a particular level and so decreases the probability of very cold days. This simple reasoning is in agreement with model results which suggest an increase in heat waves in summer and a decline in the incidence of frosts in many regions, although occasional cold winters will continue to occur.

Figure 6.20 is based on a simple shift in the distribution, but the shape of the distribution also can change. Indeed, some studies suggest that the future climate will be more variable in some regions (corresponding to a wider distribution in Figure 6.20 and thus even more frequent extremes). For instance, a drying of the

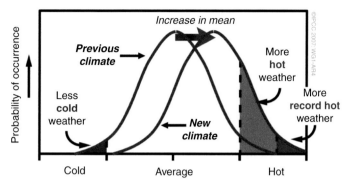

©IPCC 2007 WG1-AR4

**Fig. 6.20** Schematic diagram showing the effect of a mean temperature increase on extreme temperatures for a normal temperature distribution. (Source: Solomon et al. 2007.)

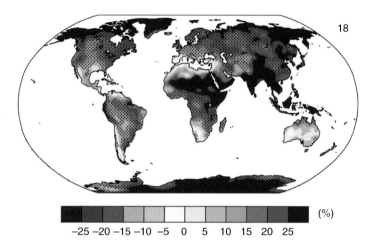

**Fig. 6.21** Projected per cent changes in the annual maximum five-day precipitation accumulation over the 2081–2100 period relative to 1981–2000 in the RCP8.5 scenario from CMIP5. (Source: Collins et al. 2013.)

soils which leads to reduced latent heat loss at the surface and thus higher temperatures can contribute to an increase in the probability of occurrence of very warm days (temperature–soil moisture feedback; see Section 4.2.4).

Although evaporation and precipitation changes are effectively energy limited and constrained by the heat budget on a global scale (Section 6.2.5), this is not the case on a local scale. Heavy rains are often associated with the rapid precipitation of most of the water available in the atmosphere. The intensity of the precipitation extreme thus is proportional to the humidity changes, and it increases at a rate of about 7% per degree Celsius, that is, at a much higher rate than mean precipitation. The impact of the warming on the frequency of extreme precipitation events is smaller. Furthermore, no systematic rule can be established between the spatial distribution of mean precipitation changes and the maximum amount of rain during extreme events, with many regions showing a decrease in mean precipitation in the future but an increase in precipitation during heavy rainfall (Figure 6.21).

## 6.2.9  Changes in the Carbon Cycle

In preceding sections we have briefly described the influence of anthropogenic forcing on climate, which is mainly due to the increase in greenhouse gas concentration. In turn, climate change has an impact on the biogeochemical cycles. This leads to modifications in the exchanges of carbon between the atmosphere, the ocean and land surface affecting the atmospheric $CO_2$ concentration and thus the **radiative forcing,** with potential feedback effects on climate (see Section 4.3.1).

As mentioned in Section 2.3.1, about half the anthropogenic $CO_2$ emitted up to the present by anthropogenic fossil fuel burning and changes in land use has stayed in the atmosphere. The remaining half is stored in the ocean and the terrestrial biosphere. However, this division of anthropogenic emissions between atmospheric, oceanic and land reservoirs will change in the future (Figure 6.22). For the RCP8.5 scenario, the fraction remaining in the atmosphere will increase as the efficiency of net carbon uptake by the ocean and land will decrease. This corresponds to a decrease in magnitude of the negative concentration-carbon feedback and an increase in the positive climate-carbon feedback. By contrast, the atmospheric fraction decreases significantly in the RCP2.6 scenario. In this scenario, emissions to the atmosphere tend to values which are very close to 0 during the twenty-first century. Moreover, the ocean continues its atmospheric carbon uptake as it takes centuries to millennia to reach its equilibrium with the current atmospheric concentration. This leads to a dominant contribution of the ocean in the storage of anthropogenic carbon during the twenty-first century in this scenario.

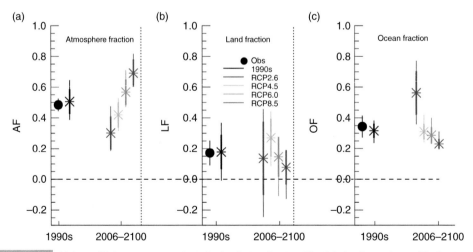

**Fig. 6.22**  Multi-model changes in atmospheric, land and ocean fraction of fossil-fuel carbon emissions. The fractions are defined as the changes in storage in each component (atmosphere, land and ocean) divided by the fossil-fuel emissions derived from each CMIP5 simulation for the four RCP scenarios. Solid circles show the observed estimates for the 1990s. Multi-model mean values are shown as star symbols, and the multi-model range (min to max) and standard deviation are shown by thin and thick vertical lines, respectively. Owing to the difficulty of estimating land-use emissions, this figure is based on fossil-fuel emissions only. (Source: Ciais et al. 2013.)

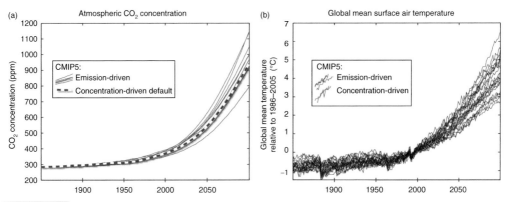

**Fig. 6.23**   Simulated changes in (a) atmospheric $CO_2$ concentration and (b) global average surface temperature (°C), as calculated for the RCP8.5 scenario when $CO_2$ emissions are prescribed to the model as external forcing (blue). Also shown (panel b, in red) is the simulated warming from the same models when directly forced by atmospheric $CO_2$ concentration (panel a, red dotted line). (Source: Collins et al. 2013.)

    The projections made by Earth system models (ESMs) including a carbon cycle driven by emissions of $CO_2$ can be compared with those directly driven by $CO_2$ concentrations (Figure 6.23). Recall here that the concentrations in the RCP scenarios (see Section 6.1.3) were obtained using a carbon-cycle model that included its own representation of climate-carbon feedbacks. Consequently, Figure 6.23 does not display the results of simulations with and without climate-carbon feedbacks but rather illustrates the impact of different representations of those feedbacks. For the eleven models selected in Figure 6.23, the simulation driven by emissions provides estimates of atmospheric $CO_2$ concentration in 2100 which are slightly higher (mean 985 ppm; full range 794–1142 ppm) than the value of 936 ppm prescribed in the RCP8.5 scenario. The resulting global temperature change between 2081–2100 and 1986–2005 in these simulations ranges between 2.5 and 5.6°C (mean 3.9°C), whilst the corresponding increase for the simulations forced by concentration is between 2.6 and 4.7°C (mean 3.7°C). This larger mean temperature increase in the emission-driven scenario is consistent with the higher $CO_2$ concentration. Furthermore, the larger range of temperature change in the simulations including carbon-cycle models illustrates clearly that the changes in carbon cycle are a key source of uncertainty in climate projections (see, e.g., Friedlingstein et al., 2014), as already mentioned in Section 4.3.1.

    In addition to the analysis of climate change as a function of time and scenario presented in Figure 6.23, it is instructive to analyse the temperature response as a function of the total amount of anthropogenic $CO_2$ released to the atmosphere (Figure 6.24). Actually, the '**transient climate response to carbon emissions**' (TCRE), defined as the ratio of global temperature change to cumulative anthropogenic emissions of $CO_2$, appears as roughly constant in time and independent of the scenario in model simulations. Its value is model dependent and estimated to be between 0.8 and 2.5°C per 1000 PgC (Collins et al. 2013). Such a quasi-linear relationship between temperature and cumulative emissions may be surprising because, among other important processes, the radiative effect of $CO_2$

is a non-linear function of its concentration [Eq. (4.1)] and the oceanic heat and carbon uptakes, which play a strong role in the transient temperature response of the system and involve complex non-linear interactions (see Sections 2.1.5 and 4.3.1). Nevertheless, those non-linearities tend to compensate for each other during the twenty-first century. The approximation appears valid for smooth future scenarios, emission of less than 2000 PgC of carbon, and holds until temperature reaches its maximum. This means that, whatever the scenario, it is possible to estimate the maximum amount of anthropogenic $CO_2$ which can be released to maintain the global mean temperature below a chosen target.

Another consequence of the flux of anthropogenic carbon from the atmosphere to the ocean is oceanic acidification [see reactions (2.42)–(2.44)]. Since the beginning of the industrial era, the surface **pH** of the global ocean decreased by about 0.1. The expected decrease at the surface by 2100 is about 0.07 for the RCP2.6 scenario and 0.3 for the RCP8.5 scenario. This ocean acidification increases the solubility of $CaCO_3$ (see Section 4.3.3); this also could be related to the reduced $CO_3^{2-}$ concentration owing to oceanic uptake of $CO_2$. This will have a clear impact on $CaCO_3$ production by corals, as well as by calcifying phytoplankton and zooplankton, and thus on their life cycles (see Section 2.3.2.2).

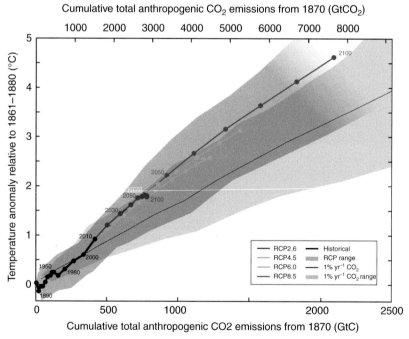

**Fig. 6.24**  Global mean surface temperature increase as a function of cumulative total global $CO_2$ emissions. Multi-model mean (straight line) and spread (coloured plume) from climate-carbon models for each RCP until 2100 are shown. The decadal means (dots) are connected by straight lines. Model results over the historical period (1860–2010) are indicated in black. The multi-model mean and range simulated by CMIP5, forced by a $CO_2$ increase of 1% per year, is given by the thin black line and grey area. For a specific amount of cumulative $CO_2$ emissions, the 1% per year $CO_2$ simulations exhibit lower warming than those driven by RCPs, which include additional non-$CO_2$ forcing. All values are given relative to the 1861–1880 base period. (Source: IPCC 2013.)

The **aragonite** will be particularly influenced by this change because it is less stable than **calcite**. For instance, it is expected that some regions of the Arctic will be under-saturated for aragonite as early as 2025, meaning that aragonite will not be stable any more in surface waters. In 2100, for the RCP8.5 scenario, the under-saturation would be widespread in the Arctic, the Southern Ocean and some coastal upwelling systems (Ciais et al. 2013).

## 6.3   Long-Term Climate Changes

### 6.3.1 The Carbon Cycle

The interactions among the atmosphere, the land biosphere and the ocean surface layer take place relatively rapidly and are predicted to play a dominant role in the changes in atmospheric $CO_2$ concentration over the twenty-first century (see Section 6.2.9). By contrast, the exchanges of $CO_2$ with the deep ocean are much slower, taking place on time scales from centuries to millennia. As the deep ocean is not in equilibrium with the surface, the carbon uptake by the deep ocean continues during the whole of the third millennium. For instance, a reduction in $CO_2$ emissions to 0 after 2300 in idealised prolongations of RCP scenarios induces a decrease in atmospheric $CO_2$ concentration after that date (Figure 6.25). The decrease in atmospheric $CO_2$ concentration is relatively slow, however. Values higher than those during pre-industrial times are found in 3000 for all the scenarios with still more than 1500 ppm in RCP8.5 700 years after cessation of the emissions.

Despite this decrease in the $CO_2$ concentration, the global mean surface temperature is more or less stable during the third millennium for RCP4.5, RCP6 and RCP8.5. The **radiative forcing** due to $CO_2$ decreases after 2300 in the scenarios, but the heat uptake by the ocean also decreases (see Section 4.1.5) as the ocean warms. The two effects nearly balance each other, leading to a simulated stabilisation of the temperature. The temperature decrease is a bit clearer in RCP2.6 after 2050 because the reduction in emissions occurs much earlier in this scenario.

The results displayed in Figure 6.25 mainly imply a long-term adjustment between the ocean and the atmosphere. However, on long time scales, the changes in acidity caused by the oceanic uptake of $CO_2$ induce dissolution of some of the $CaCO_3$ in the sediments (calcium carbonate compensation; see Section 4.3.3), modifying the ocean's **alkalinity** and allowing an additional uptake of atmospheric $CO_2$. This interaction with $CaCO_3$ in the sediments then produces a further reduction in the atmospheric concentration. However, this process is very slow, and after 10,000 years, the atmospheric $CO_2$ concentration still can be significantly higher than in pre-industrial times (Figure 6.26). On even longer time scales, an additional decrease in the atmospheric $CO_2$, eventually back to pre-industrial levels, can be achieved by the reactions of $CO_2$ with rocks and, in particular, by the negative feedback caused by weathering (see Section 4.3.4). Because of this long-term perturbation of the carbon cycle, the temperature

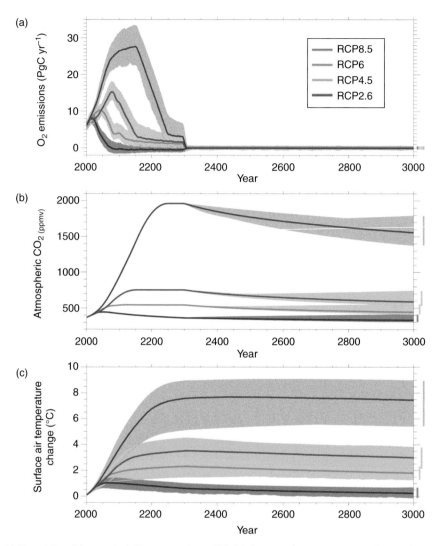

**Fig. 6.25**    (a) $CO_2$ emissions, (b) atmospheric $CO_2$ concentration and (c) global mean surface air temperature relative to the years 1986–2005 in seven intermediate-complexity models under four RCP scenarios. The thick solid lines represent the models' ensemble averages for each scenario, whilst the shading illustrates the range spanned by all the models. (Source: Zickfeld, K., M. Eby, A. J. Weaver, K. Alexander, E. Crespin, N. R. Edwards, A. V. Eliseev, G. Feulner, T. Fichefet, C. E. Forest, P. Friedlingstein, H. Goosse, P. B. Holden, F. Joos, M. Kawamiya, D. Kicklighter, H. Kienert, K. Matsumoto, I. I. Mokhov, E. Monier, S. M. Olsen, J. O. P. Pedersen, M. Perrette, G. Philippon-Berthier, A. Ridgwell, A. Schlosser, T. Schneider Von Deimling, G. Shaffer, A. Sokolov, R. Spahni, M. Steinacher, K. Tachiiri, K. S. Tokos, M. Yoshimori, N. Zeng and F. Zhao (2013). Long-term climate change commitment and reversibility: An EMIC intercomparison. *Journal of Climate* 26, 5782–809. © American Meteorological Society. Used with permission.)

remains significantly higher than in pre-industrial times during the whole period investigated in Figure 6.26, the amplitude of the temperature rise over several millennia being related to the total release of carbon at the end of the second and the beginning of the third millennium.

    This section illustrates that we cannot reliably estimate one time scale for the response of atmospheric $CO_2$ concentration to fossil fuel burning, as we can for

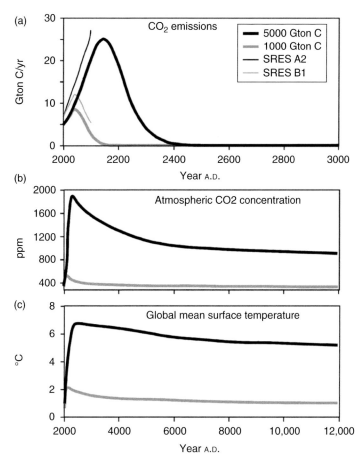

**Fig. 6.26** Response of climate model of intermediate complexity CLIMBER-2 to moderate (1000 Gton C) and large (5,000 Gton C) total fossil fuel emissions. The equilibrium climate sensitivity of the model is 2.6°C. The land carbon cycle was neglected in these simulations, whilst deep sea sediments were explicitly simulated using a sediment model. (a) Emissions scenarios and reference SRES scenarios (B1 and A2; see Figure 6.2). (b) Simulated atmospheric $CO_2$ (ppm). (c) Simulated changes in global annual mean air surface temperature (°C). (Source: Archer and Brovkin 2008.)

some other anthropogenic **forcings**, because of the wide variety of processes involved. To give an accurate representation of the time changes of atmospheric $CO_2$ concentration, several different time scales corresponding to the dominant mechanisms are required.

## 6.3.2  Sea Level and Ice Sheets

Sea-levels changes induced by climate variations have two origins. First, water can be added to the ocean from other reservoirs. The main potential contributors are the glaciers, ice caps and ice sheets and, to a smaller extent, the water stored on land. The second cause of sea-level change is related to ocean density. For a constant oceanic mass, any modification of the density affects ocean volume and thus sea level. As the density variations are mainly ruled by the water temperature, this process is often referred to as '**thermal expansion**', although salinity changes can play a non-negligible role in some regions.

For the period 1971–2010, most of the sea-level rise, estimated to be around 2 mm yr$^{-1}$, was due to the melting of glaciers and thermal expansion. The contribution of thermal expansion is slightly higher at about 0.8 mm yr$^{-1}$, whilst the contribution of glacier melting is around 0.7 mm yr$^{-1}$. The melt water flow from the Greenland and Antarctic ice sheets is hard to estimate on this time scale, but the most recent observations provide values for the years 1993–2010 of about 0.3 mm yr$^{-1}$ for each ice sheet (compared with a total change of 3 mm yr$^{-1}$ over this period). Finally, changes in storage of water over land during this period contributed to about 0.1 mm yr$^{-1}$ over the period 1971–2010 (Church et al. 2013).

Over the twenty-first century, the melting of glaciers and ice caps and thermal expansion are expected to remain the two main causes of rising sea levels. Greenland and Antarctica likely will make an additional positive contribution. However, uncertainties are large, in particular, for Antarctica, where it is not precisely known whether the net mass flow will be positive or negative. While melting is expected in some regions close to the shore, frozen water may accumulate over part of Antarctica because of additional precipitation caused by the warming (see Section 6.2.5). Indeed, temperatures in the centre of Antarctica are so low that the temperature increase estimated for the twenty-first century is far too small to produce melting there.

Depending on the scenario, the estimates of sea levels at the end of the twenty-first century, based on models representing the physics of the system, range from 30 to 80 cm higher than in the period 1986–2005 (Figure 6.27). These estimates take into account the rapid ice-flow changes which occur on relatively small scales (a kilometre or even less), which may transport ice to the ocean or to warmer areas, where it would melt relatively quickly. However, many uncertainties remain, in particular, related to the fate of **ice shelves** and the impact of ocean warming on the melting at their bases (see also Section 6.3.3). As a consequence, alternative methods have been proposed, based on simple statistical relationships between the rises in surface temperatures and sea levels. The validity of these studies is currently debated, but they suggest that a sea-level rise up to 190 cm by the end of the twenty-first century is not impossible (see, e.g., Vermeer and Rahmstorf 2009).

Even in the case of surface temperature stabilisation after 2100, the sea level is predicted to continue to rise (Figure 6.28). Firstly, the deep ocean will have to come into equilibrium with the new surface conditions, leading to warming at deeper levels and thus thermal expansion over several centuries. By contrast the contribution of glaciers will decrease with time as their total volume diminishes.

Secondly, the thermal inertia of the ice sheets is very large, taking several millennia to tens of millennia to completely melt, even when the warming is considerable. For Greenland, it has been estimated that a sustained local warming on the order of 2–4°C, which is not incompatible with the values provided by models of several scenarios, may be sufficient to induce a complete melting of the ice sheet (Church et al. 2013). The ice sheet would start to melt on its periphery and would gradually retreat to the centre of the island, to finally survive only in the eastern mountains (Figure 6.29). As the Greenland ice sheet retreats, the bedrock

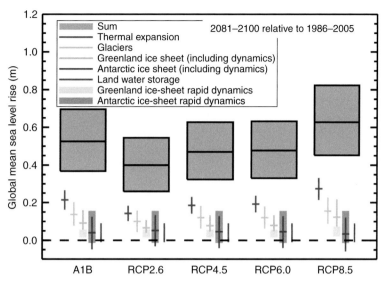

**Fig. 6.27** Projections from process-based models with median and likely range (66%) for global mean sea-level rise and its contributions in 2081–2100 relative to 1986–2005 for the four RCP scenarios and SRES scenario A1B. (Source: Church et al. 2013.)

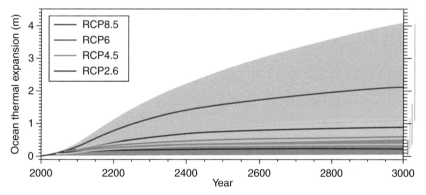

**Fig. 6.28** Changes in sea level (relative to the years 1986–2005) caused by thermal expansion, as simulated by the same seven intermediate-complexity models as in Figure 6.25 in idealised prolongations of RCP scenarios displaying a reduction in $CO_2$ emissions to 0 after 2300. The thick solid lines represent the model ensemble averages for each scenario, whilst the shading illustrates the range spanned by all the models. (Source: Zickfeld, K., M. Eby, A. J. Weaver, K. Alexander, E. Crespin, N. R. Edwards, A. V. Eliseev, G. Feulner, T. Fichefet, C. E. Forest, P. Friedlingstein, H. Goosse, P. B. Holden, F. Joos, M. Kawamiya, D. Kicklighter, H. Kienert, K. Matsumoto, I. I. Mokhov, E. Monier, S. M. Olsen, J. O. P. Pedersen, M. Perrette, G. Philippon-Berthier, A. Ridgwell, A. Schlosser, T. Schneider Von Deimling, G. Shaffer, A. Sokolov, R. Spahni, M. Steinacher, K. Tachiiri, K. S. Tokos, M. Yoshimori, N. Zeng and F. Zhao (2013). Long-term climate change commitment and reversibility: An EMIC intercomparison. *Journal of Climate* 26, 5782–809. © American Meteorological Society. Used with permission.)

will slowly rebound because of the smaller weight on the surface. This will initially cause a series of big inland lakes to appear below sea level. After 3,000 years, almost all the initially depressed areas will have risen above sea level again. Such a complete melting of the Greenland ice sheet on a millennium time scale would produce a rise in sea level of about 7 m.

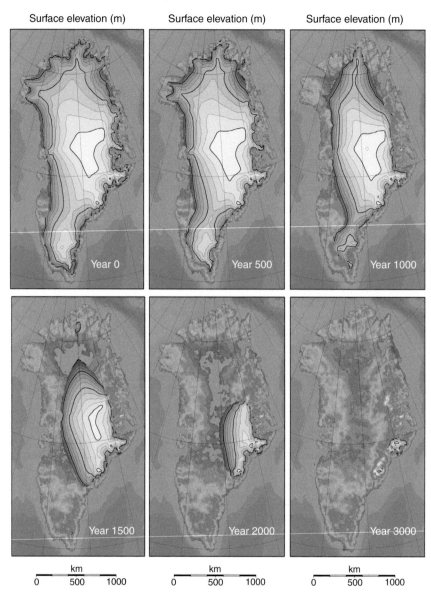

Surface elevation (m)          Surface elevation (m)          Surface elevation (m)

Year 0          Year 500          Year 1000

Year 1500          Year 2000          Year 3000

km          km          km
0    500    1000          0    500    1000          0    500    1000

**Fig. 6.29** Snapshots of the changes in the Greenland ice sheet in a scenario in which the $CO_2$ concentration is maintained at four times the pre-industrial value (four times $CO_2$ scenario) for 3,000 years. The results come from the intermediate-complexity climate model LOVECLIM. The sea and land below sea level are shown in blue, ice-free tundra in brown and green, and the ice sheet in grey. The contour intervals over the ice are 250 m, with thick lines at 1000-m intervals. (Source: Huybrechts et al. 2011.)

The melting in Antarctica will be much smaller and slower than that in Greenland because of the size of the ice sheet and the very cold temperatures there at present. However, some regions of East Antarctica may experience a significant melting on a similar time scale to that of Greenland (see also Section 6.3.3).

### 6.3.3 Abrupt Climate Changes

The time developments described in preceding sections appear to be relatively smooth. Some changes, however, may be abrupt, in particular, on regional scales. The uncertainties involved in those abrupt changes are high because there is generally no recent equivalent to such events, and only a few observations of the key processes which may be responsible for their occurrence are available. Nevertheless, because of their large impact on the climate system, they are an intense subject of research, using both theoretical and more applied approaches.

For several reasons, probably the most classical example of **abrupt change** is related to the Atlantic MOC. Firstly, there is a good theoretical understanding that the MOC can have two equilibrium states for similar boundary conditions and that a perturbation can induce an abrupt shift between those states (see Section 4.2.5). Secondly, some transitions recorded in past climates archives have been attributed to rapid changes in the MOC, indicating that such events are indeed possible (see Section 5.5.4). Thirdly, all the climate models show that the MOC can be shut down if a sufficient freshwater perturbation is applied in the North Atlantic, and many models which have been systematically tested display two equilibrium states of the MOC for certain boundary conditions. Nevertheless, there is a wide range between the different models in the amount of freshwater necessary to induce such a shutdown. Furthermore, it is not clear from model analyses or from observations whether the system is presently in a configuration in which only the mode with an active MOC is stable (mono-stable regime) or whether the shutdown state is also stable (bi-stable regime). This has implications for the irreversibility of a potential shutdown. In the mono-stable case, if the perturbation is stopped, the system will come back to its present mode of operation, whilst it could remain in a shutdown state in the bi-stable case. As discussed in Section 6.2.6, the projections do not indicate a collapse of the MOC during the twenty-first century. The situation after 2100 is more uncertain and depends on the magnitude of the warming and melting of the Greenland ice sheet (see Section 6.3.2) which could strongly reduce the MOC by bringing additional freshwater into the North Atlantic.

The vegetation cover also has experienced fast transitions in the past, both at low and at high latitudes (see Section 5.6.2). Abrupt changes thus have been postulated for the future too. In particular, some tropical forests may not be able to survive future reduction in precipitation or increased evaporation, leading to abrupt vegetation shifts. At high latitudes, droughts or fires and insect damage may increase the stress on the boreal forest and lead to higher mortality. However, although the uncertainties remain large, there is currently no evidence to suggest that the likelihood of such abrupt changes is high during the twenty-first century (Collins et al. 2013).

The sizes of both the Greenland and Antarctic ice sheets have been significantly modified over the last million years (see Section 5.5.2), but the rate of change is

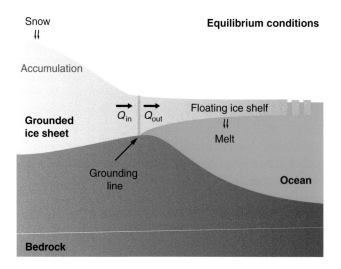

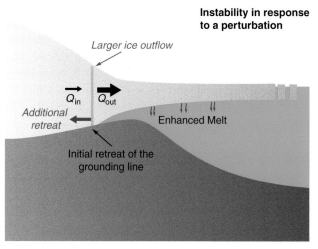

Fig. 6.30 Schematic illustration of the processes leading a potential instability at the ice-sheet margins showing first stable conditions and second the instability triggered by an initial melting which induces a retreat of the grounding line, a larger mass outflow as the ice thickness at the grounding line increases and then a subsequent retreat of the grounding line. (Source: Modified from Church et al. 2013.)

generally not precisely known. Rapid transitions may be caused by the so-called marine ice-sheet instability (Figure 6.30). In order to be active, this mechanism requires that the bedrock supporting the ice sheet is below sea level and becomes deeper as the distance to the **grounding line** increases. This configuration is widely observed today in West Antarctica as well as in some regions in Greenland and East Antarctica. At equilibrium, the ice flow crossing the grounding line $Q_{in}$ is equal to the outflow through the floating ice shelves $Q_{out}$. As a result of a perturbation, the grounding line may show an initial retreat caused by a thinning of the ice shelves. This could be due to additional melting at the surface associated with warming of the air and at the bottom because of the intrusion of warmer water. Following this retreat, the thickness of the ice shelves at the grounding line increases, leading to a larger $Q_{out}$, whilst $Q_{in}$ remains more or less the same because it must balance the net accumulation at the surface of the ice sheet. As the export of ice at the grounding line becomes larger than the import, the ice volume (thickness) decreases locally, and the inland migration of the grounding line continues, leading potentially to an instability. Furthermore, the ice shelves generally are inducing some stresses which reduce the ice flow to the ocean. Removing them because of melting can accelerate the ice discharge and thus also lead to rapid reductions of the volume of the ice sheets.

Those processes may have a rapid and strong local/regional impact but are generally too slow to have a high large-scale imprint at the decadal scale. Thus, they would not lead to abrupt changes following the CCSP (2008) definition (see Section 5.5.4). Nevertheless, they could be associated with a major shift in the system, largely irreversible on a millennial time scale, because of the time needed to grow an ice sheet from the limited amount of precipitation at high latitudes. Besides, other mechanisms potentially may stabilise the retreat, such as, for instance, the large-scale bedrock roughness, which may provide some kind of topographical pinning points. Consequently, it is not yet possible to estimate with reasonable confidence whether the West Antarctic ice sheet can be destabilised by a perturbation of the size of the one expected during the next centuries. The uncertainties thus are large on the potential relevance of marine ice-sheet instability to the future evolution of the ice sheet and the sea level.

## Review Exercises

1. An important difference between SRES scenarios and RCP scenarios is that the first step in the procedure is to estimate the future emissions for the SRES scenarios, whilst it is to choose the future changes in radiative forcing in the RCP scenarios.
   a. True
   b. False
2. In most of the scenarios, the global emissions of sulphur dioxide decrease strongly over the twenty-first century, whilst the atmospheric concentration of $CO_2$ increases over the same period.
   a. True
   b. False
3. The mean of an ensemble of simulations performed with different models in similar conditions
   a. generally reproduces well the mean state of the climate system over a reference period.
   b. provides a good estimate of the simulated response to a perturbation.
   c. provides a good estimate of climate variability.

4. The internal variability of a system is a source of uncertainty in the projections. For global mean temperature, its relative influence is larger at the end of the twenty-first century than in the next decade.
   a. True
   b. False

5. For projections, simulations are started at a date about one century back in time. A simulation then is performed up to the present day before launching simulations for the future. In predictions, the simulations covering the next decade(s) are started directly from a state which is as close as possible to the one observed at a recent time. The goal is to
   a. reduce the mean biases characteristic of the model mean state.
   b. have a better estimate of the impact of past forcings on the current state.
   c. account for the impact of the actual conditions on the future changes.

6. The projections of climate change for 2100 are characterised by a larger warming over land than over the oceans. This is explained mainly by
   a. the larger thermal inertia of the oceans compared to land surfaces.
   b. the evaporation over the ocean which is not limited by soil moisture content.

7. A larger surface warming is obtained for the Arctic than over the tropical regions in projections for the end of the twenty-first century. This polar amplification is due to
   a. the snow-and-ice–albedo feedback.
   b. a higher heat transport towards the Arctic.
   c. a lower simulated cloud cover in winter.
   d. a larger warming at the surface than at higher altitudes.

8. A dominant pattern in the simulated precipitation changes for the end of the twenty-first century is an increase in precipitation in wet regions and a decrease in dry regions. This may be explained by
   a. more evapotranspiration in a warmer world in wet regions, leading to more precipitation, whilst no additional water is available for evapotranspiration in dry regions.
   b. a larger transport of moisture by the atmospheric circulation.
   c. a reduction in the intensity of the Hadley cell.

9. A fraction of the anthropogenic $CO_2$ emitted by fossil fuel burning over the next decades will be absorbed by the oceans, but this fraction will be lower for the RCP8.5 scenario than for the last decades.
   a. True
   b. False

10. In the long term, interactions with oceanic sediments and rocks will induce a reduction in the atmospheric $CO_2$ concentration. If the atmospheric concentration reaches at least 500 ppm during the next decades, how long will it take to again reach a concentration similar to the one observed during pre-industrial times?
    a. Several decades
    b. Several centuries
    c. Several thousand years

11. Over the period 1993–2010, the largest contributor to sea-level rise was
    a. the thermal expansion.
    b. the melting of glaciers and ice caps.
    c. the melting of the Greenland ice sheet.
    d. the melting of the Antarctic ice sheet.

12. A total melting of the Greenland ice sheet likely would take several millennia. If this occurs, it would induce a sea-level rise of
    a. 2 m.
    b. 7 m.
    c. 50 m.

13. Marine ice sheet instability is possible because of
    a. snow-and-ice–albedo feedback.
    b. a warming of the ocean below the ice shelf.
    c. the bedrock is below sea level and becomes deeper as the distance to the grounding line increases.

# Concluding Remarks

Our short journey in climate science is close to its end. It has shown the diversity of the domain and has provided a sample of the large number of processes potentially responsible for climate variations on a wide range of time and spatial scales. This diversity is exciting because it allows us to create the links with pre-existing knowledge, learned in other textbooks and lectures, whilst stimulating our curiosity about many additional applications and topics not covered here. It is also frustrating because the content of an introductory textbook should be as self-sufficient as possible. Ideally, new points had to be demonstrated or justified, relying on standard skills in mathematics and physics, for instance, with a few references for additional lectures. The main mechanisms included in the preceding chapters are all explained qualitatively, often using schematics, and many of them quantitatively too. Nevertheless, determining the stability criteria of a numerical scheme or estimating precisely the magnitude of a feedback requires techniques which cannot all be described in detail here. Only the conclusions of existing studies can be given, with some general information on the method(s) applied. I hope that this will encourage many readers to go deeper into the topics that appear the most promising for them to investigate by themselves how precisely those conclusions are reached and the strong scientific arguments behind them.

Many of the cited references are recent, showing the rapid development of this knowledge. This situation also underlines the fact that uncertainties are still present on the subject addressed, justifying the current intense scientific activity. These uncertainties have been mentioned several times in this text, but it is important to recall here that many results are also well established. We have not insisted on the historical development of the field, but climatology is based on laws applied with success in mechanics, astronomy, thermodynamics, electromagnetism, chemistry, geology and numerical analysis over decades at least. Studies focussing on important topics such as the impact of changes in greenhouse gases concentrations or the astronomical theory of paleoclimates have their roots in the beginning of the twentieth century or even before [see, e.g., Arrhenius (1896) and the references therein and the discussions in Berger (1988)]. The main conclusions presented in the various chapters of this book are robust as they are based on well-understood mechanisms, have been reproduced using various techniques and have been identified for many periods in the past.

The wide range of processes which have to be accounted for in climate science is mirrored by the diversity of climate models. They cover the full spectrum from simple models with one or two linear equations to Earth system models, which are among the most sophisticated models currently developed. The complex models should be used to provide the most detailed representation of the system, but simple models are able to describe some of the main characteristics of the complex models. The 'presence of an emergent simplicity' (Held 2014) allows for clear understanding of some important processes. This also permits explanations which can convince wide audiences who probably would be lost if the full complexity of the system were required.

This understanding is a key element because decisions have to be taken on the basis of scientific arguments, in particular, to limit the impact of human activities on future climate change and to mitigate the consequences of the unavoidable effects. The uncertainties should be taken into account in any decision. The goal of scientists is to reduce them as much as possible by performing new and more precise observations, proposing new theoretical developments and improving models. Nevertheless, the current uncertainties should not be a pretext to avoid actions because clear and strong arguments based on solid science are available. This is well described by Oreskes and Conway (2010) in the conclusion of their book, *Merchants of Doubt: How a Handful of Scientists Obscured the Truth on Issues from Tobacco Smoke to Global Warming*, where they state: 'For even if modern science does not give us certainty, it does have a robust track record. We have sent men to the moon, cured diseases, figured out the internal composition of the Earth, invented new materials, and built machines to do much of our work for us – all on the basis of modern scientific knowledge. While these practical accomplishments do not prove that our scientific knowledge is true, they do suggest that modern science gives us a pretty decent basis for action.'

# Glossary

The terms defined in this glossary are highlighted in bold characters in the main text. The intention is not to provide a general definition of the selected entries but rather to give an explanation of the meaning of those terms as employed in climatology and in particular in this book. We encourage readers to refer to the suggested textbooks, online encyclopaedias and the glossary provided as annexes of the Intergovernmental Panel on Climate Change (IPCC 2007, 2013) assessment reports for complementary information.

Abrupt change    Several definitions of abrupt changes have been proposed. They all have similarities. In particular, a recent definition proposes to characterise abrupt change as rapid events leading in a few decades to substantial modifications of the state of a system which persist for at least a few decades (CCSP 2008).

Adiabatic    An adiabatic transformation is a transformation occurring without exchange of heat or matter between the system studied (e.g., a small volume of sea water or air) and the surrounding environment. An adiabatic transformation does not imply that temperature is constant as the pressure can change, leading to exchange of mechanical energy (work) between the system and the environment. In real conditions, a natural system is never isolated in such a way that the heat exchanges are totally inhibited. However, if the transformation implies large changes in height and is rapid enough, the exchange of energy is governed by the adjustment of pressure (which is much faster than heat transfers), and considering it as an adiabatic transformation is a good approximation.

Aerosols    Atmospheric aerosols are relatively small solid or liquid particles that are suspended (float) in the atmosphere. They can be produced naturally or by human activities. Their size typically ranges from a few hundredths of a micrometre to several micrometres. Aerosols have an influence on the radiative balance of the Earth.

Albedo    The albedo $\alpha$ is the ratio between reflected and incoming **radiative flux**. It varies between 0 for a perfect **black body** which absorbs all the incoming radiation to 1 for a surface which reflects it all. It depends on the wavelength, but the general term usually refers to some appropriate average across the spectrum of visible light or across the whole spectrum of solar radiation.

| | |
|---|---|
| Alkalinity | The total alkalinity is defined as the excess of bases over acid in sea water. |
| Anomaly | The anomaly of a variable (e.g., temperature) is the difference between the value under consideration and the long-term mean over a reference period. Anomalies can be computed for the annual mean, a season, a particular month or all the months together, then removing for each month the corresponding mean over the reference period. |
| Aphelion | The aphelion is the point in the Earth's orbit that is furthest from the Sun. |
| Aragonite | Aragonite is a form of calcium carbonate ($CaCO_3$). It has a different crystal lattice and crystal shape than **calcite**. |
| Ascendance | An ascendance is an upward movement of air in the atmosphere. |
| Astronomical parameters | In climatology, the characteristics of the Earth's movement are determined by three parameters, called the 'astronomical' or 'orbital parameters': the **obliquity** $\varepsilon_{obl}$, the **eccentricity** $ecc$ and the **climatic precession** $ecc \sin \omega$. |
| Atmospheric boundary layer | The atmospheric boundary layer is the lowest part of the atmosphere which is in direct contact with the Earth's surface. The properties of this layer are influenced directly by the presence of the surface, and in turn, they influence the exchanges between the surface and the atmosphere. Vertical mixing is usually strong in this layer because of the relatively intense turbulent motion. |
| Austral | Austral is a synonym for 'southern'. |
| Baroclinic instability | The baroclinic instability is an instability of the flow in the presence of vertical density gradients leading to the formation of meanders and **eddies**. For more details, see, for instance, Cushman-Roisin and Beckers (2011). |
| Bathymetry | Bathymetry is the topography of the floor of the ocean. It also refers to the measurement of the depth of the oceans. |
| Biogeochemical feedbacks | Biogeochemical feedbacks are **feedbacks** which involve interactions between climate, biological activity and the biogeochemical cycles on Earth. |

| | |
|---|---|
| Biogeophysical feedbacks | Biogeophysical feedbacks are **feedbacks** which involve the interactions between climate and some physical characteristics of the surface which are influenced by biological activity. |
| Biomes | Biomes are regions with distinctive large-scale vegetation systems. |
| Biosphere | 'The biosphere is the part of the Earth System comprising all the living organisms in the atmosphere, on land (terrestrial biosphere) and in the ocean (marine biosphere), including derived dead organic matter, such as litter, soil organic matter and oceanic detritus' (IPCC 2007). |
| Black body | A black body is an object or system that absorbs all **electromagnetic radiation** incident on it. |
| Blooms | Phytoplankton blooms are rapid increases in the mass of phytoplankton. |
| Boreal | Boreal is a synonym for 'northern'. |
| Calcite | Calcite is a form of calcium carbonate ($CaCO_3$). It is one of the most widely distributed minerals on Earth and is a constituent of sedimentary rocks, in particular, **limestone**. |
| Calendar month | The length of months and seasons is a function of the astronomical parameters and thus changes with time. Different options are available to compute time averages. The first method is based on the **true longitude**, the year being divided, for instance, into twelve parts of 30° each (see, e.g., Figure 5.24). Alternatively, the length of the months can be maintained at the values for present-day conditions (see, e.g., Figure 5.32). When comparing the insolation during two periods, the two approaches give qualitatively similar results but with quantitative changes (see, e.g., Joussaume and Braconnot 1997). |
| Calibration | The calibration of a climate model is the adjustment of the numerical or physical parameters to improve the agreement between the results of the model and observations. |
| Canopy | The canopy is the above-ground portion of a plant community formed by plant crowns. |
| Celestial equator | The celestial equator is the projection of the Earth's equator onto the celestial sphere (Figure G.1). |
| Celestial sphere | The celestial sphere is an imaginary sphere with a very large radius whose centre is the centre of the Earth. |

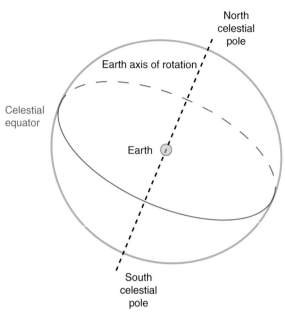

North
celestial
pole

Earth axis of rotation

Celestial
equator

Earth

South
celestial
pole

**Fig. G.1**    The celestial equator.

Chlorofluorocarbons   Chlorofluorocarbons (CFCs) are gases derived from alkanes (e.g., methane or ethane) in which all the hydrogen atoms have been replaced by chlorine or fluorine. They are a subset of the **halocarbons**. CFCs have been widely used in refrigerators, insulation and aerosol spray cans. However, because they have been shown to contribute to **stratospheric ozone** depletion, their use is now banned.

Clausius-Clapeyron    The Clausius-Clapeyron equation gives the relationship
equation               between the **latent heat** associated with a transition from phase 1 to phase 2 $[L_{1\rightarrow2}(T)]$ at the equilibrium temperature $T$, the volume $V_1(T, P)$ of the matter in phase 1, the volume $V_2(T, P)$ of the matter in phase 2 and the slope of the line separating the two phases in a $T$-$P$ diagram (i.e., $dP/dT$):

$$L_{1\rightarrow2}(T) = T\left[V_2(T,P) - V_1(T,P)\right]\frac{dP}{dT}$$

For the transition of water between the liquid and vapour phases in the atmosphere, the Clausius-Clapeyron equation can be written as

$$L_v(T) = T\left(V_{\text{vapour}} - V_{\text{liquid}}\right)\frac{de_s}{dT}$$

where $V_{\text{vapour}}$ and $V_{\text{liquid}}$ are the volumes of the water in the vapour and liquid phases, and $e_s$ is the **saturation vapour pressure**. This relationship can be used to compute the variation of $e_s$ as a function of temperature:

$$\frac{de_s}{dT} = \frac{L_v(T)}{T(V_{\text{vapour}} - V_{\text{liquid}})}$$

If we assume that the water vapour is an ideal gas and express $V_{\text{vapour}}$ using the **equation of state**, that the volume of the vapour is much larger than that of the liquid and that $L_v$ is a constant ($L_v = 2.5 \times 10^6$ J kg$^{-1}$, which is a strong approximation), we can express $e_s$ as a function of the temperature $T$ by integrating this equation between 273.15 K (for which $e_s = 611$ Pa) and the temperature $T$:

$$e_s = 611\exp\left[\frac{L_v}{R_V}\left(\frac{1}{273.15} - \frac{1}{T}\right)\right]$$

where $R_v$ is the gas constant for water vapour (461.39 J kg$^{-1}$ K$^{-1}$). This relationship can be used to compute the **specific humidity** at saturation $q_{\text{sat}}$ using the relationship between the saturation vapour pressure and humidity and knowing the air pressure $p$ ($q_{\text{sat}} \approx 0.622\, e_{\text{sat}}/p$). For more details, see, for instance, Wallace and Hobbs (2006).

| | |
|---|---|
| Climate sensitivity | See **equilibrium climate sensitivity**. |
| Climate system | The climate system consists of five major components: the atmosphere, the **hydrosphere**, the **cryosphere**, the land surface and the **biosphere**. |
| Climatic precession | Climatic precession ($ecc\sin\omega$) is related to the distance between the Earth and the Sun at the summer solstice. |
| Climatological | The climatological value of a variable is its mean over a reference period (generally thirty years). |
| Cloud microphysics | Cloud microphysics describes the physical processes which occur in clouds at scales smaller than a few centimetres. |
| Continental shelf | The continental shelf is a part of the continental plate which is below sea level. The corresponding oceanic regions, close to the coast, are shallow, with depth between a few metres to a few hundreds of metres. The connection between the continental shelf and the abyssal plain at depths of several kilometres comprises the steep continental slope and continental rise. |

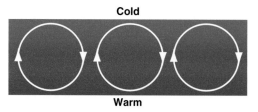

**Fig. G.2** Schematic representation of convection in a fluid heated from below and cooled from above.

Continentality    The continentality is related to the effect of the distribution of land masses and oceans on the local climate. In particular, the amplitude of the seasonal cycle tends to be larger and precipitation generally is lower inside continents compared to oceanic or coastal areas. Nevertheless, the degree of continentality is not just a function of the distance to the shore, as it depends on the wind direction, the topography, and so on.

Convection    In thermal convection, a fluid (liquid or gas) receives heat from below (cooled from above). The parcels included in the bottom layer are warmer, less dense and thus rise. They are continuously replaced by colder fluid parcels which are subsequently heated and then also rise, forming a convection loop or convection current. Convection stops when the temperature differences between different parts of the fluid are too small to create any movement (Figure G.2).

Coriolis force    The Coriolis force is an inertial or fictitious force which causes an apparent deflection of moving objects towards the right in the northern hemisphere and towards the left in the southern hemisphere when viewed from a frame of reference attached to the Earth (or, equivalently, when viewed by an observer who is standing on the Earth). This occurs because of the Earth's rotation, and the Coriolis effect is actually present whenever a rotating frame of reference is used.

Correlation    A correlation is a measure of the strength of a relationship or dependence between two variables. It is often estimated by the Pearson correlation coefficient $r$ (see also **covariance**).

Cosmic rays    Cosmic rays are high-energy particles (among which about 90% are protons) coming from outer space.

Cosmogenic isotopes    Cosmogenic isotopes are created when elements in the atmosphere or on Earth are bombarded by **cosmic rays.**

Covariance    The covariance between two variables is a measure of the link between the two variables. The covariance divided by the square root of the variance (i.e., the standard deviation) of the two variables gives a correlation coefficient. If many variables are analysed together and listed in a vector, a covariance matrix can be built. Each element $i, j$ of the matrix is equal to the covariance between the element $i$ and the element $j$ of the vector. A covariance matrix thus is symmetrical, and its diagonal elements are the variance of the elements of the vector.

| | |
|---|---|
| Cryosphere | The cryosphere is the portion of the Earth's surface where water is in solid form (**sea ice**, lake and river ice, snow cover, **glaciers**, **ice caps** and **ice sheets**). |
| Cyclones | A cyclone is a low-pressure system in the atmosphere. |
| Deep-water formation | Deep-water formation is the process through which water acquires the properties characteristic of a water mass (in particular, the temperature and salinity) close to the surface, where the heat and freshwater fluxes are large, before sinking to great depths. In the deep ocean, the temperature and salinity generally change very slowly. As a consequence, the properties of the waters found at great depths in the ocean can be traced over very large distances to their origins (their 'formation') at the surface. See also **water-mass formation** and **thermohaline circulation**. |
| Diapycnal | The diapycnal direction lies at right angles to the local **isopycnal** surface. Consequently, the angle between the diapycnal direction and the vertical is usually very small. |
| Dissolution | Dissolution is the process by which a solid or liquid forms a homogeneous mixture with a solvent (in climatology, the solvent is generally water). During this process, the crystal lattice of the solid is broken down into individual ions, atoms or molecules. |
| Dissolved inorganic carbon | Dissolved inorganic carbon (DIC) is the sum of the concentration of the three forms of inorganic carbon present in the ocean (i.e., carbonic acid, $H_2CO_3$, bicarbonate, $HCO_3^-$, and carbonate ions, $CO_3^{2-}$). |
| Downwelling | A downwelling is a downward movement of water in the ocean. |
| Dry air | Dry air is air without water vapour. Air (or moist air) is composed of dry air plus water vapour. |
| Earth system | The Earth system can be divided in five spheres: the atmosphere (gaseous envelope), the **hydrosphere** (liquid water), the **cryosphere** (solid water, i.e., ice), the lithosphere (solid Earth) and the **biosphere** (life). The Earth system |

is generally considered to be broader than the **climate system** because it explicitly includes some processes (e.g., some geological processes) which do not influence climate. Descriptions of the Earth system also generally take human activities into account.

**Earth system model**

An Earth system model is a model which includes a representation of several components of the Earth system (atmosphere, ocean, sea ice, land surface, land vegetation, carbon cycle, ice sheet, etc.), whilst the term 'climate model' is generally used for models which include representations of the atmosphere, sea ice, ocean and land surface only.

**Earth system sensitivity**

The Earth system sensitivity is the change in the global mean surface temperature after the Earth system has reached a new equilibrium in response to a perturbation. In contrast to the **equilibrium climate sensitivity**, the Earth system sensitivity implies the equilibrium of all the components of the Earth system, including, for instance, ice sheets, carbon cycle and vegetation, and thus corresponds to adjustments taking several millennia.

**Easterlies**

Easterlies are winds coming from the east which are typically found in the tropics.

**Eccentricity**

Eccentricity *ecc* describes the shape of an ellipse (e.g., that described by the Earth's orbit around the Sun) with a semi-major axis *a* and a semi-minor axis *b*. It is defined as

$$ecc = \frac{\sqrt{a^2 - b^2}}{a}$$

**Ecliptic plane**

The ecliptic plane is the geometrical plane containing the mean orbit of the Earth around the Sun. The ecliptic is the intersection of the **celestial sphere** with the ecliptic plane, and it corresponds to the apparent path that the Sun traces out in the sky.

**Eddies**

Eddies are whirlpool-like transient features in the ocean and atmosphere. Their spatial extent is smaller than that of the general circulation. Eddies in the atmosphere are called '**cyclones**' or 'anti-cyclones'. Meso-scale eddies in the ocean have a typical size of 10–100 km.

**Effective radiative forcing**

The effective radiative forcing (ERF) is the variation in radiative flux at the top of the atmosphere in response to a perturbation after allowing the atmospheric conditions (i.e., temperature, humidity and cloud cover) to adjust to the perturbation whilst keeping the surface temperature fixed over the whole surface of the globe or a fraction of it. In this framework, for most estimates, the temperature is maintained fixed over the oceans.

| | |
|---|---|
| Ekman transport | Because of the Earth's rotation, the surface ocean velocity induced by the wind is directed (outside the equatorial regions) to the right of the prevailing wind in the northern hemisphere and to the left in the southern hemisphere. Integrated over the vertical, the ocean transport caused by the wind is perpendicular to the wind direction. This is called the 'Ekman transport'. |
| Electromagnetic radiation | See **Electromagnetic spectrum.** |
| Electromagnetic spectrum | Electromagnetic radiation is classified by wavelength into radio, microwave, infrared (IR), visible, ultraviolet (UV), X-rays and gamma rays. We perceive the radiation in the visible region as light. |
| Emissivity | The emissivity of an object $\varepsilon$ is the ratio of energy radiated by the object to the energy radiated by a **black body** at the same temperature. It is a measure of a material's ability to absorb and radiate energy. A true black body would have an $\varepsilon = 1$, whilst any real object has $\varepsilon < 1$. |
| Ensemble of simulations | An ensemble of simulations is a group of simulations performed in the same framework. We can distinguish between (1) multi-model ensembles, in which different models are used, (2) perturbed physics ensembles, in which all the simulations are performed with the same model, but each member of the ensemble uses different parameters and/or different representations of some physical processes, and (3) ensembles which are obtained with a single model but using different initial conditions to sample the model's **internal variability**. When the ensemble includes a combination of these three cases, it is often referred to as a 'super-ensemble'. |
| Ensemble mean | Mean over an **ensemble of simulations**. |
| Ensemble member | Individual simulation of an ensemble of simulations. |
| Entrainment | In the context of the mixed-layer dynamics, entrainment is the transfer of water from the **thermocline** to the **oceanic mixed layer** because of **upwelling** or a deepening of the mixed layer. |
| Emipirical orthogonal function (EOF) | See **principal component**. |

| | |
|---|---|
| Equation of state | The equation of state is a law, characteristic of a material in a specific phase, which relates **state variables**. In climatology, it is often expressed as the density being a function of other state variables such as pressure, temperature and composition (i.e., salinity for sea water). |
| Equilibrium climate sensitivity | The equilibrium climate sensitivity is generally defined as the change in the global mean surface temperature after the climate system (i.e., atmosphere, ocean, sea ice or land surface) has reached a new equilibrium in response to a doubling of the $CO_2$ concentration in the atmosphere. |
| Equinox | The equinoxes are the moments when the Sun is positioned on the plane of the Earth's equator and, by extension, the apparent position of the Sun over the Earth's equator at that moment. The equinox during which the Sun passes from south to north in its apparent movement is known as the 'vernal equinox'. |
| Evapotranspiration | Evapotransipration is the transfer of water from Earth's surface to the atmosphere. It is the sum of evaporation from the soils, leaves and so on and the transpiration of plants. |
| Feedback | A feedback tends to amplify (positive feedback) or reduce (negative feedback) the response of a system to an initial perturbation via mechanisms which are internal to the system itself. |
| Forcing | A climate forcing is a perturbation, originating in elements which are not part of the system investigated, which induces changes in the climate. For instance, a change in total solar irradiance is a forcing because it modifies the Earth's climate. Forcings can be natural or anthropogenic depending of their origin. |
| Fourier series | A Fourier series decomposes a function into a sum of sines and cosines. |
| Fourier's law | Fourier's law (also called the 'conduction law') states that the heat flux through a material $F_{cond}$ is proportional to the temperature **gradient** in the material. In one dimension (along the x-axis), this can be expressed as $$F_{cond} = -k\frac{\partial T}{\partial x}$$ where $k$ is the thermal conductivity. |
| Fractionation | See **isotopic fractionation**. |

Geometric series   A geometric series is a sum in which each term is equal to the previous one multiplied by the constant number. For instance, $\left(1 + a + a^2 + a^3 + \cdots\right) = \sum_{n=0}^{\infty} a^n$ is a geometrical series in which the ratio between successive terms is $a$.

Geostrophic equilibrium   In the atmosphere and ocean, on a large scale, away from the boundaries (i.e., surface, coast) and the equator, the dominant terms in the horizontal equation of motion are the **Coriolis force** and the force due to the horizontal pressure gradient. The geostrophic balance, which assumes a balance between these two forces, is thus a reasonable approximation:

$$fv_g = \frac{1}{\rho}\frac{\partial p}{\partial x}$$
$$-fu_g = \frac{1}{\rho}\frac{\partial p}{\partial y}$$

In this equation $p$ is the pressure, $\rho$ is the density, $f$ is the Coriolis parameter and $u_g$ and $v_g$ are the components of the (geostrophic) velocity in the two horizontal directions. $f$ equals $2\Omega\sin\phi$, where $\Omega$ is the Earth's angular velocity, and $\phi$ is the latitude. $f$ is positive in the northern hemisphere and negative in the southern hemisphere. When this balance is achieved, the fluid is said to be in geostrophic equilibrium, and knowing the horizontal pressure distribution, the horizontal velocity can be computed. The geostrophic equilibrium explains why the flow is clockwise around a high-pressure system in the northern hemisphere and anti-clockwise around a low-pressure system (Figure G.3).

Glacial inception   The glacial inception is the start of a glacial period characterised by an increase in the volume of the **ice sheets**.

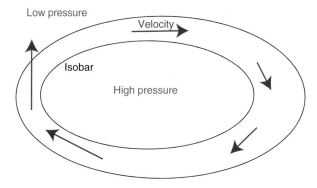

Low pressure

Velocity

Isobar

High pressure

**Fig. G.3**   Geostrophic flow around closed isobars in the northern hemisphere.

| | |
|---|---|
| Glacier | A glacier is a mass of ice which originates on land (see also **ice caps** and **ice sheets**). |
| Gradient | The gradient of a scalar field $f(x, y, z)$ is a vector which points in the direction of the greatest increase in the scalar field and whose magnitude is proportional to the rate of change. It is denoted $\mathrm{grad}(f)$, $\vec{\nabla}f$ or $\nabla f$ and is defined by |

$$\vec{\nabla}f = \left( \frac{\partial f}{\partial x}, \frac{\partial f}{\partial y}, \frac{\partial f}{\partial z} \right)$$

| | |
|---|---|
| | The projection of the gradient over one direction is often called the 'gradient' in this direction. For instance, $\partial f / \partial z$ is often called the 'vertical gradient of $f$'. |
| Greenhouse gas | A greenhouse gas is a gas which has an impact on the radiative properties of the atmosphere by its ability to absorb radiation in specific infrared wavelengths, leading to the greenhouse effect. |
| Grid | The numerical resolution of the equations governing the dynamics of the **climate system** generally requires the definition of a grid, whose nodes correspond to the locations where the model variables are computed. The values computed at these nodes provide enough information to reconstruct, over the whole domain, an approximation of the corresponding field (e.g., the temperature). An important characteristic of a grid is its resolution, which is related to the distance between two different values computed by the model. |
| Grounding line | See **ice shelf**. |
| Gulf Stream | The Gulf Stream is a strong current found along the southeast coast of the United States. The current is mainly wind driven and forms the western boundary of the subtropical **gyre** in the Atlantic. To the general public, the Gulf Stream often means the whole northern branch of the subtropical gyre, including the North Atlantic Drift. As the Gulf Stream and North Atlantic Drift transport warm water northward, their path is associated with relatively high temperatures compared with other oceanic regions at the same latitude. However, rather than stressing the climatic role of the Gulf Stream, it is more appropriate to analyse the oceanic heat transport associated with the wind-driven and **thermohaline circulations**, both of which contribute to the Gulf Stream mass transport. In particular, the thermohaline circulation in the Atlantic contributes to the relatively mild conditions found in Europe. Nevertheless, the main reason |

for the different winter temperatures in eastern Canada and western Europe is the atmospheric circulation, which brings relatively warm air of oceanic origin to Europe.

Gyres
Gyres are large-scale quasi-circular patterns of circulation in the ocean. For instance, the subtropical gyres are almost closed loops of ocean currents between roughly 15 and 45°.

Hadley cell
Hadley cells are thermally driven cells with rising air near the equator in the **intertropical convergence zone** (ITCZ), poleward flow in the upper **troposphere**, subsiding air in the subtropics at around 30° and return flow from the subtropics to the equatorial regions as part of the **trade winds**.

Halocarbons
Halocarbons are chemical compounds in which one or more carbon atoms are linked with one or more halogen atoms (e.g., fluorine, chlorine, bromine or iodine).

Hour angle
The hour angle $HA$ indicates the time since the Sun was at its local meridian, measured from the observer's meridian westward. $HA$ thus is 0 at the local solar noon. It is generally measured in radians or hours ($2\pi$ rad $= 24$ h).

Humidity
Atmospheric humidity is the amount of water vapour in the air. Different definitions are available based on the mass ratio of water vapour compared to that of air or the partial pressure of the vapour. See also **specific humidity**, **relative humidity** and **saturation vapour pressure.**

Humus
Humus is soil organic matter formed of relatively stable organic compounds resulting from the decomposition of plant litter.

Hydrates
Clathrates are lattice structures formed by a host molecule which traps a guest molecule. When the structure is made of water which traps a gas, the compound resembling ice is termed a 'gas hydrate'.

Hydrosphere
The hydrosphere is the water on and underneath the Earth's surface (i.e., oceans, seas, rivers, lakes and underground water).

Hydrostatic equilibrium
On a large scale, in the atmosphere and the ocean, the dominant terms in the vertical equation of motion are gravity $m\vec{g}$ and the force due to the vertical pressure gradient. The hydrostatic balance, which assumes that these two forces balance each other, thus holds to a very good approximation:

$$\frac{\partial p}{\partial z} = -\rho g$$

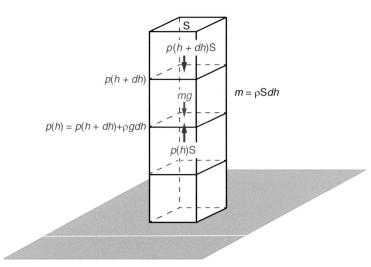

Balance of gravity and pressure forces on a small element of surface S and height $dh$. The increase in pressure between $h + dh$ and $h$ is equal to the weight of the element divided by its surface. Consequently, the sea-level pressure depends on the mass of the whole air column above the surface.

In this equation, $p$ is the pressure, $\rho$ is the density and $g$ is the gravitational acceleration. When this balance is achieved, the fluid is said to be in hydrostatic equilibrium, and knowing the density, the pressure can be computed by integrating the equation along the vertical. In the ocean, $\rho$ can be considered a constant, $\rho = \rho_0$. Integration of the preceding equation then indicates that the pressure increases linearly with depth (be cautious that $z$ is 0 at the surface and is negative at depth); thus,

$$p = p_0 - \rho_0 gz$$

In an **ideal gas**, $\rho = p/R_g T$, If $T$ can be considered a constant, then the pressure decreases exponentially with height (Figure G.4):

$$p = p_0 e^{-gz/R_g T}$$

| | |
|---|---|
| Iceberg | An iceberg is a large piece of ice, which originates on land, floating in open water. |
| Ice cap | An ice cap is a mass of ice, usually dome shaped, which covers less than 50,000 km² of land. |
| Ice sheet | An ice sheet is a mass of ice which covers more than 50,000 km² of land. |

| Ice shelf | An ice shelf is a thick platform of floating ice, originating on land, which has flowed across the coastline onto the sea. The boundary between the grounded ice which rests on bedrock and the floating ice shelf is called the 'grounding line'. |
|---|---|

Ideal gas law    The ideal gas law, also referred to as the 'perfect gas law', generally is given by

$$pV = nR^*T$$

where $p$ is the pressure of the gas, $V$ is the volume occupied by the gas, $n$ is the number of moles of gas, $R^*$ is the universal gas constant ($R^* = 8.3143$ J K$^{-1}$ mol$^{-1}$) and $T$ is the absolute temperature of the gas. This equation can be expressed as a function of $\rho$, the density, instead of the volume:

$$p = \frac{n}{V} R^*T = \frac{nm_g}{V} \frac{R^*}{m_g} T = \frac{m}{V} \frac{R^*}{m_g} T = \rho \frac{R^*}{m_g} T = \rho R_g T$$

in which $m_g$ is the molar mass of the gas, $m$ is the mass of the gas in the volume $V$ ($\rho = m/V$) and $R_g$ is the specific gas constant for this gas.

Insolation    The instantaneous insolation is the energy received per unit time on 1 m$^2$ of a horizontal plane at the top of the atmosphere (or, equivalently, on a horizontal plane at the Earth's surface if we neglect the influence of the atmosphere). It is measured in watts per square metre (W m$^{-2}$). It thus has the same unit as irradiance. The daily insolation is the total insolation received during one day (J/m$^2$).

Internal energy    The internal energy of a system is the sum of the energy of all the particles in the system, measured by reference to the centre of mass of the system. For a perfect gas (a good approximation for the atmosphere), a solid and an incompressible fluid (a good approximation for the ocean), it can be considered as a function of the temperature and mass of the system.

Internal variability    The internal variability of the climate system is the variability which is due to internal processes only and would be present in the absence of any change in the **forcing**.

Interglacial    An interglacial is a relatively warm period between two glacial periods (ice ages).

| Intertropical convergence zone | The intertropical convergence zone (ITCZ) is a band close to the equator where the trade winds of the two hemispheres meet, resulting in a convergence, rising air and heavy precipitation. |

Isopycnal
An isopycnal is a surface of equal potential density in the ocean. The adjective 'isopycnal' refers to changes or processes which take place along surfaces of equal potential density. Isopycnals are generally very close to the horizontal, but small deviations from the horizontal may have a large impact on ocean dynamics and on the representation of some processes (e.g., diffusion) in ocean models.

Isotope
Isotopes are atoms whose nuclei contain the same number of protons (and are therefore the same element) but a different number of neutrons. Isotopes have very similar chemical properties but different masses and different physical properties (some of which have an influence on chemical reactions). Isotopes can be divided into stable and unstable (radioactive) varieties. Radioactive isotopes decay, and their abundance decreases with time, unless new isotopes are produced.

Isotopic fractionation
Because of the different properties of the various **isotopes**, they can be partially separated during chemical reactions, phase changes or exchanges between different media, resulting in variations of the isotope ratio in different substances or phases.

Laplacian
The Laplacian operator $\nabla^2$ is a differential operator which can be written in Cartesian coordinates in three dimensions, $x$, $y$ and $z$, as

$$\nabla^2 = \frac{\partial^2}{\partial x^2} + \frac{\partial^2}{\partial y^2} + \frac{\partial^2}{\partial z^2}$$

Lapse rate
The temperature lapse rate $\Gamma$ is the negative of the vertical **gradient** of temperature.

Latent heat
The latent heat is the energy released or absorbed by a substance during a change of phase. More formally, it is the variation in enthalpy associated with a phase transition at a constant temperature $T$. The latent heat of fusion $L_f$ is the energy associated with changes between the solid and liquid states, whilst the latent heat of vaporisation $L_v$ is associated with transitions between the liquid and gaseous states. See **Clausius-Clapeyron equation**.

| Lead | A lead is an elongated area of open water inside the sea ice pack. |

| Leaf-area index | The leaf-area index (unitless) is the cumulated surface of the green leaves of vegetation per unit of ground surface. One side of the leaf is taken into account for broad leaves, but the estimate is more complex for needle leaves, the projected needle leaf area being generally used. Its value is generally between 0 (bare land) and 10 (see, e.g., http://ldas.gsfc.nasa.gov/gldas/GLDASlaigreen.php). |

| Least-squares method | A classical example of least-squares methods is linear **regression**. The goal is to find a linear function which minimises the sum of the squares of the difference between this function and a set of observations. For more details, see Wilks (2011) and van Storch and Zwiers (1999). |

| Legendre polynomials | Legendre polynomials $P_n(\mu)$ are polynomial of degree $n$ defined as |

$$P_n(\mu) = \frac{1}{2^n n!} \frac{d^n (\mu^2 - 1)^n}{d\mu^n}$$

The first four Legendre polynomials thus are

$$P_0(u) = 1$$
$$P_1(u) = \mu$$
$$P_2(u) = \frac{1}{2}(3\mu^2 - 1)$$
$$P_3(u) = \frac{1}{2}(5\mu^3 - 3\mu)$$

| Limestone | Limestone is a sedimentary rock mainly composed of **calcite**. |

| Lithosphere | The lithosphere is the outermost part of the solid Earth. It includes the Earth's crust and the upper part of the mantle. |

| Longwave radiation | The thermal radiation emitted by the Earth in the infrared part of the **electromagnetic spectrum** is often referred to as 'longwave radiation'. |

| Melt ponds | Melt ponds are pools of water which form at the surface of sea ice in spring and summer, mainly in the northern hemisphere. They occur because of ice and snow melting. They have a lower albedo than ice and have a significant influence on the surface heat balance (Figure G.5). |

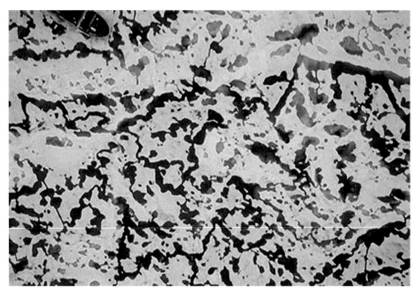

Fig. G.5   Melt ponds at the surface of Arctic sea ice. The boat in the upper left corner provides a rough scale to the figure.
(Source: Photo from D. Perovich. Reproduced with permission.)

| | |
|---|---|
| Member of an ensemble | Individual simulation of an **ensemble of simulations** (= **ensemble member**). |
| Meridional | The adjective 'meridional' refers to the North–South direction. The meridional transport (of mass or heat, for instance) is a net transport from one latitudinal band to another, either northward or southward. |
| Meridional overturning circulation | The **meridional overturning circulation** (MOC) of the ocean is a circulation traditionally described in the latitude-depth plane. It can be represented by a stream function obtained as the integral of the velocity between the East–West boundaries of the oceanic basin and from the surface to the depth considered. The MOC in the Atlantic is often related to the **thermohaline circulation**, but the difference should be kept in mind because the MOC also includes shallow wind-driven cells such as the one observed in equatorial regions (Figure G.6). |
| Metamorphism | Metamorphism involves changes in solid rocks primarily caused by variations in temperature and pressure. |
| Mitigation | Mitigation refers to the actions taken to reduce the anthropogenic emission of greenhouse gases or to stimulate their removal from the atmosphere in order to limit the magnitude of future climate changes. |

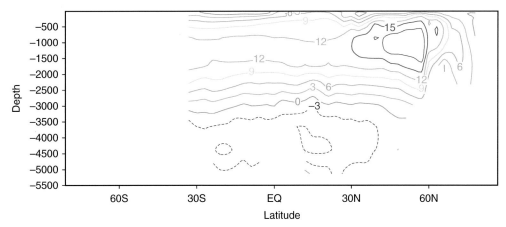

**Fig. G.6** Meridional overturning stream function (in Sv; $1 \text{ Sv} = 10^6 \text{ m}^3 \text{ s}^{-1}$) in the Atlantic for present-day conditions, as given by the climate model LOVECLIM. By convention, solid contours show clockwise flows, and dashed contours (corresponding to negative values) show anti-clockwise flows. The value plotted at a particular location represents the mass (actually the volume) transported between the surface and the point considered. For instance, there is a northward mass transport with a maximum value of about 13 Sv at 30°S over the top 1400 m of the Atlantic. Below this depth, the transport is southward, whilst close to the bottom, the transport is northward again. This figure also shows a large sinking (downward mass transport) in the North Atlantic.

| | |
|---|---|
| Monsoon | Monsoons are seasonal changes in atmospheric circulation and precipitation caused by the differential heating of a land mass and its adjacent ocean. |
| Multi-model ensemble | See **ensemble of simulations** |
| Natural variability | Natural variability represents the climate changes that are due to natural processes. It includes the **internal variability** and the response to natural **forcing**. |
| Net primary production | The net primary production (NPP) is the rate of net carbon uptake related to photosynthetic activities. It is the difference between the uptake by **photosynthesis** (gross primary production) and respiration of plants or **phytoplankton**. |
| Numerical grid | See **grid**. |
| Obliquity | The obliquity ($\varepsilon_{obl}$) is the angle between the equator and the **ecliptic**. It corresponds to the angle between the axis of rotation of the Earth and the perpendicular to the **ecliptic plane**. |
| Oceanic mixed layer | The oceanic mixed layer is the upper part of the ocean which is in direct contact with the surface. The properties of this layer are influenced by the presence of the surface and, in turn, influence the exchanges between the ocean and the |

atmosphere. Vertical mixing is generally strong in the oceanic mixed layer because of the relatively intense turbulence. Consequently, the oceanic properties (i.e., temperature, salinity, etc.) are fairly uniform in this layer.

Optical depth

The optical depth measures the fraction of light which is **scattered** or absorbed during its path through a medium, so producing a reduction in the intensity of the incident beam. If $I_0$ is the irradiance at the source (e.g., the top of the atmosphere) and $I$ is the irradiance at a particular point (usually the Earth's surface), the optical depth $\tau$ is defined by

$$\tau = -\ln(I/I_0)$$

corresponding to

$$I/I_0 = e^{-\tau}$$

In atmospheric sciences, $\tau$ is usually defined along a vertical path.

Orbital parameters

The characteristics of the Earth's movement are described by three parameters, called the 'astronomical' or 'orbital parameters'. The term **'astronomical parameter'** is usually preferred because the three parameters characterise more than the orbit of the Earth around the Sun.

Overturning circulation

An overturning circulation describes a flow characterised by sinking in one region, upward movement in another and horizontal transports between those regions to form a closed circulation. It thus can be represented schematically as a loop in a vertical plane. See also **meridional overturning circulation**.

Ozone

Ozone is a molecule consisting of three oxygen atoms ($O_3$). Its presence in the **stratosphere** protects the Earth's surface from dangerous ultraviolet radiation. In the lower **troposphere**, it is a dangerous, strongly irritating pollutant.

Parameterisation

Some processes are not explicitly included in models because of simplifications, lack of knowledge of the mechanisms involved or because the spatial resolution of the model is not high enough to include them. To take the first-order effects of these processes into account, they are represented by parameterisations.

Partial differential equations

Partial differential equations (PDEs) are equations involving an unknown function of several independent variables and its partial derivatives with respect to these variables.

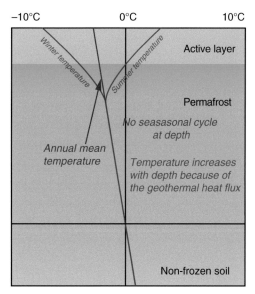

−10°C                    0°C                    10°C

*Winter temperature*

*Summer temperature*

Active layer

Permafrost

No seasasonal cycle
at depth

*Annual mean
temperature*

*Temperature increases
with depth because of
the geothermal heat flux*

Non-frozen soil

**Fig. G.7** Schematic representation of the vertical profile of temperature in a permafrost region. The vertical line at 0°C allows delimiting of the active layer, in which maximum summer temperature is above the freezing point, the permafrost region which is below the freezing point all year long and the frost-free soil at greater depth.

| | |
|---|---|
| Perihelion | The perihelion is the point in the Earth's orbit which is closest to the Sun. |
| Permafrost | Permafrost is a layer of soil or rock beneath the surface which remains below 0°C throughout the year. This occurs when the summer warming is insufficient to reach the bottom of the layer of frozen ground. The top layer which is frozen in winter but has temperatures above 0°C in summer is called the 'active layer'. Its thickness ranges from a few centimetres in very cold regions to more than 1 m in warmer regions. Permafrost can include ground ice or simply soil or rock at temperatures below 0°C (dry permafrost) (Figure G.7). |
| pH | pH is a measure of the acidity of a solution, for instance, sea water, generally estimated by the logarithm of its concentration of hydrogen ions ($H^+$): |

$$pH = -\log_{10}[H^+]$$

An acidic solution has a pH lower than 7, a neutral solution has a pH of 7 and an alkaline (basic) solution has a pH higher than 7.

| | |
|---|---|
| Phytoplankton | Phytoplankton is a type of **plankton** which produces complex organic compounds from simple inorganic molecules. This can be achieved by using energy from |

light (via photosynthesis) or through inorganic chemical reactions.

| | |
|---|---|
| Planetesimals | Planetesimals are small bodies (much smaller than a planet) in the solar system. |
| Plankton | Plankton consists of organisms (mostly microscopic plants and animals) which drift in the seas or in freshwater bodies. See **phytoplankton** and **zooplankton**. |
| Plant functional type | Plant functional types (PFTs) are groups of plants which share common characteristics (e.g., tropical trees, deciduous temperate trees, needle-leaf boreal trees and different types of grass). |
| Polynya | A polynya is a region of open water, larger than a **lead**, inside the ice pack. |
| Pore | The porosity of a soil is defined as the ratio between the volume which is not occupied by solid particles (sometimes referred to as 'voids') and the total volume. |
| Potential temperature | The potential temperature is the temperature which a sample of seawater (or air) initially at some depth (altitude) $z$ would have if it were lifted **adiabatically** to a reference level $z_r$. |
| ppm | The relative amount of gases, such as $CO_2$, in the atmosphere is measured by the mole fraction, expressed as parts per million (ppm), corresponding to the number of molecules of the gas in 1 million molecules of dry air. |
| Principal component | One goal of principal-components analysis (PCA) is to describe a data set using only a few variables or spatial patterns, which include the largest possible part of the variability in the data set, by taking profit of its structure (e.g., the links between different locations). For instance, the **anomalies** of a field $F'(x, y, t)$, available at particular locations and specific times, can be transformed using this statistical method as the product of time series $c_i(t)$ and fixed patterns $e_i(x, y)$: |

$$F'(x,y,t) = \sum_{i=1}^{k} c_i(t) e_i(x,y)$$

The $e_i(x, y)$ are called '**empirical orthogonal functions**' (EOFs) or 'eigenvectors', and the $c_i(t)$ EOFs are called 'coefficients' or 'principal components'. This decomposition is constructed in such a way that the first principal component is associated with the largest possible fraction of the variance of the field. The next components are determined to account for as much as possible of the variability of $F'(x, y, t)$ while being orthogonal to all the preceding principal components (justifying the wording EOF). To have a complete description of the field, a large value of $k$ in the preceding equation may be required, but because the variance decreases strongly with the index $i$, keeping only a few EOFs is generally enough. For more details, see Wilks (2011) or von Storch and Zwiers (1999).

| | |
|---|---|
| Predictability | Determining the predictability of the climate system is to distinguish the part of its variations that may, in theory, be predicted in contrast to the part that is intrinsically unpredictable. The predictability can be derived from two sources. The first is the response to a known **forcing**. The second is due to the knowledge of the initial conditions which determine the future development of the system. On a time scale of a few days, the sequence of individual events can be followed, and a meteorological forecast can provide estimates of the weather conditions on a particular day in a particular region. This is sometimes referred to as a 'deterministic forecast'. However, because of the rapid growth of the errors which are unavoidable (e.g., because of the limited precision of the observations used to define the initial conditions), such forecasts have only **skill** up to 1 to 2 weeks. For seasonal or decadal **predictions**, only the statistics of the system can be determined, such as the probability of having a warm summer or positive phase of the NAO. The projections thus are probabilistic in essence. Actually, weather forecasts provide now, in addition to the best estimate of the condition at a time, the probability of alternative states and thus can be related to probabilistic forecasts. |
| Predictions | Predictions are estimates of the future evolution of a climate system based on knowledge of its current state and estimates of future **forcing**. In climatology, the predictions are often made on seasonal to decadal time scales. See also **predictability**. |
| Precession | See **climatic precession**. |
| Projection | A climate projection is an estimate of the potential future state of the climate system in response to specified changes in external **forcing**. The projections thus include an uncertainty related to the **scenario** selected. |
| Proxy data | Proxy data are indirect information on climate state collected from climate-sensitive recorders present in various archives |

|                   |                                                                                                                                                                                                                                                                                                    |
|-------------------|-----------------------------------------------------------------------------------------------------------------------------------------------------------------------------------------------------------------------------------------------------------------------------------------------------|
|                   | (e.g., tree rings, isotopic composition of ice and marine cores, lake sediment composition, historical data, etc.). |
| Radiation         | See **electromagnetic radiation**. |
| Radiative forcing | 'Radiative forcing is the change in the net downward minus upward irradiances (expressed in $W\,m^{-2}$) at the **tropopause** due to a change in an external driver of climate change such as, for example, a change in the concentration of carbon dioxide or in the output of the Sun. Radiative forcing is computed with all tropospheric properties held fixed at their unperturbed values and after allowing for stratospheric temperatures to readjust to radiative-dynamical equilibrium. Radiative forcing is called 'instantaneous' if no change in stratospheric temperature is accounted for' (IPCC 2007). |
| Reanalyses        | Weather forecasting centres reconstruct the present atmospheric configuration every day, combining previous forecasts and observations using data assimilation techniques, in order to obtain physically consistent estimates of the atmospheric state (analyses). Because of changes in the structure of the models and the procedures used, the analyses are not necessarily homogeneous over long periods. In order to reduce potential spurious trends, reanalyses are performed using the same model and the same procedure over the whole period. Nevertheless, biases are still present because the amount of data available changes over time. Similarly, estimates of the changes in past state from available data and models also have been obtained for other components of the climate system, in particular, for the ocean. |
| Regression        | Linear regression is a statistical procedure which represents a variable (the dependent variable) as a linear function of one or more other variables (the independent variables). The model parameters which link the dependent and independent variables are called the 'regression coefficients'. |
| Relative humidity | The relative humidity (RH) measures the mass of water vapour in an air parcel compared to the mass at saturation at the same temperature and pressure. Numerically, it is nearly identical to compute it as the ratio of the partial pressure of water vapour in the parcel to the **saturation vapour pressure** or as the ratio of the **specific humidity** (or mixing ratio) of the parcel and the specific humidity (mixing ratio) at saturation. At saturation (i.e., equilibrium between the liquid and vapour phases), the relative humidity is 1 (or 100%). |

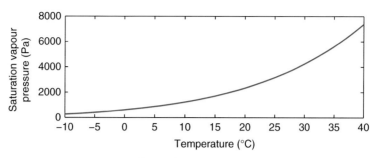

Saturation vapour pressure over a water surface as a function of temperature.

| | |
|---|---|
| Remineralisation | Remineralisation is the transformation of organic matter to an inorganic (mineral) form which can be used again by primary producers. |
| Resolution | See **grid**. |
| Saturation vapour pressure | The saturation vapour pressure $e_s$ is the partial pressure of a vapour when the vapour phase is in equilibrium with the liquid phase. When the water vapour pressure is equal to the saturation vapour pressure, the **relative humidity** is 1 (Figure G.8). |
| Scattering | Scattering describes a process in which radiation deviates from a straight trajectory, for instance, because of the presence of some particles in the gas it passes through. |
| Scenario | A climate scenario is an estimate of future changes in external forcing, emission and/or concentration of **greenhouse gases**, **aerosols**, various pollutants in the atmosphere, land use and so on. These factors can be related to estimates of future socio-economic and technological developments. |
| Sea ice | Sea ice is the ice which forms when sea water freezes. |
| Sedimentation | Sedimentation in the ocean is the tendency of particles in suspension to move downward and be deposited on the ocean floor. |
| Sensible heat | The sensible heat flux between two media or between two parcels within a medium is the energy which is transferred in the form of thermal energy (heat) and is related to the difference in temperature between the media or parcels. |
| Shortwave radiation | Shortwave radiation is another name for the radiation emitted by the Sun and received by the Earth. |
| Skill | The skill of a forecast, a **prediction** or a **projection** is its ability to provide additional information compared to |

a standard reference method. For instance, the skill of a weather forecast can be estimated by comparing its accuracy with the one of a forecast assuming constant future states identical to the initial conditions of the forecast (i.e., persistence). In climatology, the standard reference is often taken as a climate identical to the one observed presently. As projections or simulated results for past conditions are never perfect, the goal is not to describe the absolute accuracy but the relative improvement in our knowledge of the system dynamics or of its state. Various objective measures of the skill have been developed, in particular, in meteorology. See Wilks (2011) for more details.

**Solar constant**
The solar constant $S_0$ is the amount of incoming solar energy per unit area and unit time at the mean Earth–Sun distance measured on the outer surface of the Earth's atmosphere in a plane perpendicular to the rays (in $W\,m^{-2}$). As this value is not constant in time, the term '**total solar irradiance**' is preferred currently.

**Solar declination**
The solar declination $\delta$ is the angle between a line from the centre of the Earth towards the Sun and the celestial equator.

**Specific heat capacity at constant pressure**
The specific heat capacity at constant pressure of a body $c_p$ is the energy required to increase the temperature of 1 kg of the body by 1°C at a constant pressure.

**Specific heat capacity at constant volume**
The specific heat capacity at constant volume of a body $c_v$ is the energy required to increase the temperature of 1 kg of the body by 1°C at a constant volume.

**Specific humidity**
The specific **humidity** $q$ is the ratio of the mass of water vapour to the mass of dry air plus water vapour in a particular volume of air. As the mass of the water vapour is much less than the mass of the air, the specific humidity is very close to the 'mixing ratio', defined as the ratio of the mass of water vapour to the mass of dry air.

**State variable**
A state variable is a property which characterises the state of a system and depends only on the current state of the system, not of its past history. If the state of the system is perfectly known, all the state variables can be obtained, and conversely, if all the state variables are known, the state of the system is perfectly defined.

**State vector**
The state vector is a one-dimensional matrix (a vector), function of time, including all the **state variables** at all the locations at which they are computed.

| | |
|---|---|
| Statistic | A statistic is the result of applying a statistical algorithm to some data. The commonest statistics are the average and standard deviation of a range of observations. |
| Stefan-Boltzmann law | The Stefan-Boltzmann law, also known as 'Stefan's law', states that the total energy radiated per unit surface area of a **black body** in unit time is directly proportional to the fourth power of the temperature $T$: |

$$E = \sigma T^4$$

where $\sigma$ is the Stefan-Boltzmann constant ($\sigma = 5.67 \times 10^{-8}$ W m$^{-2}$ K$^{-4}$).

| | |
|---|---|
| Stomata | A stoma is an aperture, found, for instance, on leaves, which is used by plants for gas exchanges with the atmosphere. Air containing carbon dioxide and oxygen necessary for **photosynthesis** and respiration enters the plant through these openings (plural: stomata), whilst the oxygen produced by photosynthesis is expelled through them. These exchanges include a release of water vapour to the atmosphere (transpiration). |
| Storm track | The storm track is the path which **cyclones** tend to follow at middle latitudes. |
| Stratification | Stratification is a measure of vertical density **gradient**. |
| Stratosphere | The stratosphere is the layer of the atmosphere located above the **tropopause**, at an altitude between roughly 10 and 50 km. |
| Subduction | Subduction occurs in regions where two tectonic plates converge, resulting in one plate sliding underneath the other and moving down into the mantle. In subduction, lighter continental plates ride above denser oceanic plates. |
| Subsidence | Subsidence is a downward (air) motion. |
| Supercontinent | During some periods of the Earth's history, all or nearly all the continents have been clustered together, forming a single land mass referred to as a 'supercontinent'. Some recent examples are Rodinia around 800 million years ago, Pannotia around 600 million years ago and Pangea around 300 million years ago. |
| Surface boundary layer | See **atmospheric boundary layer**. |
| Taiga | Taiga is a boreal forest covered by conifers. |
| Taylor series | This is the representation of a function as an infinite power series whose coefficients are calculated from the values of |

the derivatives of the function at a single point. If the series is truncated to a finite number of terms, the resulting Taylor polynomial provides a polynomial approximation of the function around that point.

| | |
|---|---|
| Teleconnection | A teleconnection is usually indicated by the correlation between the values observed at two separate locations. This link is related to a pattern of variability, associated with wave propagation, the presence of mountains, and so on. |
| Thermal expansion | Thermal expansion is the change in volume of a constant mass (e.g., oceanic water) as a result of a change in its temperature. In the sea, a temperature increase produces an increase in the volume of water (also called 'dilation' or 'dilatation') and thus a rise in sea level if the oceanic mass remains unchanged. |
| Thermocline | A thermocline is a region in the ocean with a strong temperature gradient. Oceanographers usually make a distinction between the seasonal thermocline (which is formed at the base of the summer **mixed layer** and is absent in winter) and the permanent thermocline (which separates the surface layer from the relatively homogeneous deep ocean). |
| Thermohaline circulation | The thermohaline circulation is a large-scale circulation in the ocean which involves circulation both at the surface and at great depths. It is partly driven by the density contrasts in the ocean. |
| Time scale | The time scale of variation of a process represents the time over which significant changes can be expected to be observed. It takes millions of years for plate tectonics to induce movements of the continents which have a clear impact on the climate. This process thus is said to be important for climate on the time scale of millions of years. The time scale can represent an order of magnitude or be defined very precisely, for example, on the basis of spectral analysis or on an exponential decay. The definition of a time scale also can be related to the dominant periodicity of interest to an investigator. For instance, if someone is interested in variations on a seasonal time scale, the analysis will be mainly devoted to the differences between the various seasons. Time scales also can be daily, monthly, annual, decadal, centennial, millennial and so on. |
| Total solar irradiance | The total solar irradiance is the energy emitted by the Sun over all wavelengths which falls each second on 1-m$^2$ surface perpendicular to the Sun's rays at the mean Earth–Sun |

distance measured at the top of the Earth's atmosphere. It measures the solar energy flux in watts per square metre ($W\,m^{-2}$) and is sometimes called the '**solar constant**'.

Tracer — A tracer is a constituent which is transported by a flow. A distinction is often made between active tracers which modify the flow through their influence on density (e.g., temperature and salinity in the ocean) and passive tracers which do not influence the motion (e.g., **chlorofluorocarbons** in the ocean). In oceanography, **phytoplankton** and **zooplankton** are generally treated as tracers because their movement is mainly determined by the ocean circulation.

Trade winds — Trade winds are easterly winds (from the northeast in the northern hemisphere and from the southeast in the southern hemisphere) characteristic of tropical regions.

Transient climate response — The transient climate response (TCR) is defined as the global average of the annual mean surface temperature change averaged over years 60 to 80 in an experiment in which the $CO_2$ concentration is increased by 1% per year until year 70 (i.e., until it doubles its initial value).

Transient climate response to carbon emissions — This is the ratio of global temperature change to cumulative anthropogenic emissions (in °C per 1000 PgC).

Trophic level — Trophic levels are the various stages within food chains. Standard examples of trophic levels are the primary producers, the primary consumers (herbivores) and higher-level consumers (predators), as well as the decomposers which transform dead organisms and waste materials into nutrients available for the producers.

Tropopause — The tropopause is the boundary between the **troposphere** and the **stratosphere** located at an altitude of about 10 km.

Troposphere — The troposphere is the lowest part of the Earth's atmosphere. Its average depth is about 10 km.

True longitude — The true longitude $\lambda_t$ is the angle on the **ecliptic plane** between the position of the Earth relative to the Sun at any given time and at the vernal equinox.

Tundra — Tundra is a **biome** characteristic of regions where trees cannot grow because the temperature is too low.

Upwelling — An upwelling is an upward movement of water in the ocean.

Vernal equinox — See **equinox**.

| | |
|---|---|
| Walker circulation | The Walker circulation is a zonal **overturning circulation** in equatorial regions. In the Pacific, it is associated with **ascendance** over the warm western Pacific, **subsidence** over the cold eastern Pacific, eastward transport in the upper troposphere and westward atmospheric flow in the lower layers. |
| Water mass formation | Oceanographers talk about water mass formation when a volume of water acquires specific properties, such as temperature and salinity, because of interactions with the atmosphere and keeps them while being transported by ocean currents. See **deep-water formation** and **thermohaline circulation**. |
| Weathering | Weathering is the decomposition of rocks and soils at or near the Earth's surface. |
| Westerlies | Westerlies are winds coming from the west which are typically found at middle latitudes. |
| Wien's law | Wien's law states that the wavelength at which the radiant energy of a **black body** is greatest is only a function of temperature. For the Sun this maximum is located in the visible part of the **spectrum**, whilst for the Earth it is located in the infrared. |
| Zenith distance | The solar zenith distance is defined as the angle between the solar rays and the normal to the Earth's surface at a particular point. |
| Zonal | The adjective 'zonal' refers to the East–West direction. For instance, the zonal mean for any latitude is the average over all longitudes. |
| Zooplankton | Zooplankton is the type of **plankton** which consumes (or grazes) **phytoplankton**. |

# Cited References and Further Reading

## Books and Textbooks

Many excellent textbooks focussing on different components of the climate system or on methods applied in climatology are currently available. Here is a subjective list of the ones which have been most useful during preparation of this book and which provide complementary information to subjects covered in the various chapters at a level adapted for a general audience. A limited number of references have been given on purpose to help readers choosing material adapted to their needs. This is by no means intended to be a selection of the best books on each subject, and in many cases, alternative choices are certainly as valuable as the ones listed herein.

Archer, D. (2010). *The Global Carbon Cycle* (Princeton Primers in Climate). Princeton University Press, Princeton, NJ.

(2011). *Global Warming: Understanding the Forecast*, 2nd ed. Wiley-Blackwell, Hoboken, NJ.

Bonan, G. B. (2008). *Ecological Climatology: Concepts and Applications*, 2nd ed. Cambridge University Press.

Boucher, O. (2012). *Aérosols atmosphériques: Propriétés et impacts climatiques*. Springer, Berlin.

Bradley, R. S. (2014). *Paleoclimatology: Reconstructing Climates of the Quaternary*, 3rd ed. (International Geophysics Series 68). Academic Press, New York.

Benestad, R. E., I. Hanssen-Bauer and D. Chen (2008). *Empirical-Statistical Downscaling*. World Scientific Publishing, Hackensack, NJ.

Cronin, T. M. (2010). *Paleoclimates: Understanding Climate Change Past and Present*. Columbia University Press, New York.

Cushman-Roisin, B., and J.-M. Beckers (2011). *Introduction to Geophysical Fluid Dynamics: Physical and Numerical Aspects*, 2nd ed. (International Geophysics Series 101). Academic Press, New York.

Dijkstra, H. A. (2013). *Nonlinear Climate Dynamics*. Cambridge University Press.

Duplessy, J.-C., and G. Ramstein (eds.) (2013). *Paléoclimatologie-Trouver, dater et interpreter les indices*, Vol. 1. EDP Science/CNRS Editions, Les Ulis, France.

(eds.) (2013). *Paléoclimatologie-Enquête sur les climats anciens*, Vol. 2. EDP Science/CNRS Editions, Les Ulis, France.

Hartmann, D. L. (1994). *Global Physical Climatology* (International Geophysics Series 56). Academic Press, New York.

Houghton, J. (2009). *Global Warming: The Complete Briefing*. Cambridge University Press.

Kalnay, E. (2003). *Atmospheric Modeling: Data Assimilation and Predictability*. Cambridge University Press.

Liou, K. N. (2002). *An Introduction to Atmospheric Radiation* (International Geophysics Series 84). Academic Press, New York.

Marshall, J., and R. A. Plumb (2008). *Atmosphere, Ocean and Climate Dynamics: An Introductory Text* (International Geophysics Series 93). Academic Press, New York.

Millero, F. J. (2006). *Chemical Oceanography*, 3rd ed. Taylor and Francis, London.

McGuffie, K., and A. Henderson-Sellers (2014). *The Climate Modelling Primer*, 4th ed. Wiley-Blackwell, Hoboken, NJ.

Neelin, D. J. (2011). *Climate Change and Climate Modeling*. Cambridge University Press.

Pierrehumbert, R. Y (2010). *Principles of Planetary Climate*. Cambridge University Press.

Schlesinger, W. H., and E. S. Bernhardt (2013). *Biogeochemistry: An Analysis of Global Change*, 3rd ed. Academic Press, New York.

Sarmiento, J. L., and N. Gruber (2006). *Ocean Biogeochemical Dynamics*. Princeton University Press, Princeton, NJ.

Skinner, B. J., S. C. Porter and J. Park (2004). *Dynamic Earth: An Introduction to Physical Geology*, 5th ed. Wiley, Hoboken, NJ.

Stocker, T. (2011). *Introduction to Climate Modelling*. Springer, Berlin.

Talley, L. D., G. L. Pickard, W. J. Emery and J. H. Swift (2011). *Descriptive Physical Oceanography: An Introduction*, 6th ed. Academic Press, New York.

von Storch, H., and F. W. Zwiers (1999). *Statistical Analysis in Climate Research*. Cambridge University Press.

Wallace, J. M., and P. V. Hobbs (2006). *Atmospheric Science: An Introductory Survey*, 2nd ed. (International Geophysics Series 92). Academic Press, New York.

Wilks, D. S. (2011). *Statistical Methods in the Atmospheric Sciences* (International Geophysics Series 100). Academic Press, New York.

## Other References

Abe-Ouchi, A., F. Saito, K. Kawamura, M. E. Raymo, J. Okuno, K. Takahashi and H. Blatter (2013). Insolation-driven 100,000-year glacial cycles and hysteresis of ice-sheet volume. *Nature* 500, 190–4.

Adler, R. F., G. J. Huffman, A. Chang, R. Ferraro, P.-P. Xie, J. Janowiak, B. Rudolf, U. Schneider, S. Curtis, D. Bolvin, A. Gruber, J. Susskind, P. Arkin and E. Nelkin (2003). The version 2 Global Precipitation Climatology

Project (GPCP) monthly precipitation analysis (1979–present). *Journal of Hydrometeorolgy* 4, 1147–67.

Allen, M. R., and P. A. Stott (2003). Estimating signal amplitudes in optimal fingerprinting: I. Theory. *Climate Dynamics* 21, 477–91.

Alley, R. B., P. A. Mayewski, T. Sowers, M. Stuiver, K. C. Taylor and, P. U. Clark (1997). Holocene climatic instability: a prominent, widespread event 8200 yr ago. *Geology* 25, 483–6.

Andrews, T., J. M. Gregory, M. J. Webb and K. E. Taylor (2012). Forcing, feedbacks and climate sensitivity in CMIP5 coupled atmosphere-ocean climate models. *Geophysical Research Letters* 39, L09712.

Annan, J. D., and J. C. Hargreaves (2013). A new global reconstruction of temperature changes at the Last Glacial Maximum. *Climate of the Past* 9, 367–76.

Arakawa, A., and V. Lamb (1977). *Computational Design of the Basic Dynamical Processes in the UCLA General Circulation Model* (Methods in Computational Physics 17). Academic Press, New York, 173–265.

Archer, D., and V. Brovkin (2008). The millennial atmospheric lifetime of anthropogenic $CO_2$. *Climatic Change* 90, 283–97.

Arneth, A., S. P. Harrison, S. Zaehle, K. Tsigaridis, S. Menon, P. J. Bartlein, J. Feichter, A. Korhola, M. Kulmala, D. O'Donnell, G. Schurgers, S. Sorvari and T. Vesala (2010). Terrestrial biogeochemical feedbacks in the climate system. *Nature Geosciences* 3, 525–32.

Arora, V. K., G. J. Boer, P. Friedlingstein, M. Eby, C. D. Jones, J. R. Christian, G. Bonan, L. Bopp, V. Brovkin, P. Cadule, T. Hajima, T. Ilyina, K. Lindsay, J. F. Tjiputra and T. Wu (2013). Carbon-concentration and carbon-climate feedbacks in CMIP5 Earth System Models. *Journal of Climate* 26, 5289–314.

Arrhenius, S. (1896). On the influence of carbonic acid in the air upon the temperature of the ground. *Philosophical Magazine and Journal of Science* 5(41), 237–76.

Barber, D. C., A. Dyke, C. Hillaire-Marcel, A. E. Jennings, J. T. Andrews, M. W. Kerwin, G. Bilodeau, R. McNeely, J. Southon, M. D. Morehead and J.-M. Gagnon (1999). Forcing of the cold event of 8,200 years ago by catastrophic drainage of laurentide lakes. *Nature* 400, 344–8.

Bartlein, P. J., S. P. Harrison, S. Brewer, S. Connor, B. A. S. Davis, K. Gajewski, J. Guiot, T. I. Harrison-Prentice, A. Henderson, O. Peyron, I. C. Prentice, M. Scholze, H. Seppä, B. Shuman, S. Sugita, R. S. Thompson, A. E. Viau, J. Williams and H. Wu. (2011). Pollen-based continental climate reconstructions at 6 and 21 ka: A global synthesis. *Climate Dynamics* 37, 775–802.

Berger, A. L. (1978). Long-term variations of daily insolation and quaternary climatic changes. *Journal of Atmospheric Sciences* 35, 2363–7.

Berger, A. (1988). Milankovitch theory and climate. *Reviews of Geophysics* 26(4), 624–57.

Berger, A., and M. F. Loutre (1991). Insolation values for the last 10 million years. *Quaternary Science Reviews* 10 (4), 297–317.

(2003). Climate 400,000 years ago, a key to the future? In: *Earth's Climate and Orbital Eccentricity*. The marine isotope stage 11 question. A. W. Droxler, R. Z. Poore, L. H. Burckle (Eds). Geophysical Monograph 137. AGU, Washington, DC, 17–26.

Berger, A., J. L. Mélice and M. F. Loutre (2005). On the origin of the 100-kyr cycles in the astronomical forcing. *Paleoceanography* 20, PA4019.

Berner, R. A. (2006). GEOCARBSULF: A combined model for Phanerozoic atmospheric $O_2$ and $CO_2$. *Geochimica et Cosmochimica Acta* 70, 5653–64.

Berger, A., X. S. Li and M. F. Loutre (1999). Modelling northern hemisphere ice volume over the last 3 Ma. *Quaternary Science Reviews* 18, 1–11.

Bindoff, N. L., P. A. Stott, K. M. AchutaRao, M. R. Allen, N. Gillett, D. Gutzler, K. Hansingo, G. Hegerl, Y. Hu, S. Jain, I. I. Mokhov, J. Overland, J. Perlwitz, R. Sebbari and X. Zhang (2013). Detection and attribution of climate change: From global to regional. In: T. F. Stocker, D. Qin, G.-K. Plattner, M. Tignor, S. K. Allen, J. Boschung, A. Nauels, Y. Xia, V. Bex and P. M. Midgley (eds.), *Climate Change 2013: The Physical Science Basis* (Contribution of Working Group I to the Fifth Assessment Report of the Intergovernmental Panel on Climate Change). Cambridge University Press.

Bitz, C. M., and G. H. Roe (2004). A mechanism for the high rate of sea-ice thinning in the Arctic Ocean. *Journal of Climate* 17, 3623–32.

Bony, S., R. Colman, V. M. Kattsov, R. P. Allan, C. S. Bretherton, J. L. Dufresne, A. Hall, S. Hallegatte, M. M. Holland, W. Ingram, D. A. Randall, B. J. Soden, G. Tselioudis and M. J. Webb (2006). How well do we understand and evaluate climate change feedback processes? *Journal of Climate* 19, 3445–82.

Booth, B. B. B., N. J. Dunstone, P. R. Halloran, T. Andrews and N. Bellouin (2012). Aerosols implicated as a prime driver of twentieth-century North Atlantic climate variability. *Nature* 484, 228–32.

Boucher, O., D. Randall, P. Artaxo, C. Bretherton, G. Feingold, P. Forster, V.-M. Kerminen, Y. Kondo, H. Liao, U. Lohmann, P. Rasch, S. K. Satheesh, S. Sherwood, B. Stevens and X. Y. Zhang (2013). Clouds and aerosols. In: T. F. Stocker, D. Qin, G.-K. Plattner, M. Tignor, S. K. Allen, J. Boschung, A. Nauels, Y. Xia, V. Bex and P. M. Midgley (eds.), *Climate Change 2013: The Physical Science Basis* (Contribution of Working Group I to the Fifth Assessment Report of the Intergovernmental Panel on Climate Change). Cambridge University Press.

Braconnot, P., B. Otto-Bliesner, S. Harrison, S. Joussaume, J.-Y. Peterchmitt, A. Abe-Ouchi, M. Crucifix, E. Driesschaert, Th. Fichefet, C. D. Hewitt, M. Kageyama, A. Kitoh, A. Laîné, M.-F. Loutre, O. Marti, U. Merkel, G. Ramstein, P. Valdes, S. L. Weber, Y. Yu and Y. Zhao (2007). Results of PMIP2 coupled simulations of the Mid-Holocene and Last Glacial Maximum: 1. Experiments and large-scale features. *Climate of the Past* 3, 261–77.

Braconnot P., S. P. Harrison, B. O.-Bliesner, A. Abe-Ouchi, J. Jungclaus and J.-Y. Peterschmitt (2011). The Paleoclimate Modeling Intercomparison Project contribution to CMIP5. *CLIVAR Exchanges* 56, 16(2), 15–19.

Broecker, W. S., and T.-H. Peng (1987). The role of $CaCO_3$ compensation in the glacial to interglacial atmospheric $CO_2$ change. *Global Biogeochemical Cycles* 1, 15–29.

Brohan, P., J. J. Kennedy, I. Harris, S. F. B. Tett and P. D. Jones (2006). Uncertainty estimates in regional and global observed temperature changes: a new data set from 1850. *Journal of Geophysical Research* 111 (D12), D12106.

Brovkin, V., A. Ganopolski and Y. Svirezhev (1997). A continuous climate-vegetation classification for use in climate-biosphere studies. *Ecological Modelling* 101, 251–61.

Budyko, M. I. (1969). The effect of solar radiation variations on the climate of the Earth. *Tellus* 21, 611–19.

Cabré, A., I. Marinov and S. Leung (2015). Consistent global responses of marine ecosystems to future climate change across the IPCC AR5 earth system models. *Climate Dynamics* (in press).

Cai, W., S. Borlace, M. Lengaigne, P. van Rensch, M. Collins, G. Vecchi, A. Timmermann, A. Santoso, M. J. McPhaden, L. Wu, M. H. England, G. Wang, E. Guilyardi and F.-F. Jin (2014). Increasing frequency of extreme El Niño events due to greenhouse warming. *Nature Climate Change* 4, 111–16.

Catling, D. C., and M. W. Claire (2005). How Earth's atmosphere evolved to an oxic state: a status report. *Earth and Planetary Science Letters* 237, 1–20.

CCSP (2008). *Abrupt Climate Change: A Report by the U.S. Climate Change Science Program and the Subcommittee on Global Change Research*, P. U. Clark, A. J. Weaver (coordinating lead authors), E. Brook, E. R. Cook, T. L. Delworth and K. Steffen (chapter lead authors). U.S. Geological Survey, Reston, VA.

Christensen, J. H., K. Krishna Kumar, E. Aldrian, S.-I. An, I. F. A. Cavalcanti, M. de Castro, W. Dong, P. Goswami, A. Hall, J. K. Kanyanga, A. Kitoh, J. Kossin, N.-C. Lau, J. Renwick, D. B. Stephenson, S.-P. Xie and T. Zhou (2013). Climate phenomena and their relevance for future regional climate change. In: T. F. Stocker, D. Qin, G.-K. Plattner, M. Tignor, S. K. Allen, J. Boschung, A. Nauels, Y. Xia, V. Bex and P. M. Midgley (eds.), *Climate Change 2013: The Physical Science Basis* (Contribution of Working Group I to the Fifth Assessment Report of the Intergovernmental Panel on Climate Change). Cambridge University Press.

Church, J. A., P. U. Clark, A. Cazenave, J. M. Gregory, S. Jevrejeva, A. Levermann, M. A. Merrifield, G. A. Milne, R. S. Nerem, P. D. Nunn, A. J. Payne, W. T. Pfeffer, D. Stammer and A. S. Unnikrishnan (2013). Sea level change. In: T. F. Stocker, D. Qin, G.-K. Plattner, M. Tignor, S. K. Allen, J. Boschung, A. Nauels, Y. Xia, V. Bex and P. M. Midgley (eds.), *Climate Change 2013: The Physical Science Basis* (Contribution of Working Group I to the Fifth Assessment Report of the Intergovernmental Panel on Climate Change). Cambridge University Press.

Ciais, P., and J. Jouzel (1994). Deuterium and oxygen 18 in precipitation: isotopic model, including mixed cloud processes. *Journal of Geophysical Research: Atmospheres* 99 (D8), 16793–803.

Ciais, P., C. Sabine, G. Bala, L. Bopp, V. Brovkin, J. Canadell, A. Chhabra, R. DeFries, J. Galloway, M. Heimann, C. Jones, C. Le Quéré, R. B. Myneni, S. Piao and P. Thornton (2013). Carbon and other biogeochemical cycles. In: T. F. Stocker, D. Qin, G.-K. Plattner, M. Tignor, S. K. Allen, J. Boschung, A. Nauels, Y. Xia, V. Bex and P. M. Midgley (eds.), *Climate Change 2013: The Physical Science Basis* (Contribution of Working Group I to the Fifth Assessment Report of the Intergovernmental Panel on Climate Change). Cambridge University Press.

Claussen, M. (2009). Late quaternary vegetation: climate feedbacks. *Climate of the Past* 5, 203–16.

Claussen, M., L. A. Mysak, A. J. Weaver, M. Crucifix, T. Fichefet, M. F. Loutre, S. L. Weber, J. Alcamo, V. A. Alexeev, A. Berger, R. Calov, A. Ganopolski, H. Goosse, G. Lohman, F. Lunkeit, I. I. Mohkov, V. Petoukhov, P. Stone and Z. Wang (2002). Earth system models of intermediate complexity: closing the gap in the spectrum of climate system models. *Climate Dynamics* 18, 579–86.

Clement, A. C., and L. C. Peterson (2008). Mechanisms of abrupt climate change of the last glacial period. *Review of Geophysics*, 46, RG4002.

*Climate and Cryosphere (CliC) Project Science and Co-ordination Plan* (2001). Edited by I. Allison, R. G. Barry and B. E. Goodison. WCRP-114 WMO/TD No. 1053.

Clough, S. A., M. W. Shephard, E. J. Mlawer, J. S. Delamere, M. J. Iacono, K. Cady-Pereira, S. Boukabara and P. D. Brown (2005). Atmospheric radiative transfer modeling: a summary of the AER codes. *Journal of Quantitative Spectroscopy and Radiative Transfer* 91, 233–44.

Collins, M., R. Knutti, J. Arblaster, J.-L. Dufresne, T. Fichefet, P. Friedlingstein, X. Gao, W. J. Gutowski, T. Johns, G. Krinner, M. Shongwe, C. Tebaldi, A. J. Weaver and M. Wehner (2013). Long-term climate change: projections, commitments and irreversibility. In: T. F. Stocker, D. Qin, G.-K. Plattner, M. Tignor, S. K. Allen, J. Boschung, A. Nauels, Y. Xia, V. Bex and P. M. Midgley (eds.), *Climate Change 2013: The Physical Science Basis* (Contribution of Working Group I to the Fifth Assessment Report of the Intergovernmental Panel on Climate Change). Cambridge University Press.

Collins, M., S.-I. An, W. Cai, A. Ganachaud, E. Guilyardi, F.-F. Jin, M. Jochum, M. Lengaigne, S. Power, A. Timmermann, G. Vecchi and A. Wittenberg (2010). The impact of global warming on the tropical Pacific Ocean and El Niño. *Nature Geosciences* 3, 391–7.

Comiso, J. C., and F. Nishio (2008). Trends in the sea ice cover using enhanced and compatible AMSR-E, SSM/I, and SMMR data. *Journal of Geophysical Research* 113, C02S07.

Cook, B. I., J. E. Smerdon, R. Seager and E. R. Cook (2014). Pan-continental droughts in North America over the last millennium. *Journal of Climate* 27, 383–97.

Cook, E. R., R. Seager, M. A. Cane and D. W. Stahle (2007). North American drought: reconstructions, causes, and consequences. *Earth Science Reviews* 81, 93–134.

Cox, M. D. (1987). Isopycnal diffusion in a z-coordinate ocean model. *Ocean Modelling* 74, 1–5.

Crespin, E., H. Goosse, T. Fichefet, A. Mairesse and Y. Sallaz-Damaz (2013). Arctic climate over the past millennium: annual and seasonal responses to external forcings. *The Holocene* 23(3) 321–9.

Crockford, R. H., and D. P. Richardson (2000). Partitioning of rainfall into throughfall, stemflow and interception: effect of forest type, ground cover and climate. *Hydrological Processes* 14, 2903–20.

Crucifix, M. (2012a). Traditional and novel approaches to palaeoclimate modelling. *Quaternary Science Reviews* 57 1–16.

(2012b). Oscillators and relaxation phenomena in Pleistocene climate theory. *Philosophical Transactionsof the Royal Society A* 370, 1140–65.

Crucifix, M., M. F. Loutre, P. Tulkens, T. Fichefet and A. Berger (2002). Climate evolution during the Holocene: a study with an Earth system model of intermediate complexity. *Climate Dynamics* 19, 43–60.

Dansgaard, W., S. J. Johnsen, H. B. Clausen, D. Dahl-Jensen, N. S. Gundestrup, C. U. Hammer, C. S. Hvidberg, J. P. Steffensen, A. E. Sveinbjörnsdottir, J. Jouzel and G. Bond (1993). Evidence for general instability of past climate from a 250 kyr ice-core record. *Nature* 364, 218–20.

DeConto, R. M., and D. Pollard (2003). Rapid Cenozoic glaciation of Antarctica induced by declining atmospheric $CO_2$. *Nature* 421, 245–9.

DeConto, R. M., S. Galeotti, M. Pagani, D. Tracy, K. Schaefer, T. Zhang, D. Pollard and D. J. Beerling (2010). Past extreme warming events linked to massive carbon release from thawing permafrost. *Nature*, 484, 87–92.

Dee, D. P., S. M. Uppala, A. J. Simmons, P. Berrisford, P. Poli, S. Kobayashi, U. Andrae, M. A. Balmaseda, G. Balsamo, P. Bauer, P. Bechtold, A. C. M. Beljaars, L. van de Berg, J. Bidlot, N. Bormann, C. Delsol, R. Dragani, M. Fuentes, A. J. Geer, L. Haimberger, S. B. Healy, H. Hersbach, E. V. Hólm, L. Isaksen, P. Kållberg, M. Köhler, M. Matricardi, A. P. McNally, B. M. Monge-Sanz, J.-J. Morcrette, B.-K. Park, C. Peubey, P. de Rosnay, C. Tavolato, J.-N. Thépaut and F. Vitart (2011). The ERA-interim reanalysis: configuration and performance of the data assimilation system. *Quarterly Journal of the Royal Meteorological Society* 137, 553–97.

Delworth, T. L., and M. E. Mann (2000). Observed and simulated multidecadal variability in the northern hemisphere. *Climate Dynamics* 16, 661–76.

Denman, K. L., G. Brasseur, A. Chidthaisong, P. Ciais, P. M. Cox, R. E. Dickinson, D. Hauglustaine, C. Heinze, E. Holland, D. Jacob, U. Lohmann, S Ramachandran, P. L. da Silva Dias, S. C. Wofsy and X. Zhang (2007). Couplings between changes in the climate system and biogeochemistry. In: T. F. Stocker, D. Qin, G.-K. Plattner, M. Tignor, S. K. Allen, J. Boschung, A. Nauels, Y. Xia, V. Bex and P. M. Midgley (eds.), *Climate Change 2007: The Physical Science Basis* (Contribution of Working Group I to the Fourth Assessment Report of the Intergovernmental Panel on Climate Change). Cambridge University Press.

Dlugokencky, E. J., P. M. Lang, A. M. Crotwell, K. A. Masarie and M. J. Crotwell (2013). Atmospheric methane dry air mole fractions from the

NOAA ESRL Carbon Cycle Cooperative Global Air Sampling Network, 1983–2012, Version 2013-08-28; available at: ftp://aftp.cmdl.noaa.gov/data/trace_gases/ch4/flask/surface/.

Donnadieu, Y., Y. Goddéris, G. Ramstein, A. Nédélec and J. Meert (2004). A 'snowball Earth' climate triggered by continental break-up through changes in runoff. *Nature* 428, 303–6.

Donnadieu, Y., Y. Goddéris, R. Pierrehumbert, G. Dromart, F. Fluteau and R. Jacob (2006). A GEOCLIM simulation of climatic and biogeochemical consequences of Pangea breakup. *Geochemistry, Geophysics, Geosystems* 7(11), Q11019.

Elderfield, H., P. Ferretti, M. Greaves, S. Crowhurst, I. N. McCave, D. Hodell and A. M. Piotrowski (2012). Evolution of ocean temperature and ice volume through the mid-Pleistocene climate transition. *Science* 337, 704–9.

Emanuel, K. A. (1991). A scheme for representing cumulus convection in large-scale models. *Journal of the Atmospheric Sciences* 48, 2313–35.

Enfield, D. B., A. M. Mestas-Nuñez and P. J. Trimble (2001). The Atlantic multidecadal oscillation and its relation to rainfall and river flows in the continental US. *Geophysical Research Letters* 28, 2077–80.

Evans, M. N., S. E. Tolwinski-Ward, D. M. Thompson and K. J. Anchukaitis (2013). Applications of proxy system modeling in high resolution paleoclimatology. *Quaternary Science Reviews* 76, 16–28.

Eyring, V., J. M. Arblaster, I. Cionni, J. Sedláček, J. Perlwitz, P. J. Young, S. Bekki, D. Bergmann, P. Cameron-Smith, W. J. Collins, G. Faluvegi, K.-D. Gottschaldt, L. W. Horowitz, D. E. Kinnison, J.-F. Lamarque, D. R. Marsh, D. Saint-Martin, D. T. Shindell, K. Sudo, S. Szopa and S. Watanabe (2013). Long-term ozone changes and associated climate impacts in CMIP5 simulations. *Journal of Geophysical Research: Atmospheres* 118, 5029–60.

Fasullo, J. T., and K. E. Trenberth (2008). The annual cycle of the energy budget: II. Meridional structures and poleward transports. *Journal of Climate* 21, 2313–25.

Feldl, N., and G. H. Roe (2013). The nonlinear and nonlocal nature of climate feedbacks. *Journal of Climate*, 26, 8289–304.

Feulner, G. (2012). The faint young Sun problem. *Review of Geophysics* 50, RG2006.

Feulner, G., S. Rahmstorf, A. Levermann and S. Volkwardt (2013). On the origin of the surface air temperature difference between the hemispheres in Earth's present-day climate. *Journal of Climate* 26, 7136–50.

Flato, G., J. Marotzke, B. Abiodun, P. Braconnot, S. C. Chou, W. Collins, P. Cox, F. Driouech, S. Emori, V. Eyring, C. Forest, P. Gleckler, E. Guilyardi, C. Jakob, V. Kattsov, C. Reason and M. Rummukainen (2013). Evaluation of climate models. In: T. F. Stocker, D. Qin, G.-K. Plattner, M. Tignor, S. K. Allen, J. Boschung, A. Nauels, Y. Xia, V. Bex and P. M. Midgley (eds.), *Climate Change 2013: The Physical Science Basis* (Contribution of Working Group I to the Fifth Assessment Report of the Intergovernmental Panel on Climate Change). Cambridge University Press.

Forster, P., V. Ramaswamy, P. Artaxo, T. Berntsen, R. Betts, D. W. Fahey, J. Haywood, J. Lean, D. C. Lowe, G. Myhre, J. Nganga, R. Prinn, G. Raga, M. Schulz and R. Van Dorland (2007). Changes in atmospheric constituents and in radiative forcing. In: S. Solomon, D. Qin, M. Manning, Z. Chen, M. Marquis, K. B. Averyt, M. Tignor and H. L. Miller (eds.), *Climate Change 2007: The Physical Science Basis* (Contribution of Working Group I to the Fourth Assessment Report of the Intergovernmental Panel on Climate Change). Cambridge University Press.

Forster, P. M., T. Andrews, P. Good, J. M. Gregory, L. S. Jackson and M. Zelinka (2013). Evaluating adjusted forcing and model spread for historical and future scenarios in the CMIP5 generation of climate models. *Journal of Geophysical Research: Atmospheres* 118, 1139–50.

Frank, D. C., J. Esper, C. C. Raible, U. Buntgen, V. Trouet, B. Stocker and F. Joos (2010). Ensemble reconstruction constraints on the global carbon cycle sensitivity to climate. *Nature*, 463, 527–32.

Friedlingstein, P., M. Meinshausen, V. K. Arora, C. D. Jones, A. Anav, S. K. Liddicoat and R. Knutti (2014). Uncertainties in CMIP5 climate projections due to carbon cycle feedbacks. *Journal of Climate* 27, 511–26.

Friedlingstein, P., P. Cox, R. Betts, L. Bopp, W. von Bloh, V. Brovkin, P. Cadule, S. Doney, M. Eby, I. Fung, G. Bala, J. John, C. Jones, F. Joos, T. Kato, M. Kawamiya, W. Knorr, K. Lindsay, H. D. Matthews, T. Raddatz, P. Rayner, C. Reick, E. Roeckner, K.-G. Schnitzler, R. Schnur, K. Strassmann, A. J. Weaver, C. Yoshikawa and N. Zeng (2006). Climate-carbon cycle feedback analysis: results from the C4MIP model intercomparison. *Journal of Climate* 19, 3337–53.

Ganopolski, A., and R. Calov (2011). The role of orbital forcing, carbon dioxide and regolith in 100 kyr glacial cycles. *Climate of the Past* 7, 1415–25.

Ganopolski, A., and S. Rahmstorf (2001). Rapid changes of glacial climate simulated in a coupled climate model. *Nature* 409, 153–8.

Gao, C., A. Robock and C. Ammann (2008). Volcanic forcing of climate over the past 1500 years: an improved ice core-based index for climate models. *Journal of Geophysical Research* 113, D23111.

Gent, P. R., and J. C. McWilliams (1990). Isopycnal mixing in ocean general circulation models. *Journal of Physical Oceanography* 20, 150–5.

Gent, P. R., J. Willebrand, T. J. McDougall and J. C. McWilliams (1995). Parameterizing eddy-induced tracer transports in ocean circulation models. *Journal of Physical Oceanography* 25, 463–74.

Gleckler, P. J., K. E. Taylor and C. Doutriaux (2008). Performance metrics for climate models. *Journal of Geophysical Research* 113, D06104.

Gómez-Navarro, J. J., J. P. Montávez, S. Jerez, P. Jiménez-Guerrero, R. Lorente-Plazas, J. F. González-Rouco and E. Zorita (2011). A regional climate simulation over the Iberian Peninsula for the last millennium. *Climate of the Past* 7, 451–72.

Grabowski, W. W., and P. K. Smolarkiewicz (1999). CRCP: a cloud resolving convection parameterization for modeling the tropical convecting atmosphere. *Physica D* 133, 171–8.

Gregory, J. M., C. D. Jones, P. Cadule and P. Friedlingstein (2009). Quantifying carbon cycle feedbacks. *Journal of Climate* 22, 5232–50.

Gregory, J. M., and P. M. Forster (2008). Transient climate response estimated from radiative forcing and observed temperature change. *Journal of Geophysical Research* 113, D23105.

Gregory, J. M., W. J. Ingram, M. A. Palmer, G. S. Jones, P. A. Stott, R. B. Thorpe, J. A. Lowe, T. C. Johns and K. D. Williams (2004). A new method for diagnosing radiative forcing and climate sensitivity. *Geophysical Research Letters* 31, L03205.

Guiot, J., and A. de Vernal (2007). Transfer functions: methods for quantitative paleoceanography based on microfossils. In: C. Hillaire-Marcel and A. de Vernal (eds.), *Proxies in Late Cenozoic Paleoceanography*, Elsevier, New York, 523–63.

Hanert, E., D. Y. L. Roux, V. Legat and E. Deleersnijder (2005). An efficient Eulerian finite element method for the shallow water equations. *Ocean Modelling* 10, 115–36.

Hansen, J. E., A. Lacis, D. Rind, G. Russell, P. Stone, I. Fung, M. Ruedy and J. Lerner (1984) Climate sensitivity: analysis of feedback mechanisms. In: J. E. Hansen and T. Takahashi (eds.), *Climate Processes and Climate Sensitivity* (Geophysical Monograph 29). American Geophysical Union, Washington, DC, 130–63.

Hansen, J., G. Russell, A. Lacis, I. Fung, D. Rind and P. Stone (1985). Climate response times: dependence on climate sensitivity and ocean mixing. *Science* 229, 857–9.

Hansen, J., M. Sato and R. Ruedy (1997). Radiative forcing and climate response. *Journal of Geophysical Research* 102(D6), 6831–84.

Hansen, J., M. Sato, R. Ruedy, L. Nazarenko, A. Lacis, G. A. Schmidt, G. Russell, I. Aleinov, M. Bauer, S. Bauer, N. Bell, B. Cairns, V. Canuto, M. Chandler, Y. Cheng, A. Del Genio, G. Faluvegi, E. Fleming, A. Friend, T. Hall, C. Jackman, M. Kelley, N. Kiang, D. Koch, J. Lean, J. Lerner, K. Lo, S. Menon, R. Miller, P. Minnis, T. Novakov, V. Oinas, Ja. Perlwitz, Ju. Perlwitz, D. Rind, A. Romanou, D. Shindell, P. Stone, S. Sun, N. Tausnev, D. Thresher, B. Wielicki, T. Wong, M. Yao and S. Zhang (2005). Efficacy of climate forcings. *Journal Geophysical Research* 110, D18104.

Hansen, J., M. Sato, P. Kharecha, D. Beerling, R. Berner, V. Masson-Delmotte, M. Pagani, M. Raymo, D. L. Royer and J. C. Zachos (2008). Target atmospheric $CO_2$: where should humanity aim? *Open Atmospheric Science Journal* 2, 217–31.

Hansen, J., R. Ruedy, M. Sato and K. Lo (2010). Global surface temperature change. *Reviews of Geophysics*, 48, RG4004.

Hartmann, D. L., A. M. G. Klein Tank, M. Rusticucci, L. V. Alexander, S. Brönnimann, Y. Charabi, F. J. Dentener, E. J. Dlugokencky, D. R. Easterling, A. Kaplan, B. J. Soden, P. W. Thorne, M. Wild and P. M. Zhai (2013). Observations: atmosphere and surface. In: T. F. Stocker, D. Qin, G.-K. Plattner, M. Tignor, S. K. Allen, J. Boschung, A. Nauels, Y. Xia, V. Bex and P. M. Midgley (eds.), *Climate Change 2013: The Physical Science Basis*

(Contribution of Working Group I to the Fifth Assessment Report of the Intergovernmental Panel on Climate Change). Cambridge University Press.

Hasselmann, K. (1993). Optimal fingerprints for the detection of time-dependent climate change. *Journal of Climate* 6, 1957–69.

Hawkins, E., and R. Sutton (2009). The potential to narrow uncertainty in regional climate predictions. *Bulletin of the American Meteorological Society* 90, 1095–1107.

   (2012). Time of emergence of climate signals. *Geophysical Research Letters* 39, L01702.

Hays, J. D., J. Imbrie and N. J. Shackleton (1976). Variations in the Earth's orbit: pacemaker of the ice ages. *Science* 194, 1121–32.

Hegerl, G., and F. Zwiers (2011). Use of models in detection and attribution of climate change. *WIREs Climate Change* 2, 570–91.

Heinrich, H. (1988). Origin and consequences of cyclic ice rafting in the northeast Atlantic Ocean during the past 130,000 years. *Quaternary Research* 29, 142–52.

Held, I. (2014). Simplicity amid complexity. *Science* 343, 1206–7.

Held, I., and B. J. Soden (2006). Robust responses of the hydrological cycle to global warming. *Journal of Climate* 19, 5686–99.

Held, I. M., and K. M. Shell (2012). Using relative humidity as a state variable in climate feedback analysis. *Journal of Climate* 25, 2578–82.

Hemming, S. (2004). Heinrich events: massive late Pleistocene detritus layers of the North Atlantic and their global climate imprint. *Review of Geophysics* 42, RG1005.

Hibler, W. D., III (1979). A dynamic-thermodynamic sea ice model. *Journal of Physical Oceanography* 9, 815–46.

Ho, C. K., D. B. Stephenson, M. Collins, C. A. T. Ferro and S. J. Brown (2012). Calibration strategies: a source of additional uncertainty in climate change projections. *Bulletin of the American Meteorological Society* 93, 21–6.

Hoffman, P. F., A. J. Kaufman, G. P. Halverson and D. P. Schrag (1998). A neoproterozoic snowball Earth. *Science* 281, 1342–6.

Hourdin, F., J.-Y. Grandpeix, C. Rio, S. Bony, A. Jam, F. Cheruy, N. Rochetin, L. Fairhead, A. Idelkadi, I. Musat, J.-L. Dufresne, A. Lahellec, M.-P. Lefebvre and R. Roehrig. (2013). LMDZ5B: the atmospheric component of the IPSL climate model with revisited parameterizations for clouds and convection. *Climate Dynamics* 40, 2193–2222.

Huang, P., S.-P. Xie, K. Hu, G. Huang and R. Huang (2013). Patterns of the seasonal response of tropical rainfall to global warming. *Nature Geoscience* 6, 357–61.

Hurrell, J. W., Y. Kushnir, M. Visbeck and G. Ottersen (2003). An overview of the North Atlantic Oscillation. In: J. W. Hurrell, Y. Kushnir, G. Ottersen and M. Visbeck (eds.), *The North Atlantic Oscillation: Climate Significance and Environmental Impact* (Geophysical Monograph 134). American Geophysical Union, Washington, DC, 1–35.

Huybers, P., and C. Wunsch (2005). Obliquity pacing of the late Pleistocene glacial terminations. *Nature* 434, 491–4.

Huybrechts, P., H. Goelzer, I. Janssens, E. Driesschaert, T. Fichefet, H. Goosse and M.-F. Loutre (2011). Response of the Greenland and Antarctic ice sheets to multi-millennial greenhouse warming in the Earth system model of intermediate complexity LOVECLIM. *Surveys in Geophysics* 32, 397–416.

Ilyina, T., K. D. Six, J. Segschneider, E. Maier-Reimer, H. Li and I. Núñez-Riboni (2013). Global ocean biogeochemistry model HAMOCC: model architecture and performance as component of the MPI-Earth system model in different CMIP5 experimental realizations. *Journal of Advances in Modeling Earth Systems* 5, 287–315.

Imbers, J., A. Lopez, C. Huntingford and M. R. Allen (2013). Testing the robustness of the anthropogenic climate change detection statements using different empirical models. *Journal of Geophysical Research: Atmospheres* 118, 3192–9.

Ingram, W. (2013). Some implications of a new approach to the water vapour feedback. *Climate Dynamics* 40, 925–33.

IOC, SCOR and IAPSO (2010). *The International Thermodynamic Equation of Seawater 2010: Calculation and Use of Thermodynamic Properties* (Manual and Guides 56). Intergovernmental Oceanographic Commission, New York.

IPCC (2007). *Climate Change 2007: The Physical Basis* (Contribution of Working Group I to the Fourth Assessment Report of the Intergovernmental Panel on Climate Change), S. Solomon, D. Qin, M. Manning, Z. Chen, M. Marquis, K. B. Averyt, M. Tignor and H. L. Miller (eds.). Cambridge University Press.

IPCC (2013). Summary for Policymakers. In: T. F. Stocker, D. Qin, G.-K. Plattner, M. Tignor, S. K. Allen, J. Boschung, A. Nauels, Y. Xia, V. Bex and P. M. Midgley (eds.), *Climate Change 2013: The Physical Science Basis* (Contribution of Working Group I to the Fifth Assessment Report of the Intergovernmental Panel on Climate Change). Cambridge University Press.

Jin, F.-F. (1997). An equatorial ocean recharge paradigm for ENSO: I. Conceptual model. *Journal of the Atmospheric Sciences* 54, 811–29.

Jones, P. D., K. R. Briffa, T. J. Osborn, J. M. Lough, T. D. van Ommen, B. M. Vinther, J. Luterbacher, E. R. Wahl, F. W. Zwiers, M. E. Mann, G. A. Schmidt, C. M. Ammann, B. M. Buckley, K. M. Cobb, J. Esper, H. Goosse, N. Graham, E. Jansen, T. Kiefer, C. Kull, M. Küttel, E. Mosley-Thompson, J. T. Overpeck, N. Riedwyl, M. Schulz, A. W. Tudhope, R. Villalba, H. Wanner, E. Wolff and E. Xoplaki (2009). High-resolution paleoclimatology of the last millennium: a review of the current status and future prospects. *The Holocene* 19, 3–49.

Jones, G. S., P. A. Stott and N. Christidis (2013). Attribution of observed historical near-surface temperature variations to anthropogenic and natural causes using CMIP5 simulations, *Journal Geophysical Research: Atmospheres* 118, 4001–24.

Joshi, M. M., F. H. Lambert and M. J. Webb (2013). An explanation for the difference between twentieth and twenty-first century land–sea warming ratio in climate models. *Climate Dynamics* 41, 1853–69.

Joussaume, S., and P. Braconnot (1997). Sensitivity of paleoclimate simulation results to season definitions. *Journal of Geophysical Research* 102 (D2) 1943–56.

Jouzel, J., V. Masson-Delmotte, O. Cattani, G. Dreyfus, S. Falourd, G. Hoffmann, B. Minster, J. Nouet, J. M. Barnola, J. Chappellaz, H. Fischer, J. C. Gallet, S. Johnsen, M. Leuenberger, L. Loulergue, D. Luethi, H. Oerter, F. Parrenin, G. Raisbeck, D. Raynaud, A. Schilt, J. Schwander, E. Selmo, R. Souchez, R. Spahni, B. Stauffer, J. P. Steffensen, B. Stenni, T. F. Stocker, J. L. Tison, M. Werner and E. W. Wolff. (2007). Orbital and millennial Antarctic climate variability over the past 800,000 years. *Science* 317, 793–6.

Kalnay, E., M. Kanamitsu, R. Kistler, W. Collins, D. Deaven, L. Gandin, M. Iredell, S. Saha, G. White, J. Woollen, Y. Zhu, A. Leetmaa, R. Reynolds, M. Chelliah, W. Ebisuzaki, W. Higgins, J. Janowiak, K. C. Mo, C. Ropelewski, J. Wang, R. Jenne and D. Joseph (1996). The NCEP/NCAR 40-year reanalysis project. *Bulletin of the American Meteorological Society* 77, 437–71.

Kerkhoff, C., H. R. Künsch and C. Schär (2014). Assessment of bias assumptions for climate models. *Journal of Climate* 27, 6799–818.

Kharin, V. V., G. J. Boer, W. J. Merryfield, J. F. Scinocca and W.-S. Lee (2012). Statistical adjustment of decadal predictions in a changing climate. *Geophysical Research Letters* 39, L19705.

Kiehl, J. T., and K. E. Trenberth (1997). Earth's annual global mean energy budget. *Bulletin of the American Meteorological Society* 78, 197–208.

Kirtman, B., S. B. Power, J. A. Adedoyin, G. J. Boer, R. Bojariu, I. Camilloni, F. J. Doblas-Reyes, A. M. Fiore, M. Kimoto, G. A. Meehl, M. Prather, A. Sarr, C. Schär, R. Sutton, G. J. van Oldenborgh, G. Vecchi and H. J. Wang (2013). Near-term climate change: projections and predictability. In: T. F. Stocker, D. Qin, G.-K. Plattner, M. Tignor, S. K. Allen, J. Boschung, A. Nauels, Y. Xia, V. Bex and P. M. Midgley (eds.), *Climate Change 2013: The Physical Science Basis* (Contribution of Working Group I to the Fifth Assessment Report of the Intergovernmental Panel on Climate Change). Cambridge University Press.

Knutson, T. R., F. Zeng and A. T. Wittenberg (2013). Multimodel assessment of regional surface temperature trends: CMIP3 and CMIP5 twentieth-century simulations. *Journal of Climate* 26, 8709–43.

Knutti, R., R. Furrer, C. Tebaldi, J. Cermak and G. A. Meehl (2010). Challenges in combining projections from multiple climate models. *Journal of Climate* 23, 2739–58.

Koch, D., S. Menon, A. Del Genio, R. Ruedy, I. A. Alienov and G. A. Schmidt (2009). Distinguishing aerosol impacts on climate over the past century. *Journal of Climate* 22, 2659–77.

Kopp, G., and J. L. Lean (2011). A new, lower value of total solar irradiance: evidence and climate significance. *Geophysical Research Letters* 38, L01706.

Kuhlbrodt, T., A. Griesel, M. Montoya, A. Levermann, M. Hofmann and S. Rahmstorf (2007). On the driving processes of the Atlantic meridional overturning circulation. *Reviews of Geophysics* 45, RG2001.

Kutzbach, J. E., and B. L. Otto-Bliesner (1982). The sensitivity of the African-Asian monsoonal climate to orbital parameter changes for 9000 years BP in a low-resolution general-circulation model. *Journal of the Atmospheric Sciences* 39, 1177–88.

Lamarque, J.-F., D. T. Shindell, B. Josse, P. J. Young, I. Cionni, V. Eyring, D. Bergmann, P. Cameron-Smith, W. J. Collins, R. Doherty, S. Dalsoren, G. Faluvegi, G. Folberth, S. J. Ghan, L.W. Horowitz, Y. H. Lee, I. A. MacKenzie, T. Nagashima, V. Naik, D. Plummer, M. Righi, S. T. Rumbold, M. Schulz, R. B. Skeie, D. S. Stevenson, S. Strode, K. Sudo, S. Szopa, A. Voulgarakis and G. Zeng 2014. The Atmospheric Chemistry and Climate Model Intercomparison Project (ACCMIP): overview and description of models, simulations and climate diagnostics. *Geoscientific Model Development* 6, 179–206.

Lean, J. L., and D. H. Rind (2008). How natural and anthropogenic influences alter global and regional surface temperatures: 1889 to 2006. *Geophysical Research Letters* 35, L18701.

Lenton, T., H. Held, E. Kriegler, J. Hall, W. Lucht, S. Rahmstorf and H. J. Schellnhuber (2008). Tipping elements in the Earth's climate system. *Proceedings of the National Academy of Sciences of the United States of America* 105, 1786–93.

Levitus, S., M. E. Conkright, T. P. Boyer, T. O'Brien, J. I. Antonov, C. Stephens, L. Stathoplos, D. Johnson and R. Gelfeld (1998). *World Ocean Database 1998,* Vol. 1: *Introduction. NOAA Atlas NESDIS 18*. U.S. Government Printing Office, Washington, DC.

Lisiecki, L. E., and M. E. Raymo (2005). A Pliocene-Pleistocene stack of 57 globally distributed benthic $\delta^{18}O$ records. *Paleoceanography* 20, PA1003.

Liu, Z., J. Zhu, Y. Rosenthal, X. Zhang, B. L. Otto-Bliesner, A. Timmermann, R. S. Smith, G. Lohmann, W. Zheng and O. Elison Timm (2014). The Holocene temperature conundrum. *Proceedings of the National Academy of Sciences of the United States of America* 111, E3501–5.

Loeb, N. G., B. A. Wielicki, D. R. Doelling, G. L. Smith, D. F. Keyes, S. Kato, N. Manalo-Smith, and T. Wong (2009). Toward optimal closure of the Earth's top-of-atmosphere radiation budget. *Journal of Climate* 22, 748–66.

Lorenz, E. N. (1963). Deterministic nonperiodic flow. *Journal of the Atmospheric Sciences* 20, 130–41.

   (1984). Formulation of a low-order model of a moist general circulation. *Journal of the Atmospheric Sciences* 41, 1933–45.

Lumpkin, R., and Z. Garraffo (2005). Evaluating the decomposition of tropical Atlantic drifter observations. *Journal of Atmospheric and Oceanic Technology* 22, 1403–15.

Lyons, T. W., C. T. Reinhard and N. J. Planavsky (2014). The rise of oxygen in Earth's early ocean and atmosphere. *Nature* 506, 307–15.

Manabe, S., and K. Bryan (1969). Climate calculations with a combined ocean-atmosphere model. *Journal of the Atmospheric Sciences* 26, 786–9.

Mann, G. W., K. S. Carslaw, C. L. Reddington, K. J. Pringle, M. Schulz, A. Asmi, D. V. Spracklen, D. A. Ridley, M. T. Woodhouse, L. A. Lee, K. Zhang,

S. J. Ghan, R. C. Easter, X. Liu, P. Stier, Y. H. Lee, P. J. Adams, H. Tost, J. Lelieveld, S. E. Bauer, K. Tsigaridis, T. P. C. van Noije, A. Strunk, E. Vignati, N. Bellouin, M. Dalvi, C. E. Johnson, T. Bergman, H. Kokkola, K. von Salzen, F. Yu, G. Luo, A. Petzold, J. Heintzenberg, A. Clarke, J. A. Ogren, J. Gras, U. Baltensperger, U. Kaminski, S. G. Jennings, C. D. O'Dowd, R. M. Harrison, D. C. S. Beddows, M. Kulmala, Y. Viisanen, V. Ulevicius, N. Mihalopoulos, V. Zdimal, M. Fiebig, H.-C. Hansson, E. Swietlicki and J. S. Henzing (2014). Intercomparison and evaluation of global aerosol microphysical properties among AeroCom models of a range of complexity. *Atmospheric Chemistry and Physics* 14, 4679–713.

Mann, M. E, Z. H. Zhang, S. Rutherford, R. S. Bradley, M. K. Hughes, D. Shindell, C. Ammann, G. Faluvegi and F. B. Ni (2009). Global signatures and dynamical origins of the little ice age and medieval climate anomaly. *Science*, 326, 1256–60.

Mantua, N. J., S. R. Hare, Y. Zhang, J. M. Wallace and R. C. Francis (1997). A Pacific interdecadal climate oscillation with impacts on salmon production. *Bulletin of the American Meteorological Society* 78, 1069–79.

Maraun, D., F. Wetterhall, A. M. Ireson, R. E. Chandler, E. J. Kendon, M. Widmann, S. Brienen, H. W. Rust, T. Sauter, M. Themeßl, V. K. C. Venema, K. P. Chun, C. M. Goodess, R. G. Jones, C. Onof, M. Vrac and I. Thiele-Eich (2010). Precipitation downscaling under climate change: recent developments to bridge the gap between dynamical models and the end user. *Reviews of Geophysics* 48, RG3003.

Markovic, M., H. Lin and K. Winger (2010). Simulating global and North American climate using the Global Environmental Multiscale Model with a variable-resolution modeling approach. *Monthly Weather Reviews* 138, 3967–87.

Marotzke, J. (2000). Abrupt climate change and thermohaline circulation: Mechanisms and predictability. *Proceedings of the National Academy of Sciences of the United States of America* 97, 1347–50.

Marshall, J., A. Donohoe, D. Ferreira and D. McGee (2014). The ocean's role in setting the mean position of the Inter-Tropical Convergence Zone. *Climate Dynamics* 42, 1967–79.

Masson, D., and R. Knutti (2011). Spatial-scale dependence of climate model performance in the CMIP3 ensemble. *Journal of Climate* 24, 2680–92.

Masson-Delmotte, V., M. Schulz, A. Abe-Ouchi, J. Beer, A. Ganopolski, J. F. González Rouco, E. Jansen, K. Lambeck, J. Luterbacher, T. Naish, T. Osborn, B. Otto-Bliesner, T. Quinn, R. Ramesh, M. Rojas, X. Shao and A. Timmermann (2013). Information from paleoclimate archives. In: T. F. Stocker, D. Qin, G.-K. Plattner, M. Tignor, S. K. Allen, J. Boschung, A. Nauels, Y. Xia, V. Bex and P. M. Midgley (eds.), *Climate Change 2013: The Physical Science Basis* (Contribution of Working Group I to the Fifth Assessment Report of the Intergovernmental Panel on Climate Change). Cambridge University Press.

Masson-Delmotte, V., S. Hou, A. Ekaykin, J. Jouzel, A. Aristarain, R. T. Bernardo, D. Bromwich, O. Cattani, M. Delmotte, S. Falourd, M. Frezzotti,

H. Gallée, L. Genoni, E. Isaksson, A. Landais, M. M. Helsen, G. Hoffmann, J. Lopez, V. Morgan, H. Motoyama, D. Noone, H. Oerter, J. R. Petit, A. Royer, R. Uemura, G. A. Schmidt, E. Schlosser, J. C. Simões, E. J. Steig, B. Stenni, M. Stievenard, M. R. van den Broeke, R. S. W. van de Wal, W. J. van de Berg, F. Vimeux and J. W. C. White (2008). A review of Antarctic surface snow isotopic composition: observations, atmospheric circulation, and isotopic modeling. *Journal of Climate* 21, 3359–87.

Massonnet, F., T. Fichefet, H. Goosse, C. M. Bitz, G. Philippon-Berthier, M. M. Holland and P.-Y. Barriat (2012). Constraining projections of summer Arctic sea ice. *The Cryosphere* 6, 1383–94.

Maurer, J. (2007). *Atlas of the Cryosphere* (digital media). National Snow and Ice Data Center, Boulder, CO.

Mauritsen, T., B. Stevens, E. Roeckner, T. Crueger, M. Esch, M. Giorgetta, H. Haak, J. Jungclaus, D. Klocke, D. Matei, U. Mikolajewicz, D. Notz, R. Pincus, H. Schmidt and L. Tomassini (2012). Tuning the climate of a global model. *Journal of Advances in Modelling Earth Systems*, 4, M00A01.

Maykut, G. A. (1982). Large-scale heat exchange and ice production in the central Arctic. *Journal of Geophysical Research* 87(C10), 7971–84.

Maykut, G. A., and N. Untersteiner (1971). Some results from a time-dependent thermodynamic model of sea ice. *Journal of Geophysical Research* 76, 1550–75.

McInerney, F. A., and S. L. Wing (2011). The Paleocene-Eocene thermal maximum: a perturbation of carbon cycle, climate, and biosphere with implications for the future. *Annual Review of Earth and Planetary Sciences* 39, 489–516.

Meehl, G. A., L. Goddard, G. Boer, R. Burgman, G. Branstator, C. Cassou, S. Corti, G. Danabasoglu, F. Doblas-Reyes, E. Hawkins, A. Karspeck, M. Kimoto, A. Kumar, D. Matei, J. Mignot, R. Msadek, A. Navarra, H. Pohlmann, M. Rienecker, T. Rosati, E. Schneider, D. Smith, R. Sutton, H. Teng, G. Jan van Oldenborgh, G. Vecchi and S. Yeager (2014). Decadal climate prediction: an update from the trenches. *Bulletin of the American Meteorological Society* 95, 243–67.

Meehl, G. A., L. Goddard, J. Murphy, R. J. Stouffer, G. Boer, G. Danabasoglu, K. Dixon, M. A. Giorgetta, A. M. Greene, E. Hawkins, G. Hegerl, D. Karoly, N. Keenlyside, M. Kimoto, B. Kirtman, A. Navarra, R. Pulwarty, D. Smith, D. Stammer and T. Stockdale (2009). Decadal prediction, can it be skillful? *Bulletin of the American Meteorological Society* 90, 1467–85.

Mekaoui, S., and S. Dewitte (2008). Total solar irradiance measurement and modelling during cycle 23. *Solar Physics* 247, 203–16.

Menviel, L., A. Timmermann, T. Friedrich and M. H. England (2014). Hindcasting the continuum of Dansgaard-Oeschger variability: mechanisms, patterns and timing. *Climate of the Past* 10, 63–77.

Mo, K., and J. N. Paegle (2001). The Pacific–South American modes and their downstream effects. *International Journal of Climatology*, 21, 1211–29.

Moody, E. G., M. D. King, C. B. Schaaf and S. Platnick (2008). MODIS-derived spatially complete surface albedo products: spatial and temporal pixel distribution and zonal averages. *Journal of Applied Meteorology and Climatology* 47, 2879–94.

Morice, C. P., J. J. Kennedy, N. A. Rayner and P. D. Jones (2012). Quantifying uncertainties in global and regional temperature change using an ensemble of observational estimates: the HadCRUT4 data set. *Journal of Geophysical Research-Atmospheres* 117(22), D08101.

Morrison, H., and A. Gettelman (2008). A new two-moment bulk stratiform cloud microphysics scheme in the community atmosphere model, version 3 (CAM3): I. Description and numerical tests. *Journal of Climate* 21, 3642–59.

Moss, R. H., J. A. Edmonds, K. A. Hibbard, M. R. Manning, S. K. Rose, D. P. van Vuuren, T. R. Carter, S. Emori, M. Kainuma, T. Kram, G. A. Meehl, J. F. B. Mitchell, N. Nakicenovic, K. Riahi, S. J. Smith, R. J. Stouffer, A. M. Thomson, J. P. Weyant and T. J. Wilbanks (2010). The next generation of scenarios for climate change research and assessment. *Nature* 463, 747–56.

Moss, R., M. Babiker, S. Brinkman, E. Calvo, T. Carter, J. Edmonds, I. Elgizouli, S. Emori, L. Erda, K. Hibbard, R. Jones, M. Kainuma, J. Kelleher, J.-F. Lamarque, M. Manning, B. Matthews, J. Meehl, L. Meyer, J. Mitchell, N. Nakicenovic, B. O'Neill, R. Pichs, K. Riahi, S. Rose, P. Runci, R. Stouffer, D. van Vuuren, J. Weyant, T. Wilbanks, J. P. van Ypersele and M. Zurek (2007). *Towards New Scenarios for Analysis of Emissions, Climate Change, Impacts, and Response Strategies* (IPCC expert meeting report), 19–21 September, 2007, Noordwijkerhout, The Netherlands.

Myhre, G., E. J. Highwood, K. P. Shine and F. Stordal (1998). New estimates of radiative forcing due to well mixed greenhouse gases. *Geophysical Research Letters*, 25, 2715–18.

Myhre, G., D. Shindell, F.-M. Bréon, W. Collins, J. Fuglestvedt, J. Huang, D. Koch, J.-F. Lamarque, D. Lee, B. Mendoza, T. Nakajima, A. Robock, G. Stephens, T. Takemura and H. Zhang (2013). Anthropogenic and natural radiative forcing. In: T. F. Stocker, D. Qin, G.-K. Plattner, M. Tignor, S. K. Allen, J. Boschung, A. Nauels, Y. Xia, V. Bex and P. M. Midgley (eds.), *Climate Change 2013: The Physical Science Basis* (Contribution of Working Group I to the Fifth Assessment Report of the Intergovernmental Panel on Climate Change). Cambridge University Press.

Nakicenovic, N., and R. Swart (eds.) (2000). *IPCC Special Report on Emission Scenarios*. Cambridge University Press.

Nan, S., and J. P. Li (2003). The relationship between the summer precipitation in the Yangtze River Valley and the boreal spring southern hemisphere annular mode. *Geophysical Research Letters* 30(24), 2266.

NRC (2002). *Abrupt Climate Change: Inevitable Surprises*. Committee on Abrupt Climate Change, National Research Council. National Academy Press, Washington, DC.

Oberkampf, W. L., T. G. Trucano and C. Hirsch (2002). Verification, validation, and predictive capability in computational engineering and physics. Paper presented at the Foundation for Verification and Validation in the 21st Century workshop, 22–23 October 2002, Johns Hopkins University/Applied Physics Laboratory, Laurel, MD.

O'Connor, F. M., O. Boucher, N. Gedney, C. D. Jones, G. A. Folberth, R. Coppell, P. Friedlingstein, W. J. Collins, J. Chappellaz, J. Ridley and C. E. Johnson (2010). Possible role of wetlands, permafrost, and methane hydrates in the methane cycle under future climate change: A review. *Reviews of Geophysics* 48(4) RG4005.

Oldenburg, G. J., van, M. A. Balmaseda, L. Ferranti, T. N. Stockdale and D. L. T. Anderson (2005). Evaluation of atmospheric fields from the ECMWF seasonal forecasts over a 15-year period. *Journal of Climate* 18, 3250–69.

Oreskes, N., and E. W. Conway (2010). *Merchants of Doubt: How a Handful of Scientists Obscured the Truth on Issues from Tobacco Smoke to Global Warming*. Bloomsbury Press, London.

Oreskes, N., K. Shrader-Frechette and K. Belitz (1994). Verification, validation, and confirmation of numerical models in the Earth sciences. *Science* 263: 641–6.

Oschlies, A., and V. Garçon (1999). An eddy-permitting coupled physical-biological model of the North Atlantic: 1. Sensitivity to advection numerics and mixed layer physics. *Global Biogeochemical Cycles* 13, 135–60.

PAGES2K Consortium: M. Ahmed, K. Anchukaitis, A. Asrat, H. Borgaonkar, M. Braida, B. Buckley, U. Büntgen, B. Chase, D. Christie, E. Cook, M. Curran, H. Diaz, J. Esper, Z. X. Fan, N. Gaire, Q. Ge, J. Gergis, J. F. Gonzalez-Rouco, H. Goosse, S. Grab, N. Graham, R. Graham, M. Grosjean, S. Hanhijärvi, D. Kaufman, T. Kiefer,. K. Kimura, A. Korhola, P. Krusic, A. Lara, A. M. Lézine, F. Ljungqvist, A. Lorrey,. J. Luterbacher, V. Masson-Delmotte, D. McCarroll, J. McConnell, N. McKay, M. Morales, A. Moy, R. Mulvaney, I. Mundo, T. Nakatsuka, D. Nash, R. Neukom, S. Nicholson, H. Oerter, J. Palmer, S. Phipps, M. Prieto, A. Rivera, M. Sano, M. Severi, T. Shanahan, X. Shao, F. Shi, M. Sigl, J. Smerdon, O. Solomina, E. Steig, B. Stenni, M. Thamban, V. Trouet, C. Turney, M. Umer, T. van Ommen, D. Verschuren, A. Viau, R. Villalba, B. Vinther, L. von Gunten, S. Wagner, E. Wahl, H. Wanner, J. Werner, J. White, K. Yasue and E. Zorita (2013). Continental-scale temperature variability during the past two millennia. *Nature Geoscience* 6, 339–46.

Paillard, D. (2015). Quaternary glaciations: from observations to theories. *Quaternary Science Reviews* 107, 11–24.

Paleosens project members: E. J. Rohling, A. Sluijs, H. A. Dijkstra, P. Köhler, R. S. W. van de Wal, A. S. von der Heydt, D. J. Beerling, A. Berger, P. K. Bijl, M. Crucifix, R. DeConto, S. S. Drijfhout, A. Fedorov, G. L. Foster, A. Ganopolski, J. Hansen, B. Hönisch, H. Hooghiemstra, M. Huber, P. Huybers, R. Knutti, D. W. Lea, L. J. Lourens, D. Lunt, V. Masson-Demotte, M. Medina-Elizalde, B. Otto-Bliesner, M. Pagani, H. Pälike, H. Renssen, D. L. Royer, M. Siddall, P. Valdes, J. C. Zachos and R. E. Zeebe (2012). Making sense of palaeoclimate sensitivity. *Nature* 491, 683–91.

Parrenin, F., V. Masson-Delmotte, P. Köhler, D. Raynaud, D. Paillard, J. Schwander, C. Barbante, A. Landais, A. Wegner and J. Jouzel (2013). Synchronous change of atmospheric $CO_2$ and Antarctic temperature during the last deglacial warming. *Science* 339, 1060–3.

Pavlova, T. V., V. M. Kattsov and V. A. Govorkova (2011). Sea ice in CMIP5 models: closer to reality? *Trudy GGO ( MGO Proc. )* 564, 7–18 (in Russian).

Petit, J. R., J. Jouzel, D. Raynaud, N. I. Barkov, J. M. Barnola, I. Basile, M. Bender, J. Chapellaz, M. Davis, G. Delaygue, M. Delmotte, V. M. Kotlyakov, M. Legrand, V. Y. Lipenkov, C. Lorius, L. Pépin, C. Ritz, E. Saltzman and M. Stievenard (1999). Climate and atmospheric history of the past 420,000 years from Vostok ice core, Antarctica. *Nature* 399, 429–36.

Pongratz, J., C. Reick, T. Raddatz and M. Claussen (2008). A reconstruction of global agricultural areas and land cover for the last millennium. *Global Biogeochemical Cycles* 22, GB3018.

Rahmstorf, S. (1996). On the freshwater forcing and transport of the Atlantic thermohaline circulation. *Climate Dynamics* 12, 799–811.

  (2002). Ocean circulation and climate during the past 120,000 years. *Nature* 419, 207–14.

Rahmstorf, S., M. Crucifix, A. Ganopolski, H. Goosse, I. Kamenkovich, R. Knutti, G. Lohmann, R. Marsh, L.A. Mysak, Z. Wang and A.J. Weaver (2005). Thermohaline circulation hysteresis: a model intercomparison. *Geophysical Research Letters* 32, L23605.

Ramanathan, V., and J. A. Coakley (1978). Climate modeling through radiative-convective models. *Reviews of Geophysics and Space Physics* 16, 465–89.

Ramankutty, N., and J. A. Foley (1999). Estimating historical changes in global land cover: croplands from 1700 to 1992. *Global. Biogeochemical Cycles* 13, 997–1027.

Rayner, N. A., D. E. Parker, E. B. Horton, C. K. Folland, L. V. Alexander, D. P. Rowell, E. C. Kent and A. Kaplan (2003). Global analyses of sea surface temperature, sea ice, and nigh marine air temperature since the late nineteenth century. *Journal of Geophysical Research* 108(D14), 4407.

Redi, M. H. (1982). Oceanic isopycnal mixing by coordinate rotation. *Journal of Physical Oceanography* 12, 1154–8.

Reimer, P. J., E. Bard, A. Bayliss, J. W. Beck, P. G. Blackwell, C. Bronk Ramsey, C. E. Buck, H. Cheng, R. L. Edwards, M. Friedrich, P. M Grootes, T. P. Guilderson, H. Haflidason, I Hajdas, C. Hatté, T. J. Heaton, D. L. Hoffmann, A. G. Hogg, K. A. Hughen, K. F. Kaiser, B. Kromer, S. W Manning, M. Niu, R. W. Reimer, D. A. Richards, E. M. Scott, J. R. Southon, R. A. Staff, C. S. M. Turney and J. van der Plicht (2013). INTCAL13 and MARINE13 radiocarbon age calibration curves 0–50,000 years CAL BP. *Radiocarbon* 55, 1869–87.

Renssen, H., H. Seppä, X. Crosta, H. Goosse and D. M. Roche (2012). Global characterization of the Holocene thermal maximum. *Quaternary Science Reviews* 48, 7–19.

Rhein, M., S. R. Rintoul, S. Aoki, E. Campos, D. Chambers, R. A. Feely, S. Gulev, G. C. Johnson, S. A. Josey, A. Kostianoy, C. Mauritzen, D. Roemmich, L. D. Talley and F. Wang (2013). Observations: Ocean. In: T. F. Stocker, D. Qin, G.-K. Plattner, M. Tignor, S. K. Allen, J. Boschung, A. Nauels, Y. Xia, V. Bex and P. M. Midgley (eds.), *Climate Change 2013: The*

*Physical Science Basis* (Contribution of Working Group I to the Fifth Assessment Report of the Intergovernmental Panel on Climate Change). Cambridge University Press.

Ringeval, B., P. Friedlingstein, C. Koven, P. Ciais, N. de Noblet-Ducoudré, B. Decharme and P. Cadule (2011). Climate-$CH_4$ feedback from wetlands and its interaction with the climate-$CO_2$ feedback. *Biogeosciences* 8, 2137–57.

Robock, A. (2000). Volcanic eruptions and climate. *Review of Geophysics* 38, 191–219.

Roe, G. H. (2009). Feedbacks, timescales, and seeing red. *Annual Reviews of Earth and Planetary Sciences* 37, 93–115.

Ropelewski, C. F., and M. S. Halpert (1987). Global and regional precipitation associated with El Niño/Southern Oscillation. *Monthly Weather Review* 115, 1606–26.

Royer, D. L., R. A. Berner and J. Park (2007). Climate sensitivity constrained by $CO_2$ concentrations over the past 420 million years. *Nature* 446, 530–2.

Runyan, C. W., P. D'Odorico and D. Lawrence (2012). Physical and biological feedbacks of deforestation. *Review of Geophysics* 50, RG4006

Schmidt, G. A., and S. Sherwood (2014). A practical philosophy of complex climate modelling. *European Journal for Philosophy of Science*, doi:10.1007/s13194-014-0102-9.

Schmidt, G. A., G. Hoffmann, D. T. Shindell and Y. Hu (2005). Modelling atmospheric stable water isotopes and the potential for constraining cloud processes and stratosphere-troposphere water exchange. *Journal of Geophysical Research* 110, D21314.

Schmidt, G. A., R. A. Ruedy, R. L. Miller and A. A. Lacis (2010). Attribution of the present-day total greenhouse effect. *Journal of Geophysical Research* 115, D20106.

Schuur, E. A. G., J. Bockheim, J. G. Canadell, E. Euskirchen, C. B. Field, S. V. Goryachkin, S. Hagemann, P. Kuhry, P. M. Lafleur, H. Lee, G. Mazhitova, F. E. Nelson, A. Rinke, V. E. Romanovsky, N. Shiklomanov, C. Tarnocai, S. Venevsky, J. G. Vogel and S. A. Zimov (2008). Vulnerability of permafrost carbon to climate change: implications for the global carbon cycle. *BioScience* 58(8), 701–14.

Sellers, W. D. (1969). A global climatic model based on the energy balance of the Earth-atmosphere system. *Journal of Applied Meteorology* 8, 392–400.

Sellers, P. J., R. E. Dickinson, D. A. Randall, A. K. Betts, F. G. Hall, J. A. Berry, G. J. Collatz, A. S. Denning, H. A. Mooney, C. A. Nobre, N. Sato, C. B. Field and A. Henderson-Sellers (1997). Modeling the exchanges of energy, water, and carbon between continents and the atmosphere. *Science* 275, 502–9.

Seneviratne, S. I., T. Corti, E. L. Davin, M. Hirschi, E. B. Jaeger, I. Lehner, B. Orlowsky and A. J. Teuling (2010). Investigating soil moisture–climate interactions in a changing climate: A review. *Earth-Science Reviews* 99, 125–61.

Serreze, M. C., and R. G. Barry (2011). Processes and impacts of Arctic amplification: a research synthesis. *Global and Planetary Change* 77, 85–96.

Shine, K. P., and P. M. Forster (1999). The effect of human activity on radiative forcing of climate change: a review of recent developments. *Global and Planetary Change* 20, 205–25.

Shakun, J. D., P. U. Clark, F. He, S. A. Marcott, A. C. Mix, Z. Liu, B. Otto-Bliesner, A. Schmittner and E. Bard (2012). Global warming preceded by increasing carbon dioxide concentrations during the last deglaciation. *Nature* 484, 49–55.

Siddall, M., E. J. Rohling, W. G. Thompson and C. Waelbroeck (2008). Marine isotope stage 3 sea level fluctuations: data synthesis and new outlook. *Review of Geophysics* 46, RG4003.

Smith, D. M., A. A. Scaife, G. J. Boer, M. Caian, F. J. Doblas-Reyes, V. Guemas, E. Hawkins, W. Hazeleger, L. Hermanson, C. Kit Ho, M. Ishii, V. Kharin, M. Kimoto, B. Kirtman, J. Lean, D. Matei, W. J. Merryfield, W. A. Müller, H. Pohlmann, A. Rosati, B. Wouters and K. Wyser (2013). Real-time multi-model decadal climate predictions. *Climate Dynamics* 41, 2875–88.

Soden, B. F., and I. M. Held (2006). An assessment of climate feedbacks in coupled ocean-atmosphere models. *Journal of Climate* 19, 3354–60.

Soden, B. F., I. M. Held, R. Colman, K. M. Shell, J. T. Kiehl and C. A. Shields (2008). Quantifying climate feedbacks using radiative kernels. *Journal of Climate* 21, 3504–20.

Solomon, S., D. Qin, M. Manning, R. B. Alley, T. Berntsen, N. L. Bindoff, Z. Chen, A. Chidthaisong, J. M. Gregory, G. C. Hegerl, M. Heimann, B. Hewitson, B. J. Hoskins, F. Joos, J. Jouzel, V. Kattsov, U. Lohmann, T. Matsuno, M. Molina, N. Nicholls, J. Overpeck, G. Raga, V. Ramaswamy, J. Ren, M. Rusticucci, R. Somerville, T. F. Stocker, P. Whetton, R. A. Wood and D. Wratt (2007). Technical summary. In: S. Solomon, D. Qin, M. Manning, Z. Chen, M. Marquis, K. B. Avery, M. Tignor and H. L. Miller (eds.), *Climate Change 2007: The Physical Science Basis* (Contribution of Working Group I to the Fourth Assessment Report of the Intergovernmental Panel on Climate Change). (eds.)]. Cambridge University Press.

Sterl, A., G. J. van Oldenborgh, W. Hazeleger and G. Burgers (2007). On the robustness of ENSO teleconnections. *Climate Dynamics* 29, 469–85.

Stier, P., J. Feichter, S. Kinne, S. Kloster, E. Vignati, J. Wilson, L. Ganzeveld, I. Tegen, M. Werner, Y. Balkanski, M. Schulz, O. Boucher, A. Minikin and A. Petzold (2005). The aerosol-climate model ECHAM5-HAM. *Atmospheric Chemistry and Physics* 5, 1125–56.

Stocker, T. F., and S. J. Johnsen (2003). A minimum thermodynamic model for the bipolar seesaw. *Paleoceanography* 18, 1087.

Stocker, T. F., D. Qin, G.-K. Plattner, L. V. Alexander, S. K. Allen, N. L. Bindoff, F.-M. Bréon, J. A. Church, U. Cubasch, S. Emori, P. Forster, P. Friedlingstein, N. Gillett, J. M. Gregory, D. L. Hartmann, E. Jansen, B. Kirtman, R. Knutti, K. Krishna Kumar, P. Lemke, J. Marotzke, V. Masson-Delmotte, G. A. Meehl, I. I. Mokhov, S. Piao, V. Ramaswamy, D. Randall, M. Rhein, M. Rojas, C. Sabine, D. Shindell, L. D. Talley, D. G. Vaughan and S.-P. Xie (2013). Technical summary. In: T. F. Stocker, D. Qin, G.-K. Plattner, M. Tignor, S. K. Allen, J. Boschung, A. Nauels, Y. Xia, V. Bex and P. M. Midgley (eds.), *Climate Change 2013: The Physical Science Basis* (Contribution of Working Group I to the Fifth Assessment Report of the Intergovernmental Panel on Climate Change). Cambridge University Press.

Stommel, H. (1961). Thermohaline convection with two stable regimes of flow. *Tellus* 13, 224–30.

Tao, W.-K., W. Lau, J. Simpson., J.-D. Chern, R. Atlas, D. Randall, M. Khairoutdinov, J.-L. Li, D. E. Waliser, J. Jiang, A. Hou, X. Lin, and C. Peters-Lidard (2009). A multiscale modeling system: developments, applications, and critical issues. *Bulletin of the American Meteorological Society* 90, 515–34.

Talagrand, O. (1997). Assimilation of observations, an introduction. *Journal of the Meteorological Society of Japan Special Issue* 75(1B), 191–209.

Tarnocai, C., J. G. Canadell, E. A. G. Schuur, P. Kuhry, G. Mazhitova and S. Zimov (2009). Soil organic carbon pools in the northern circumpolar permafrost region. *Global Biogeochemical Cycles* 23, GB2023.

Taylor, K. E., R. J. Stouffer and G. A. Meehl (2012). An overview of CMIP5 and the experimental design. *Bulletin and the American Meteorological Society* 93, 485–98.

Tiedtke, M. (1989). A comprehensive mass flux scheme for cumulus parameterization in large-scale models. *Monthly Weather Review* 117, 1779–1800.

Thompson, D. W. J. and J. M. Wallace (2000). Annular modes in the extratropical circulation: I. Month-to-month variability. *Journal of Climate* 13, 1000–16.

Thorndike, A. S., D. A. Rothrock, G. A. Maykut and R. Colony (1975). The thickness distribution of sea ice. *Journal of Geophysical Research* 80, 4501–13.

Trenberth, K. E., and D. P. Stepaniak (2003). Seamless poleward atmospheric energy transports and implications for the Hadley circulation. *Journal of Climate* 16, 3705–21.

Trenberth, K. E., and D. J. Shea (2006). Atlantic hurricanes and natural variability in 2005. *Geophysical Research Letters* 33, L12704.

Trenberth, K. E., and J. M. Caron (2001). Estimates of meridional atmosphere and ocean heat transports. *Journal of Climate* 14, 3433–43.

Trenberth, K. E., L. Smith, T. T. Qian, A. G. Dai and J. Fasullo (2007). Estimates of the global water budget and its annual cycle using observational and model data. *Journal of Hydrometeorology* 8, 758–69.

Trenberth, K. E., J. T. Fasullo and J. Kiehl (2009). Earth's global energy budget. *Bulletin of the American Meteorological Society* 90, 311–23.

van der Werf, G. R., J. T. Randerson, L. Giglio, G. J. Collatz, M. Mu, P. S. Kasibhatla, D. C. Morton, R. S. DeFries, Y. Jin, and T. T. van Leeuwen (2010). Global fire emissions and the contribution of deforestation, savanna, forest, agricultural, and peat fires (1997–2009). *Atmospheric Chemistry and Physics* 10, 11707–35.

van Leeuwen, P. J. (2009). Particle filtering in geophysical systems. *Monthly Weather Review* 137, 4089–114.

Vannitsem, S., and L. De Cruz (2014). A 24-variable low-order coupled ocean-atmosphere model: OA-QG-WS v2. *Geoscientific Model Development* 7, 649–62.

Vaughan, D. G., J. C. Comiso, I. Allison, J. Carrasco, G. Kaser, R. Kwok, P. Mote, T. Murray, F. Paul, J. Ren, E. Rignot, O. Solomina, K. Steffen and T. Zhang (2013). Observations: Cryosphere. In: T. F. Stocker, D. Qin, G.-K. Plattner, M. Tignor, S. K. Allen, J. Boschung, A. Nauels, Y. Xia, V. Bex and P. M. Midgley (eds.), *Climate Change 2013: The Physical Science Basis* (Contribution of Working Group I to the Fifth Assessment Report of the Intergovernmental Panel on Climate Change). Cambridge University Press.

Vial, J., J.-L. Dufresne and S. Bony (2013). On the interpretation of inter-model spread in CMIP5 climate sensitivity estimates. *Climate Dynamics* 41, 3339–62.

Vermeer, M., and S. Rahmstorf (2009). Global sea level linked to global temperature. *Proceedings of the National Academy of Sciences of the United States of America* 106, 21527–32.

Vose, R. S., D. Arndt, V. F. Banzon, D. R. Easterling, B. Gleason, B. Huang, E. Kearns, J. H. Lawrimore, M. J. Menne, T. C. Peterson, R. W. Reynolds, T. M. Smith, C. N. Williams, Jr., and D. B. Wuertz (2012). NOAA's merged land-ocean surface temperature analysis. *Bulletin of the American Meteorological Society* 93, 1677–85.

Walker, J. C. G., P. B. Hays and J. F. Kasting (1981). A negative feedback mechanism for the long-term stabilization of Earth's surface temperature. *Journal of Geophysical Research: Oceans* 86(C10), 9776–82.

Wallace, J. M., and D. S. Gutzler (1981). Teleconnections in the geopotential height field during the Northern Hemisphere winter. *Monthly Weather Review* 109, 784–812.

Wang, Q., S. Danilov, D. Sidorenko, R. Timmermann, C. Wekerle, X. Wang, T. Jung, and J. Schröter (2014). The finite element sea ice-ocean model (FESOM), version 1.4: Formulation of an ocean general circulation model. *Geoscientific Model Development* 7, 663–93.

Wang, Y.-M., J. L. Lean and N. R. Sheeley, Jr. (2005). Modeling the Sun's magnetic field and irradiance since 1713. *Astrophysical Journal* 625, 522–38.

Wanner, H., S. Brönnimann, C. Casty, D. Gyalistras, J. Luterbacher, C. Schmutz, D. B. Stephenson and E. Xoplaki (2001). North Atlantic Oscillation: concept and studies. *Surveys in Geophysics* 22, 321–82.

Watterson, I. G. (2000). Interpretation of simulated global warming using a simple model. *Journal of Climate* 13, 202–15.

Winkelmann, R., M. A. Martin, M. Haseloff, T. Albrecht, E. Bueler, C. Khroulev and A. Levermann (2011). The Potsdam parallel ice sheet model (PISM-PIK): 1. Model description. *The Cryosphere* 5, 715–26.

Wild, M., D. Folini, C. Schär, N. Loeb, E. G. Dutton and G. König-Langlo (2013). The global energy balance from a surface perspective. *Climate Dynamics* 40, 3107–34.

Whiticar, M. J., E. Faber and M. Schoell (1986). Biogenic methane formation in marine and freshwater environments: $CO_2$ reduction vs. acetate fermentation—isotope evidence. *Geochimica et Cosmochimica Acta* 50, 693–709.

WMO (2014). Scientific assessment of ozone depletion: 2014 Global Ozone Research and Monitoring Project (Rep. No. 55). World Meteorological Organization, Geneva, Switzerland.

Wolff, E. W., J. Chappellaz, T. Blunier, S. O. Rasmussen and A. Svensson (2010). Millennial-scale variability during the last glacial: the ice core record. *Quaternary Science Reviews* 29, 2828–38.

Xie, P., and P. A. Arkin (1997). Global precipitation: a 17-year monthly analysis based on gauge observations, satellite estimates, and numerical model outputs. *Bulletin of the American Meteorological Society* 78, 2539–58.

Yin, Q. Z., and A. Berger (2012). Individual contribution of insolation and $CO_2$ to the interglacial climates of the past 800,000 years. *Climate Dynamics* 38, 709–24.

Yokoyama, Y., and T. M. Esat (2011). Global climate and sea level: enduring variability and rapid fluctuations over the past 150,000 years. *Oceanography* 24, 54–69.

Zachos, J. C., G. R. Dickens and R. E. Zeebe (2008). An early Cenozoic perspective on greenhouse warming and carbon-cycle dynamics. *Nature* 451, 279–83.

Zachos, J., M. Pagani, L. Sloan, E. Thomas and K. Billups (2001). Trends, rhythms, and aberrations in global climate 65 Ma to present. *Science* 292, 686–93.

Zeebe, R. E. (2013). What caused the long duration of the Paleocene-Eocene thermal maximum? *Paleoceanography* 28, 440–52.

Zelinka, M. D., and D. L. Hartmann (2010). Why is longwave cloud feedback positive? *Journal of Geophysical Research* 115, D16117.

(2012). Climate feedbacks and their implications for poleward energy flux changes in a warming climate. *Journal of Climate* 25, 608–24.

Zhao, Y., and S. P. Harrison (2012). Mid-Holocene monsoons: a multi-model analysis of the inter-hemispheric differences in the responses to orbital forcing and ocean feedbacks. *Climate Dynamics* 39, 1457–87.

Zhang, X., G. Lohmann, G. Knorr and C. Purcell (2014). Abrupt glacial climate shifts controlled by ice sheet changes. *Nature* 512, 290–4.

Zickfeld, K., M. Eby, A. J. Weaver, K. Alexander, E. Crespin, N. R. Edwards, A. V. Eliseev, G. Feulner, T. Fichefet, C. E. Forest, P. Friedlingstein, H. Goosse, P. B. Holden, F. Joos, M. Kawamiya, D. Kicklighter, H. Kienert, K. Matsumoto, I. I. Mokhov, E. Monier, S. M. Olsen, J. O. P. Pedersen, M. Perrette, G. Philippon-Berthier, A. Ridgwell, A. Schlosser, T. Schneider Von Deimling, G. Shaffer, A. Sokolov, R. Spahni, M. Steinacher, K. Tachiiri, K. S. Tokos, M. Yoshimori, N. Zeng and F. Zhao (2013). Long-term climate change commitment and reversibility: an EMIC intercomparison. *Journal of Climate* 26, 5782–809.

# Solutions of the Review Exercises

## Chapter 1

1. a. No. In order to define climate, it is important to characterise the variability at all time scales, not only the seasonal cycle.
   b. Yes. c is also valid and provides a wider definition.
   c. Yes. b is also valid.
2. a. No.
   b. Yes.
   c. No.
3. a. No. Such a situation corresponds to stable conditions, but it is not the limit for stability.
   b. Yes. As the air rises, it expands because of the decrease in pressure, modifying its density during a transformation which is close to an adiabatic. The air column is unstable if the density of the air parcel moving upward, accounting for this transformation, is lower than the density of its environment.
4. b. False. The downward branch of the Hadley cells occurs at a latitude of about 30°, not at the pole, because of the Earth's rotation.
5. a. True.
6. b. False. The thermohaline circulation, which is driven mainly by density contrasts, also has a near-surface branch.
7. b.
8. d.

## Chapter 2

1. b.
2. a. No. The albedo effect is already taken into account in the computation of the emission temperature.
   b. No. This flux is a very small element of the Earth's energy balance.
   c. Yes. This is the so-called greenhouse effect.
3. a. No. The Earth–Sun distance changes with time, influencing the amount of energy received at the surface of the Earth, but it is not the cause of the seasonal cycle.
   b. Yes. Because of the inclination of the Earth's rotation axis on the ecliptic plane, the northern hemisphere receives more energy in boreal spring and summer.

4. a. No. The imbalance reaches about 60 W m$^{-2}$.
   b. Yes.
   c. Heat storage is not able to compensate for the imbalance on annual mean.
5. a. True.
6. b. False. The atmosphere transports much more heat poleward than the ocean at middle latitudes, whilst in tropical regions between 20°S and 20°N the ocean plays a larger role.
7. b. False. Because of deep-water formation in the North Atlantic and not in the North Pacific, the ocean transports much more heat northward in the North Atlantic.
8. a. True. The evaporation of water requires a large amount of energy. When water is available (such as over the oceanic surfaces or on relatively wet land areas), the latent heat flux thus is generally a dominant term is the surface heat balance.
9. b. False. Evaporation is large over the ocean, but the net transfer from ocean to land corresponds to about 35% of the total precipitation over land.
10. b. False. Strong imbalances in the freshwater fluxes at the surface can be compensated by ocean transport. Over land, the imbalance is generally smaller and compensated by river runoff.
11. b. False. In many continental regions, this is indeed the case, but evaporation requires energy, and in some wet regions, this is the limiting factor (as over the ocean where the water availability is never limiting).
12. a. No. The solubility of $CO_2$ is not much higher than that for other gases, such as oxygen, for instance, which do not have such a large stock in the ocean.
    b. Yes. The carbonic acid, which has to equilibrate directly with atmospheric $CO_2$, represents only 0.5% of the dissolved inorganic carbon (DIC). About 90% is in the bicarbonate form and around 10% in the carbonate form.
13. a. No. Photosynthesis on land explains the seasonal cycle of the atmospheric $CO_2$ concentration, but the regional differences over the ocean are much lower than those caused by oceanic processes.
    b. Yes. The variations in $CO_2$ flux at the ocean-atmosphere interface are mainly due to oceanic processes.
14. a. No. The equilibrium relations among carbonic acid, bicarbonate and carbonate are pushed towards the production of less carbonic acid because of the lower acidity of the water, and thus $p^{CO_2}$ in the ocean decreases.
    b. No. See above.
    c. Yes.
15. a. Yes. The surface DIC decreases as a result of photosynthesis because a fraction of the organic matter is exported out of the surface layer.
    b. No. Most of the organic matter produced in the surface layer is indeed decomposed locally by remineralisation and respiration, but a fraction of this organic matter is exported out of the surface layer, leading to a reduction of the DIC there. This is the soft-tissue pump.

16. a. No. $CaCO_3$ formation induces a decrease in alkalinity and thus an increase in $p^{CO_2}$.
    b. No. See above.
    c. Yes.
17. a. Yes. At high temperature, calcium carbonate reacts with silicon dioxide to produce calcium-silicate rocks, inducing a $CO_2$ release.
    b. No. See above.
    c. No. See above.
18. a. Yes. Calcium-silicate rocks react with carbonate acid to form calcium carbonate, and this reaction induces a carbon uptake from the atmosphere.
    b. No. See above.
19. a. True. This is the end of the long inorganic carbon cycle which takes place over millions of years.
20. a. Yes. These are the main natural sources of methane.
    b. No. Rice production over a flooded surface and emissions from livestock and fossil fuels are the main sources of methane related to human activities.
    c. No. The oxidation of methane requires oxygen, but the rate of the reaction is not influenced by oxygen concentration at the present level in the atmosphere.
    d. Yes. Methane oxidation requires the presence of highly reactive constituents such as the hydroxyl radical.

## Chapter 3

1. a. Yes. When a process is not well understood, it is better to account for its effect as simply as possible. The other answers are also correct.
    b. Yes. It is often sufficient to reproduce only the first-order effects of some processes, so parameterisations can be used instead of a detailed representation. The other answers are also correct.
    c. Yes. Even for climate models with the highest resolution presently affordable, it is not possible to represent small-scale processes such as turbulence and the interactions of the circulation with small-scale topography features, thunderstorms, cloud micro physics and so on. Their effect at larger scale therefore must be accounted for through parameterisations. The other answers are also correct.
2. a. No. The greenhouse effect influences mainly the infrared flux, whilst albedo affects solar radiation.
    b. No. Those sophisticated treatments of the radiative transfer are not included in energy-balance models (EBMs) but are present in some Earth models of intermediate complexity (EMICs) and general circulation models (GCMs).
    c. Yes. This allows taking into account the first-order effect of the greenhouse gases on the radiative balance of the atmosphere.

3. a. True. It is not yet possible to use GCMs to make simulations longer than a few hundred to a few thousand years.

4. a. True. The topography has to be averaged at the scale of the grid, leading to a strong simplification of complex terrains in GCMs. b is also correct.

   b. True. Some circulation features, which have a clear regional impact, have a scale of few tenths of kilometres that can be accounted for in regional climate models (RCMs) but not directly in GCMs. a is also correct.

   c. False. A higher resolution tends to reduce the truncation error, but the improvement is generally not coming from the numerical scheme itself because similar schemes can be used in GCMs and RCMs.

5. b. False. Many additional elements are required, such as a radiative scheme, parameterisations of the dissipation and mixing and so on.

6. a. Yes. Because of the very different horizontal and vertical scales, the horizontal heat diffusion can be safely neglected.

   b. No. This could be used to justify that a mean velocity at the size of the grid box could be computed, but not here.

7. a. False. These processes are included in bucket models, but indeed in a simplified way which does not include the influence of roots and the storage of water at different levels in the soil.

   b. True. In standard bucket models, the excess of water is immediately transferred to the ocean.

8. b. No. Because these properties only ensure the validity of the scheme at the limit of the time and spatial steps tending to zero. It is also necessary to estimate the accuracy of the scheme when time and spatial steps have realistic values and to verify that the scheme conserves some important properties such mass and energy, if required by the problem.

9. b. False. Validation is a continuous process which should be carried out each time new observations are available. Furthermore, validation is always partial, as it is impossible to perform it on all cases.

10. b. False. There is always a risk of having good results because the selection of some parameters compensates for model error on processes not directly related to those parameters. The calibration thus should be performed with care, but it is allowed to select the adequate value of unknown parameters inside their range of uncertainty.

11. a. No. This is a nice by-product, but large repositories of model results are also useful even if the experiments are not co-ordinated.

    b. Yes. As the experimental design is controlled and similar in all models, the difference between model results should be found in the models themselves.

12. a. Yes, but not only.

    b. Yes, but not only.

    c. In theory, the comparison of model results with past observations can be used to correct any model biases, but from a practical point of view, not enough observations are available in conditions similar to those of the simulation which has to be corrected. Consequently, it is not possible to propose robust corrections in many cases.

13. b. False. Except for very simple methods, data assimilation techniques are able, on the basis of the link between different variables derived from model results, to update variables that are not observed using observations of another variable. This is, of course, valid only if those two variables have a clear enough physical link.

## Chapter 4

1. a. No. Adjustment of the temperature of the stratosphere is fast compared with lower levels. This is the reason why stratospheric temperature is allowed to adjust for both the radiative forcing (RF) and the effective radiative forcing (ERF) to arrive at an estimate of the radiative imbalance of the Earth on time scales longer than a few months, which are generally the ones of interest in climatology.
   b. Yes. Some changes in the atmosphere in response to a forcing are fast and independent of surface changes. For similar reasons that the stratosphere temperature is allowed to adjust for RF, tropospheric temperature also may change in ERF to get a useful estimate of the modification of the radiative budget imposed by the perturbation.
   c. No. If the temperature adjusts everywhere, a new equilibrium is reached, and the net radiative flux at the tropopause or the top of the atmosphere is always zero.

2. a. No. Some saturation effects occur.
   b. Yes. The radiative forcing caused by an increase in the $CO_2$ concentration is larger at low concentrations than at higher ones.

3. b. Ozone depletion in the stratosphere induced a global mean radiative forcing of about $-0.05$ W m$^{-2}$ between 1750 and 2011. This is much less than the greenhouse gas forcing and tends to cool the climate. The ozone hole has a local impact on climate in Antarctica but is not the main cause of the recent global warming.

4. a. Negative. The forcing is negative because it implies the scattering of a significant part of the incoming solar radiation back to space.
   b. Positive. As more solar radiation is absorbed, the forcing is positive.
   c. Positive. This reduces the surface albedo and thus increases the fraction of the incoming solar radiation which is absorbed at the surface.
   d. Negative. A high concentration of aerosols leads to smaller water droplets in clouds, a higher albedo of clouds and thus a negative effective radiative forcing.

5. b. By definition, at equilibrium, the net radiative flux should be equal to zero in order to ensure a radiative balance of the Earth. When the radiative forcing is applied, an initial imbalance occurs, but the system adjusts, and a new balance is reached.

6. b.

7. a.

8. c. Both of them are important. For a large heat capacity and a large equilibrium climate sensitivity, the response time is longer.

9. b.

10. a. A positive feedback because the temperature in the upper troposphere increases less than in the case of a homogeneous temperature change, leading to a lower increase in the upward longwave flux.

11. b. Yes. The albedo of snow and ice is much higher than that of the ocean or land.

12. c. Negative. As they have a relatively high albedo, they reflect a significant fraction of incoming shortwave radiations. Moreover, because of their high temperature, they do not modify enough of the longwave radiation of the Earth to compensate for their shortwave effect.

13. a. This effect of soil moisture on heat capacity exists but is generally not dominant.

    b. This is the dominant effect in regions where evaporation is controlled by the soil moisture content.

14. a. No. This is a negative feedback, which tends to stabilise the system, as the cooling increases the surface density in regions of deep-water formation.

    b. Yes. This is positive feedback, as the freshening decreases the surface density at high latitudes.

15. Both answers are correct.

16. a. Yes. The warming reduces the oceanic solubility of $CO_2$ and thus the oceanic uptake of $CO_2$.

    b. Yes. This reduces the exchanges between the surface layers containing a larger amount of carbon because of the perturbation and the deeper layers which do not yet contain much carbon originating in this perturbation.

    c. Yes. A warming tends to accelerate the decomposition in soils.

    d. Both signs are possible because climate change generally enhances primary production is some regions, inducing an uptake of $CO_2$, and reduces it in some others.

17. a. Yes, the albedo of trees is lower than that of grass, in particular, when snow is present, leading to an additional warming.

    b. No. Evapotranspiration is generally not controlled by moisture availability at high latitudes and thus is not strongly influenced by the ability of plants to use soil water. At low latitudes, the presence of trees generally increases the evapotranspiration, leading to surface cooling because, among other effects, the roots of the trees are able to extract water more efficiently from the soil than grass.

18. a. No. The inflow of calcium carbonate from rivers does not have a direct influence on biological production and thus on the carbonate pump because the upper layers are oversaturated with respect to calcium carbonate.

    b. No. The inflow of calcium carbonate from rivers does not have a direct influence on biological production and thus on the soft-tissue pump, and the soft-tissue pump is not associated with a direct export of calcium carbonate.

    c.  Yes. This is the calcium carbonate compensation mechanism.

19. b.  A negative feedback because weathering is associated with a reduction in atmospheric $CO_2$ concentration which tends to limit the warming.

# Chapter 5

1. b.  False. Natural variability is generally associated with temperature changes of different signs in different regions which tend to partially cancel each other when looking at the global mean surface temperature.

2. b.  False. A model should indeed be able to reproduce the observed statistics of the internal variability of the system, such as the variance of the surface temperature in a region or the correlation of the sea-level pressure between the two centres of action of the North Atlantic oscillation. However, the time evolution of the internal modes of variability is so sensitive to small perturbations that there is no chance that a model fits the observed time series perfectly, except if the model is constrained to follow the observations. Similarly, there is no chance that two GCMs with slightly different physics or initial conditions systematically display the same events at the same time over a long period if those events are not due to a specific forcing.

3. c.

4. a.  True. The sea surface temperature (SST) gradient drives changes in sea-level pressure (SLP) which drive the easterlies.

5. a.  Yes.
    b.  Yes.
    c.  Yes.

6. a.  Yes. The larger meridional gradient in SLP induces stronger winds.
    b.  No. The stronger westerlies induce a larger transport of warm and moist air from the ocean to the continents and thus a warming over northern Europe and Asia.
    c.  Yes. The stronger winds, with a clear northerly component, bring colder air to this area.

7. b.  False. Statistical techniques are very useful tools to obtain quantitative reconstructions, but the law derived from those statistical methods must be consistent with at least a minimum understanding of the way the climatic signal is recorded in the archive. Furthermore, in the so-called forward approach, the variable measured in the archive can be estimated directly from model results, for instance, without the need to first estimate climate variables from the proxy records using a statistical method.

8.   a and b are both correct.

9. a.  No. This probably played a role during the first hundred million years of Earth's history, but this could not alone explain the paradox.
    b.  No. Information on the albedo of Earth at that time is scarce, but the changes are probably too low to play a dominant role.

c. Yes. The atmosphere was very different from today, with much higher $CO_2$ and methane concentrations, leading to a much stronger greenhouse effect.

10. a. No. Molecular oxygen has a low radiative effect. The radiative influence of ozone is larger, but this does not have a dominant influence on surface temperature on these time scales.

b. Yes. Methane is a very efficient greenhouse gas. Modifying its concentration thus has a large influence on surface temperature.

11. a. Yes.

b. No. The recycling associated with this process is too rapid to have a long-term impact (except through an indirect influence on points c and d).

c. Yes

d. Yes

e. No, as in b.

12. a. Change in solar irradiance is not the dominant forcing on these time scales.

b. Changes in orbital parameters have a dominant influence on shorter-term fluctuations but not on this long-term cooling

c. The $CO_2$ concentration has strongly decreased during this period, inducing a cooling.

13. a. True. The annual mean energy received by the Earth is inversely proportional to the square root of one minus the square of the eccentricity.

14. a. No. Increasing obliquity also has an influence on the annual mean insolation, increasing it by a few watts per square metre at high latitudes and decreasing it at the equator.

b. Yes.

15. a. Yes

b. Yes

c. Yes.

16. a. No. Insolation changes may have an indirect effect on atmospheric circulation, but such circulation changes are not the dominant effect.

b. No. Insolation changes may have an indirect effect on oceanic circulation, but such circulation changes are not the dominant effect.

c. Yes. If summer insolation is too low, the summer is cold, and only a fraction of the snow which has fallen at high latitudes over land during the winter melts in summer, leading to the formation of ice sheets. By contrast, if summer insolation is high enough, snow melting over the existing ice sheet may be larger than winter accumulation, leading to ice-sheet shrinkage.

17. a. No. The changes in continental biosphere tend to increase $CO_2$ concentration during glacial periods.

b. No. Geological processes are too slow to account for many of the observed changes.

c. Yes. The changes in solubility contribute to the decrease.

d. Yes. This is likely a dominant contributor, in particular, the changes in the Southern Ocean.

18. a. Yes. The glacial period and the deglaciation display millennium-scale variability, generally attributed to changes in the oceanic circulation and the oceanic heat transport, which is much weaker during the Holocene.
   b. No.
19. a. Yes. This is due to the direct radiative effect.
   b. No. The summer monsoons have weakened because of a weaker temperature contrast between land and ocean induced by changes in insolation.
   c. Yes. The weakening of summer monsoons has induced a decrease in precipitation over the Sahara and thus a desertification of the region.
20. a. Yes.
   b. No. Dust forcing has a significant influence when analysing the climate of the last glacial maximum but not for the last millennium.
   c. Probably, but the amplitude of the signal is uncertain and hard to formally detect.
   d. The orbital forcing plays a dominant role on longer time scales, but its influence is weaker in annual mean on a global scale for the last millennium. Nevertheless, it is still a significant contributor on the regional scale for some seasons over this period.
21. b. False. The changes in external forcing play a role on a regional scale, but the effect of internal climate variability, purely related to the intrinsic dynamics of the system, can completely mask the influence of the forcing in some periods.
22. a. No. Without taking into account the anthropogenic forcings, it is not possible to explain the observed warming.
   b. Yes. The large changes in climate observed recently appear to be outside the range of natural variability, but these changes are compatible with those predicted when taking into account anthropogenic forcings.

## Chapter 6

1. a. True
2. a. True
3. a. True. If the model biases are random, averaging several simulations performed with different models improves the performance compared with any individual simulation. If the bias is systematic and common to most models, no improvement is expected.
   b. True. As the internal variability tends to be uncorrelated between the different members of an ensemble, averaging them strongly reduces the magnitude of this internal variability, providing a good estimate of the forced response which is common to all the simulations.
   c. False. As the averaging of several simulations strongly reduces the internal variability compared with individual simulations, the total variability is underestimated by the ensemble mean.

4. b. False. At the end of the twenty-first century, for global mean surface temperature, the two main sources of uncertainties are the uncertainty in future emissions of greenhouse gases (scenario uncertainty) and the uncertainty in the physical processes represented in the models. The internal variability plays a significant role only during a few decades, the natural fluctuations of global mean temperature at decadal to centennial time scales having an amplitude which is much smaller than the changes expected in 2100.

5. a. No. As soon as the simulation is started, the model drifts towards its mean state, and this drift has to be corrected. To avoid this problem, some models initialise the predictions using anomalies.

   b. No. The effect of the past forcings is included in the projections. Reconstruction of the history of past forcings and the model response may be uncertain, but correcting for this is not the main goal of the initialisation from a state close to observations (see also point a).

   c. Yes. Knowing the current state provides information on its future evolution over the next months to years, at least for some regions.

6. a. No. This plays a role in some regions, but it is not the dominant effect. In particular, the land-sea contrast is still present at equilibrium when thermal inertia does not contribute anymore.

   b. Yes. A larger evaporation reduces the surface temperature and influences the modification of the lapse rate in response to the warming.

7. a. Yes. This is often given as a dominant cause of the polar amplification, but other processes also play an important role.

   b. Yes. Many models simulate a larger heat transport towards the Arctic, in particular, because of a larger transport of moisture and thus of latent heat in a warmer atmosphere.

   c. No. This is the opposite. More clouds in winter reduce the longwave radiative heat losses and contribute to the polar amplification.

   d. Yes. Because of the vertical structure of the atmosphere, the warming is 'trapped' close to the surface.

8. a. No. Evaporation-precipitation interactions are complex and cannot explain the wet-get-wetter pattern. Moreover, this pattern is clearer over oceans, where evaporation is not limited by moisture availability.

   b. Yes. A warmer atmosphere is able to transport more water to regions where there is a moisture convergence and strong precipitation in current conditions.

   c. No. A reduction in the intensity of the Hadley cell would induce opposite changes and would tend to compensate for the effect described in b.

9. a. True. Among other mechanisms, the warming reduces the solubility of $CO_2$ in sea water, and the past oceanic uptake of $CO_2$ has decreased the availability of carbonates in the ocean. A larger fraction of the DIC thus will remain in the future as $H_2CO_3$, increasing the partial pressure of carbon dioxide in the ocean and thus reducing the oceanic uptake of anthropogenic carbon.

10. a. No.
    b. No.
    c. Yes. These processes are relatively slow compared to the rate of current emissions of $CO_2$.
11. a. Yes. Thermal expansion is estimated to be the main contributor for the twenty-first century too.
    b. No. This is the second largest contributor.
    c. No. The time scale of the response of ice sheets is long, and their contribution for the twentieth and twenty-first centuries is expected to be smaller than those of thermal expansion and glaciers.
    d. No. The contribution of melting of the Antarctic ice sheet in the sea-level rise during the twentieth and twenty-first centuries is expected to be modest.
12. b.
13. a. No. This feedback may amplify the surface changes but is not the cause of the instability.
    b. No. The melting below the ice shelf can trigger the instability by a reduction in the thickness of the ice shelf and a retreat of the grounding line, but it is not the cause of the instability itself.
    c. Yes. As the bedrock becomes deeper, the ice thickness becomes larger when the distance to the grounding line increases, leading to a larger mass outflow as the grounding line retreats, potentially producing an instability.

# Index

The terms defined in the glossary are in bold type.

Printed in the United States
By Bookmasters